Collected Papers

Volume III

Springer
New York
Berlin
Heidelberg
Barcelona
Hong Kong
London
Milan
Paris
Singapore
Tokyo

Photo by Yale Photographic Services

Serge Lang
(ca. early 1970s)

Serge Lang

Collected Papers

Volume III

1978–1990

Springer

Serge Lang
Department of Mathematics
Yale University
10 Hillhouse Avenue
PO Box 208283
New Haven, CT 06520-8283
USA

Mathematics Subject Classification (1991): 11Gxx, 14Kxx, 11Fxx, 11Jxx, 14H

Library of Congress Cataloging-in-Publication Data
Lang, Serge, 1927–
Collected papers / Serge Lang.
p. cm.
Includes bibliographical references.
Contents: v. 1. 1952–1970 — v. 2. 1971–1977 — v. 3. 1978–1990.
ISBN 0-387-98802-5 (v. 1 : hardcover : alk. paper) — ISBN 0-387-98803-3
(v. 2 : hardcover : alk. paper) — ISBN 0-387-98800-9 (v. 3 : hardcover : alk. paper)
1. Mathematics. I. Title.
QA7.L288 2000
510—dc21 99-17359

Printed on acid-free paper.

Production managed by Michael Koy; manufacturing supervised by Joe Quatela.
Photocomposed copy prepared from the author's original manuscripts.
Printed and bound by Sheridan Books, Inc., Ann Arbor, MI.
Printed in the United States of America.

9 8 7 6 5 4 3 2 1

ISBN 0-387-98800-9 Springer-Verlag New York Berlin Heidelberg SPIN 10715720

Contents

Bibliography (through 1999)

(Boldface items are books or Lecture Notes.)

[1952a] On quasi algebraic closure, *Ann. of Math.* **55** No. 2 (1952) pp. 373–390.

[1952b] Hilbert's nullstellensatz in infinite dimensional space, *Proc. AMS* **3** No. 3 (1952) pp. 407–410.

[1952c] (with J. TATE) On Chevalley's proof of Luroth's theorem, *Proc. AMS* **3** No. 4 (1952) pp. 621–624.

[1953] The theory of real places, *Ann. of Math.* **57** No. 2 (1953) pp. 378–391.

[1954a] Some applications of the local uniformization theorem, *Am. J. Math.* **76** No. 2 (1954) pp. 362–374.

[1954b] (with A. WEIL) Number of points of varieties in finite fields, *Am. J. Math.* **76** No. 4 (1954) pp. 819–827.

[1955] Abelian varieties over finite fields, *Proc. NAS* **41** No. 3 (1955) pp. 174–176.

[1956a] Unramified class field theory over function fields in several variables, *Ann. of Math.* **64** No. 2 (1956) pp. 285–325.

[1956b] On the Lefschetz principle, *Ann. of Math.* **64** No. 2 (1956) pp. 326–327.

[1956c] L-series of a covering, *Proc. NAS* **42** No. 7 (1956) pp. 422–424.

[1956d] Sur les séries L d'une variété algébrique, *Bull. Soc. Math. France* **84** (1956) pp. 385–407.

[1956e] Algebraic groups over finite fields, *Am. J. Math.* **78** (1956) pp. 555–563.

[1957a] (with J.-P. SERRE) Sur les revêtements non ramifiés des variétés algébriques, *Am. J. Math.* **79**, No. 2 (1957) pp. 319–330.

[1957b] (with W.-L. CHOW) On the birational equivalence of curves under specialization, *Am. J. Math.* **79**, No. 3 (1957) pp. 649–652.

[1957c] Divisors and endomorphisms on an abelian variety, *Am. J. Math.* **79** No. 4 (1957) pp. 761–777.

[1957d] Families algébriques de Jacobiennes (d'après IGUSA), *Séminaire Bourbaki* No. 155, 1957/1958.

[1958a] Reciprocity and correspondences, *Am. J. Math.* **80** No. 2 (1957) pp. 431–440.

[1958b] (with J. TATE) Principal homogeneous spaces over abelian varieties, *Am. J. Math.* **80** No. 3 (1958) pp. 659–684.

[1958c] (with E. KOLCHIN), Algebraic groups and the Galois theory of differential fields, *Am. J. Math.* **80** No. 1 (1958) pp. 103–110.

[1958d] *Introduction to algebraic geometry*, Wiley-Interscience, 1958.

[1959a] (with A. NÉRON) Rational points of abelian varieties over function fields, *Am. J. Math.* **81** No. 1 (1959) pp. 95–118.

[1959b] Le théorème d'irreductibilité de Hilbert, *Séminaire Bourbaki* No. 201, 1959/1960.

[1959c] *Abelian varieties*, Wiley-Interscience, 1959; Springer-Verlag, 1983.

[1960a] (with E. KOLCHIN) Existence of invariant bases, *Proc. AMS* **11** No. 1 (1960) pp. 140–148.

[1960b] Integral points on curves, *Pub. IHES* No. **6** (1960) pp. 27–43.

[1960c] Some theorems and conjectures in diophantine equations, *Bull. AMS* **66** No. 4 (1960) pp. 240–249.

[1960d] On a theorem of Mahler, *Mathematika* **7** (1960) pp. 139–140.

[1960e] L'équivalence homotopique tangentielle (d'apres MAZUR), *Séminaire Bourbaki* No. 222, 1960/1961.

[1961] Review: Elements de géométrie algébrique (A. Grothendieck). *Bull. AMS* **67** No. 3 (1961) pp. 239–246.

[1962a] A transcendence measure for E-functions, *Mathematika* **9** (1962) pp. 157–161.

[1962b] Transcendental points on group varieties, *Topology* **1** (1962) pp. 313–318.

[1962c] Fonctions implicities et plongements Riemanniens, *Séminaire Bourbaki* 1961/1962, No. 237, May 1962.

[1962d] *Introduction to Differential Manifolds*, Addison Wesley, 1962.

[1962e] *Diophantine Geometry*, Wiley-Interscience, 1962.

[1963] *Transzendente Zahlen*, Bonn Math. Schr. No. 21 (1963).

[1964a] Diophantine approximations on toruses, *Am. J. Math.* **86** No. 3 (1964) pp. 521–533.

[1964b] Les formes bilinéaires de Néron et Tate, *Séminaire Bourbaki* 1963/64 Fasc. 3 Exposé 274, Paris 1964.

[1964c] *First Course in Calculus*, Addison Wesley 1964; Fifth edition by Springer-Verlag, 1986.

[1964d] *Algebraic and Abelian Functions*, W.A. Benjamin Lecture Notes, 1964. See also [1982c].

[1964e] *Algebraic Numbers*, Addison Wesley, 1964; superceded by [1970d].

[1965a] Report on diophantine approximations, *Bulletin Soc. Math. France* **93** (1965) pp. 177–192.

[1965b] Division points on curves, *Annali Mat. pura ed applicata*, Serie IV **70** (1965) pp. 229–234.

[1965c] Algebraic values of meromorphic functions, *Topology* **3** (1965) pp. 183–191.

[1965d] Asymptotic approximations to quadratic irrationalities I, *Am. J. Math.* **87** No. 2 (1965) pp. 481–487.

[1965e] Asymptotic approximations to quadratic irrationalities II, *Am. J. Math.* **87** No. 2 (1965) pp. 488–496.

[1965f] (with W. ADAMS) Some computations in diophantine approximations, *J. reine angew. Math.* Band **220** Heft 3/4 (1965) pp. 163–173.

[1965g] Corps de fonctions méromorphes sur une surface de Riemann (d'après ISS'SA), *Séminaire Bourbaki* No. 292, 1964/65.

[1965h] *Algebra*, Addison Wesley, 1965; second edition 1984; third edition 1993.

[1966a] Algebraic values of meromorphic functions II, *Topology* **5** (1960) pp. 363–370.

[1966b] Asymptotic diophantine approximations, *Proc. NAS* **55** No. 1 (1966) pp. 31–34.

[1966c] *Introduction to transcendental numbers*, Addison Wesley, 1966.

[1966d] *Introduction to Diophantine Approximations*, Addison Wesley 1966; see [1995d].

[1966e] *Rapport sur la cohomologie des groupes*. Benjamin, 1966.

[1967] *Algebraic structures*, Addison Wesley 1967.

[1968] *Analysis I,* Addison Wesley, 1968; (superceded by [1983c]).

[1969] *Analysis II*, Addison Wesley, 1969; (superceded by [1993c]).

[1970a] (with E. BOMBIERI) Analytic subgroups of group varieties, *Inventiones Math.* **11** (1970) pp. 1–14.

[1970b] Review: L.J. Mordell's *Diophantine Equations*, *Bull. AMS* **76** (1970) pp. 1230–1234.

[1970c] *Introduction to Linear Algebra*, Addison Wesley 1970; see also [1986b].

[1970d] *Algebraic Number Theory*, Addison Wesley 1970; see also [1994c].

[1971a] Transcendental numbers and diophantine approximations, *Bull. AMS* **77** No. 5 (1971) pp. 635–677.

[1971b] On the zeta function of number fields, *Invent. Math.* **12** (1971) pp. 337–345.

[1971c] The group of automorphisms of the modular function field, *Invent. Math.* **14** (1971) pp. 253–254.

[1971d] *Linear Algebra*, Addison Wesley 1971; see also [1987b].

[1971e] *Basic Mathematics*, Addison Wesley 1971; Springer-Verlag 1988.

[1972a] Isogenous generic elliptic curves, *Amer. J. Math.* **94** (1972) pp. 661–674.

[1972b] (with H. TROTTER) Continued fractions for some algebraic numbers, *J. reine angew. Math.* **255** (1972) pp. 112–134.

[1972c] *Differential manifolds*, Addison Wesley, 1972.

[1972d] *Introduction to Algebraic and Abelian Functions*, Benjamin-Addison Wesley, 1972; second edition see [1982c].

[1973a] Frobenius automorphisms of modular function fields, *Amer. J. Math.* **95** (1973) pp. 165–173.

[1973b] *Calculus of Several Variables*, Addison Wesley 1973; Third edition see [1987d].

[1973c] *Elliptic functions*, Addison Wesley 1973; second edition see [1987d].

[1974a] Higher dimensional diophantine problems, *Bull. AMS* **80** No. 5 (1974) pp. 779–787.

[1974b] (with H. TROTTER) Addendum to "Continued fractions of some algebraic numbers," *J. reine angew. Math.* **267** (1974) pp. 219–220.

[1975a] Diophantine approximations on abelian varieties with complex multiplication, *Advances Math.* **17** (1975) pp. 281–336.

[1975b] Division points of elliptic curves and abelian functions over number fields, *Amer. J. Math.* **97** No. 1 (1975) pp. 124–132.

[1975c] (with D. KUBERT) Units in the modular function field I, Diophantine Applications, *Math. Ann.* **218** (1975) pp. 67–96.

[1975d] (with D. KUBERT) Units in the modular function field II, A full set of units, *Math. Ann.* **218** (1975) pp. 175–189.

[1975e] (with D. KUBERT) Units in the modular function field III, Distribution relations, *Math. Ann.* **218** (1975) pp. 273–285.

[1975f] La conjecture de Catalan d'après Tijdeman, *Séminaire Bourbaki* 1975/76 No. 29.

[1975g/85] $SL_2(\mathbf{R})$, Addison Wesley, 1975; Springer-Verlag corrected second printing, 1985.

[1976a] (with J. COATES) Diophantine approximation on abelian varieties with complex multiplication, *Invent. Math.* **34** (1976) pp. 129–133.

[1976b] (with D. KUBERT) Distribution on toroidal groups, *Math. Z.* **148** (1976) pp. 33–51.

[1976c] (with D. KUBERT) Units in the modular function field, in *Modular Functions in One Variable* V, Springer Lecture Notes **601** (Bonn Conference) 1976, pp. 247–275.

[1976d] (with H. TROTTER) *Frobenius distributions in* GL_2*-extensions*, Springer Lecture Notes **504**, Springer-Verlag 1976.

[1976e] *Introduction to Modular Forms*, Springer-Verlag, 1976.

[1977a] (with D. KUBERT) Units in the modular function field IV, The Siegel functions are generators, *Math. Ann.* **227** (1997) pp. 223–242.

[1977b] (with H. TROTTER) Primitive points on elliptic curves, *Bull. AMS* **83** No. 2 (1977) pp. 289–292.

[1977c] *Complex Analysis*, Addison Wesley; second Edition Springer-Verlag 1985; fourth edition Springer-Verlag 1999.

[1978a] (with D. KUBERT) The *p*-primary component of the cuspidal divisor class group of the modular curve *X*(*p*), *Math. Ann.* **234** (1978) pp. 25–44.

[1978b] (with D. KUBERT) Units in the modular function field V, Iwasawa theory in the modular tower, *Math. Ann.* **237** (1978) pp. 97–104.

[1978c] (with D. KUBERT) Stickelberger ideals, *Math. Ann.* **237** (1978) pp. 203–212.

[1978d] (with D. KUBERT) The index of Stickelberger ideals of order 2 and cuspidal class numbers, *Math. Ann.* **237** (1978) pp. 213–232.

[1978e] Relations de distributions et exemples classiques, *Séminaire Delange-Pisot-Poitou (Théorie des Nombres)*, 1978 No. 40 (6 pages).

[1978f] *Elliptic curves: Diophantine Analysis*, Springer-Verlag 1978.

[1978g] *Cyclotomic Fields* I, Springer-Verlag, 1978.

[1979a] (with D. KUBERT) Cartan-Bernoulli numbers as values of L-series, *Math. Ann.* **240** (1979) pp. 21–26.

[1979b] (with D. KUBERT) Independence of modular units on Tate curves, *Math. Ann.* **240** (1979) pp. 191–201.

[1979c] (with D. KUBERT) Modular units inside cyclotomic units, *Bull. Soc. Math. France* **107** (1979) pp. 161–178.

[1980] *Cyclotomic Fields* II, Springer-Verlag 1980.

[1981a] (with N. KATZ) Finiteness theorems in geometric classfield theory, *Enseignement mathématique* **27** (3–4) (1981) pp. 285–314.

[1981b] (with DAN KUBERT) *Modular Units*, Springer-Verlag 1981.

[1982a] Représentations localement algébriques dans les corps cyclotomiques, Séminaire de Théorie des Nombres 1982, Birkhauser, pp. 125–136.

[1982b] Units and class groups in number theory and algebraic geometry, *Bull. AMS* **6** No. 3 (1982) pp. 253–316.

[1982c] *Introduction to algebraic and abelian functions, Second Edition*, Springer-Verlag, 1982.

[1983a] Conjectured diophantine estimates on elliptic curves, in Volume I of *Arithmetic and Geometry*, dedicated to Shafarevich, M. Artin and J. Tate editors, Birkhauser (1983) pp. 155–171.

[1983b] *Fundamentals of Diophantine Geometry*, Springer-Verlag 1983.

[1983c] *Undergraduate Analysis*, Springer-Verlag, 1983.

[1983d] (with GENE MURROW) *Geometry: A High School Course*, Springer Verlag, 1983 Second edition 1988.

[1983e] *Complex Multiplication*, Springer-Verlag 1983.

[1984a] Vojta's conjecture, *Arbeitstagung Bonn 1984*, Springer Lecture Notes **1111** 1985, pp. 407–419.

[1984b] Variétés hyperboliques et analyse diophantienne, *Séminaire de théorie des nombres*, 1984/85, pp. 177–186.

[1985a] (with W. FULTON) *Riemann-Roch Algebra*, Springer-Verlag, 1985.

[1985b] *The Beauty of Doing Mathematics*, Springer-Verlag, 1985 originally published as articles in the *Revue du Palais de la Découverte*, Paris, 1982–1984, specifically:
Une activité vivante: faire des mathématiques, *Rev. P.D.* Vol. **11** No. **104** (1983) pp. 27–62
Que fait un mathématicien pure et pourquoi?, *Rev. P.D.* Vol. **10** No. **94** (1982) pp. 19–44
Faire des Maths: grands problèmes de géométrie et de l'espace, *Rev. P.D.* Vol. **12** No. **114** (1984) pp. 21–72.

[1985c] *Math! Encounters with High School Students*, Springer-Verlag, 1985 (French edition *Serge Lang, des Jeunes et des Maths*, Belin, 1984).

[1986a] Hyperbolic and diophantine analysis, *Bulletin AMS* **14** No. 2 (1986) pp. 159–205.

[1986b] *Introduction to Linear Algebra*, Second Edition, Springer-Verlag 1986.

[1987a] Diophantine problems in complex hyperbolic analysis, *Contemporary Mathematics AMS* **67** (1987) pp. 229–246.

[1987b] *Linear Algebra*, Third Edition, Springer-Verlag 1987.

[1987c] *Undergraduate Algebra*, Springer-Verlag 1987.

[1987d] *Elliptic functions*, Second Edition, Springer-Verlag 1987.

[1987e] *Introduction to complex hyperbolic spaces*, Springer-Verlag 1987.

[1988a] The error term in Nevanlinna theory, *Duke Math. J.* **56** No. 1 (1988) pp. 193–218.

[1988b] *Introduction to Arakelov Theory*, Springer-Verlag 1988.

[1990a] The error term in Nevanlinna Theory II, *Bull. AMS* **22** No. 1 (1990) pp. 115–125.

[1990b] Old and new conjectured diophantine inequalities. *Bull. AMS* **23** No. 1 (1990) pp. 37–75.

[1990c] *Lectures on Nevanlinna theory*, in *Topics in Nevanlinna Theory*, Springer Lecture Notes **1433** (1990) pp. 1–107.

[1990d] *Cyclotomic Fields I and II*, combined edition with an appendix by Karl Rubin, Springer-Verlag, 1990.

[1991] *Number Theory III*, Survey of Diophantine Geometry, Encyclopedia of Mathematical Sciences, Springer-Verlag 1991.

[1993a] (with J. JORGENSON) On Cramér's theorem for general Euler products with functional equations, *Math. Ann.* **297** (1993) pp. 383–416.

[1993b] (with J. JORGENSON) *Basic analysis of regularized series and products*, Springer Lecture Notes **1564** (1993).

[1993c] *Real and Functional Analysis*, Springer-Verlag, 1993.

[1993d] *Algebra, Third Edition*, Addison Wesley 1993.

[1994a] (with J. JORGENSON) Artin formalism and heat kernels, *J. reine angew. Math.* **447** (1994) pp. 165–200.

[1994b] (with J. JORGENSON) *Explicit Formulas for regularized products and series*, in Springer Lecture Notes **1593** pp. 1–134.

[1994c] *Algebraic Number Theory, Second Edition*, Springer-Verlag 1994.

[1995a] Mordell's review, Siegel's letter to Mordell, diophantine geometry, and 20th century mathematics, *Notices AMS* March 1995 pp. 339–350.

[1995b] Some history of the Shimura-Taniyama conjecture, *Notices AMS* November 1995 pp. 1301–1307.

[1995c] *Differential and Riemannian Manifolds*, Springer-Verlag 1995.

[1995d] *Introduction to Diophantine Approximations, new expanded edition*, Springer-Verlag 1995.

[1996a] La conjecture de Bateman-Horn, *Gazette des mathématiciens* January 1996 No. 67 pp. 82–84.

[1996b] Comments on Chow's works, *Notices AMS* **43** (1996) No. 10 pp. 1117–1124.

[1996c] (with J. JORGENSON) Extension of analytic number theory and the theory of regularized harmonic series from Dirichlet series to Bessel series, *Math. Ann.* **306** (1996) pp. 75–124.

[1996d] *Topics in Cohomology of groups*, Springer-Verlag, Springer Lecture Notes **1625**, 1996 (English translation and expansion of *Rapport sur la Cohomologie des Groupes*, Benjamin, 1966).

[1997] *Survey of Diophantine Geometry*, Springer-Verlag 1997 (same as *Number Theory III*, with corrections and additions).

[1998] The Kirschner article and HIV: Scientific and journalistic (ir)responsibilities.

[1999a] (with J. JORGENSON) Hilbert-Asai Eisenstein series, regularized products and heat kernels, *Nagoya Math. J.* **153** (1999) pp. 155–188.

[1999b] Response to the Steele Prize, *Notices AMS* **46** No. 4, April 1999 p. 458.

[1999c] *Fundamentals of Differential Geometry*, Springer-Verlag 1999.

[1999d] *Complex analysis*, fourth edition, Springer-Verlag 1999.

[1999e] *Math Talks for Undergraduates*, Springer-Verlag 1999.

The Zurich Lectures

I was invited by Wustholz for a decade to give talks to students in Zurich. I express here my appreciation, also to Urs Stammbach for his translations, and for his efforts in producing and publishing the articles in *Elemente der Mathematik.*

Primzahlen, *Elem. Math.* **47** (1992) pp. 49–61
Die abc-vermutung, *Elem. Math.* **48** (1993) pp. 89–99
Approximationssätze der Analysis, *Elem. Math.* **49** (1994) pp. 92–103
Die Wärmeleitung auf dem Kreis und Thetafunktionen, *Elem. Math.* **51** (1996) pp. 17–27
Globaler Integration lokal integrierbarer Vektorfelder, *Elem. Math.* **52** (1997) pp. 1–11
Bruhat-Tits-Raüme, *Elem. Math.* **54** (1999) pp. 45–63

Articles on Scientific Responsibility

Circular A-21. A history of bureaucratic encroachment, J. Society of Research Administrators, 1984
Questions de responsabilité dans le journalisme scientifique, *Revue du Palais de la Découverte* Paris February 1991 pp. 17–46
Questions of scientific responsibility: The Baltimore case, *J. Ethics and Behavior* **3(1)** (1993) pp. 3–72
The Kirschner article and HIV: Scientific and journalistic (ir)responsibilities, Refused publication by the *Notices AMS*, dated 5 January 1998

Books on Scientific Responsibility

The Scheer Campaign, W.A. Benjamin, 1966
The File, Springer-Verlag, 1981
Challenges, Springer-Verlag, 1998

Math. Ann. 234, 25—44 (1978)

The p-Primary Component of the Cuspidal Divisor Class Group on the Modular Curve $X(p)$

Daniel S. Kubert* and Serge Lang*

Department of Mathematics, Cornell University, Ithaca, NY 14853, USA
Department of Mathematics, Yale University, New Haven, Conn. 06520, USA

Let K be the cyclotomic field $\mathbf{Q}(e^{2\pi i/p})$ of p-th roots of unity. The Galois group (of order $p-1$) operates on the ideal class group, and on the unit group modulo cyclotomic units. Classical problems of number theory are concerned with the eigenspace decomposition of the p-primary part of these groups, and include results of Kummer, Herbrand, and more recently Iwasawa [I] and Ribet [Ri]. Also in recent times, such results have been the object of study in the case of complex multiplication of elliptic curves for the elliptic units, as in Robert [Ro] and Coates-Wiles [CW].

In the present paper we give a theorem analogous to these results, for the p-primary part of the divisor class group generated by the cusps on the modular curve $X(p)$. Serious difficulties involving number theory arising in the classical cases mentioned above have no counterpart in the function field case, so that our result is essentially complete. We rely heavily on the series of papers [KL], which in particular characterized all the geometric units in the function field of $X(p)$. The cuspidal class group consists of the factor group of the divisors (of degree 0) generated by the cusps, modulo the divisors of geometric units. We determine the integral Stickelberger ideal annihilating the p-primary part, and the computations actually involve Gauss sums and the second Bernoulli numbers (whereas in the strictly number theoretic cases, it is the first Bernoulli numbers which occur). After recalling terminology in §1, we give the statement of the main theorem. The role of $G=(\mathbf{Z}/p\mathbf{Z})^*$ in the number-theoretic case is here played by the Cartan group $\mathbf{F}_{p^2}^*$, the multiplicative group of the field with p^2 elements (mod ± 1), which operates in a natural way simply transitively on the cusps, and again has order prime to p. Thus we meet here again an abelian situation, with abelian characters. Ultimately, we can describe completely the structure of the χ-eigenspace for any character χ, in terms of p-divisibility of $B_{2,\bar{\chi}_Z}$, where χ_Z is the restriction of χ to $(\mathbf{Z}/p\mathbf{Z})^*$. Expressing χ as a power of the Teichmuller character reduces the non-triviality to the ordinary Bernoulli numbers B_k because of elementary congruences relating the Bernoulli numbers with characters to those without characters.

* Supported by NSF grants. Kubert is also a Sloan Fellow

0025-5831/78/0234/0025/\$04.00

In some cases when the χ-eigenspace is not zero, the formula for its order contains an extra 1. This is explained in §4 by exhibiting a special subgroup, generated by the quotients of Weierstrass forms.

When projected on $X_1(p)$, this special subgroup disappears, and we have sketched without complete proofs some corresponding results on $X_1(p)$ in a final section. Details and the evaluation of cuspidal class numbers will be carried out in a subsequent paper.

Contents

§1. Statement of the Theorems

We shall assume throughout that p is a prime $\geqq 7$. The primes 2, 3, 5 are excluded for the following reasons. The modular curves $X(2), X(3), X(5)$ have genus 0, so one knows that they have trivial divisor class group. More generally, 2, 3 are always "bad" primes, and in addition, $2 \equiv -2 \pmod{p-1}$ for the prime number $p=5$. The congruence $2 \not\equiv -2$ turns out to be important in tabulating the cases which may arise.

Although the case of the p-primary part of the cuspidal ideal class group on $X(p)$ is by far the most important case, it is even easier to determine the l-primary part for primes $l \neq p$ not dividing p^2-1. Consequently we shall recall some terminology on modular curves $X(N)$ for arbitrary N first, to get the notation straight.

The group $\mathrm{GL}_2(N) = \mathrm{GL}_2(\mathbf{Z}(N)) = \mathrm{GL}_2(\mathbf{Z}/N\mathbf{Z})$ operates on the modular function field of level N. If $\alpha \in \mathrm{GL}_2(N)$ we let σ_α be the corresponding automorphism. Let $q_\tau = e^{2\pi i\tau}$. If

$$\alpha = \begin{pmatrix} 1 & 0 \\ 0 & d \end{pmatrix}$$

with d prime to N, then σ_α is the automorphism on the cyclotomic field $\mathbf{Q}(\mu_N)$ such that

$$\sigma_\alpha : \zeta \mapsto \zeta^d,$$

and σ_α leaves $q_\tau^{1/N}$ fixed. We view the function field as embedded in $\mathbf{Q}(\mu_N)((q_\tau^{1/N}))$ (power series field in $q_\tau^{1/N}$ over the cyclotomic field), so the automorphism σ_α above operates on the coefficients of the q-expansion at the standard prime at infinity, which we denote by P_∞.

In [KL II] we defined the **Cartan group** $C(N)$. If N is a prime p, then $C(N)$ is any regular representation of the multiplicative group $\mathbf{F}_{p^2}^*$ of the field with p^2 elements, thus embedded in $\mathrm{GL}_2(\mathbf{Z}(p))$. We refer to [KL II] for the general definition. We let $C(N) = C$, and write

$$C(N)(\pm) = C(\pm) = C/\pm 1.$$

Then $C(\pm)$ operates simply transitively on the primes at infinity (also called the cusps). Let $\mathscr{D}$ be the free abelian group generated by the cusps, and let $\mathscr{D}_0$ be the subgroup of divisors of degree 0. Let $\mathscr{F}$ be the subgroup of divisors of "units", i.e. modular functions on $X(N)$ which have zeros and poles only at the cusps. If A is an abelian group we let $A^{(p)}$ be its p-primary part, i.e. the subgroup of elements annihilated by p-powers. We wish to determine the structure of $(\mathscr{D}_0/\mathscr{F})^{(p)} = \mathscr{C}^{(p)}$ when N is prime $\geqq 7$ and p does not divide N^2-1.

By the simple transitive action of $C(\pm)$ on the cusps, the group of cuspidal divisors $\mathscr{D}$ may be identified with the group ring $\mathbf{Z}[C(\pm)]$. Our problem consists in describing $\mathscr{F}$ in a suitable manner, and decomposing $(\mathscr{D}/\mathscr{F})$ according to eigenspaces of $C(\pm)$, corresponding to the even characters of C.

We let

$$T: C \to \mathbf{Z}(N)$$

be the map given by

$$T: \begin{pmatrix} a & b \\ c & d \end{pmatrix} \mapsto a.$$

We let $\mathbf{B}_2(X) = X^2 - X + \frac{1}{6}$ be the second Bernoulli polynomial.

Let a_1, a_2 be rational numbers, not both integers. Then we have the Siegel functions (cf. [KL 1] or [L 2]), having q_τ-expansion:

$$g_{a_1,a_2} = -q_\tau^{\frac{1}{2}\mathbf{B}_2(a_1)} e^{2\pi i a_2(a_1-1)/2}(1-q_z)\prod_{n=1}^{\infty}(1-q_\tau^n q_z)(1-q_\tau^n/q_z)$$

where $q_z = e^{2\pi i z}$ and $z = a_1\tau + a_2$. If we measure the order at P_∞ in terms of the local parameter $q_\tau^{1/N}$ then the expansion shows that

$$\operatorname{ord}_{P_\infty} g_{a_1,a_2} = \frac{N}{2}\mathbf{B}_2(\langle a_1\rangle),$$

where $\langle a_1\rangle$ is the smallest number $\geqq 0$ in the residue class mod $\mathbf{Z}$ of a_1.

For $\alpha \in C(\pm)$ represented by the matrix $\begin{pmatrix} a & b \\ c & d \end{pmatrix}$, we define

$$g_\alpha = g_{(1,0)\alpha/N} = g_{a/N,\,b/N}.$$

This is defined up to a constant factor, since a, b, c, d are only defined mod N. Then

$$\operatorname{ord}_{P_\infty} g_\alpha = \frac{N}{2}\mathbf{B}_2\left(\left\langle \frac{T\alpha}{N}\right\rangle\right).$$

We have for $\beta \in C(\pm)$,

$$\sigma_\beta g_\alpha = g_{\alpha\beta}\varepsilon(\alpha,\beta)$$

where $\varepsilon(\alpha,\beta)$ is a 12-th root of unity, irrelevant for us since we shall only be interested in the divisors of functions. Then we have the functorial formula

$$\operatorname{ord}_{\sigma_\beta^{-1}P_\infty} g_\alpha = \operatorname{ord}_{P_\infty}\sigma_\beta g_\alpha = \frac{N}{2}\mathbf{B}_2\left(\left\langle \frac{T(\alpha\beta)}{N}\right\rangle\right).$$

We may therefore write the divisor of g_α in the form

$$(g_\alpha) = \sum \frac{N}{2} \mathbf{B}_2\left(\left\langle \frac{T(\alpha\beta)}{N} \right\rangle\right)(\sigma_\beta^{-1} P_\infty),$$

where the sum is taken over $\beta \in C(\pm)$. In particular, we find

$$(g_\alpha) = \sigma_\alpha \theta(N), \quad \text{where} \quad \theta(N) = \sum \frac{N}{2} \mathbf{B}_2\left(\left\langle \frac{T\beta}{N} \right\rangle\right) \sigma_\beta^{-1} P_\infty,$$

and $\theta(N)$ is the "Stickelberger element". It is often convenient to identify $\theta(N)$ with an element of the group algebra over $\mathbf{Q}$,

$$\theta(N) = \sum_\beta \frac{N}{2} \mathbf{B}_2\left(\left\langle \frac{T\beta}{N} \right\rangle\right) \sigma_\beta^{-1}.$$

It will be useful to define for real x,

$$Q(x) = \mathbf{B}_2\left(\left\langle \frac{x}{N} \right\rangle\right),$$

so that $\theta = \theta(N)$ can be expressed in terms of Q.

If f, g are functions on C, we define the sum

$$S_C(f, g) = \sum_{\alpha \in C} f(\alpha) g(\alpha).$$

Let p be a prime $\geqq 7$ which does not divide the order of $C(\pm)$. In particular, if $N = p$ then $C(p)$ has order $p^2 - 1$, and p itself can be taken as the prime under consideration. We let q be a power of p such that the multiplicative group $\mathbf{F}_q^*$ of the field with q elements has order divisible by the order of $C(\pm)$. If $N = p$ then we take $q = p^2$.

By the theory of quadratic relations of Kubert [Ku], Theorem 7, and [KL IV], the module $\mathbf{Z}_p \otimes \mathscr{F}$ is generated over $\mathbf{Z}_p$ by the divisors of functions of the form

$$g_{(\alpha,\beta)} = \frac{g_{\alpha+\beta} g_{\alpha-\beta}}{g_\alpha^2 g_\beta^2},$$

where (α, β) ranges over pairs such that $\alpha, \beta, \alpha+\beta, \alpha-\beta$ lie in C. Define the divisor

$$\pi(\alpha, \beta) = (\alpha+\beta) + (\alpha-\beta) - 2(\alpha) - 2(\beta)$$

and let I_p be the ideal of $\mathbf{Z}_p[C(\pm)]$ generated by all elements $\pi(\alpha, \beta)$. Then the divisor of $g_{(\alpha,\beta)}$ is given by

$$\pi(\alpha, \beta)\theta = (g_{(\alpha,\beta)}),$$

and we have an isomorphism

$$\mathscr{C}^{(p)} \approx \mathbf{Z}_p[C(\pm)]/I_p \theta(N).$$

Let $\mathfrak{o}_q$ be the ring of integers in the unramified extension of $\mathbf{Z}_p$ whose residue class field has q elements. It will be convenient to tensor all objects with $\mathfrak{o}_q$. Using elementary divisors, we have an isomorphism

$$\mathbf{Z}_p \otimes \mathscr{C}^{(p)} \approx (\mathbf{Z}_p \otimes \mathscr{D}_0)/(\mathbf{Z}_p \otimes \mathscr{F}),$$

and hence an isomorphism

$$\mathfrak{o}_q \otimes \mathscr{C}^{(p)} \approx (\mathfrak{o}_q \otimes \mathscr{D}_0)/(\mathfrak{o}_q \otimes \mathscr{F}).$$

If, as an abelian group, $\mathscr{C}^{(p)}$ is of type $(p^{e_1}, \ldots, p^{e_m})$, then $\mathfrak{o}_q \otimes \mathscr{C}^{(p)}$ is of the same type as a module over $\mathfrak{o}_q$. Thus to determine the structure of $\mathscr{C}^{(p)}$ it suffices to determine $\mathfrak{o}_q \otimes \mathscr{C}^{(p)}$. For any non-trivial even character χ of C we let

$$\mathscr{C}^{(p)}(\chi)$$

be the χ-eigenspace of $\mathfrak{o}_q \otimes \mathscr{C}^{(p)}$. Since $\mathscr{D}/\mathscr{D}_0$ is the eigenspace for the trivial character, we have a direct sum decomposition

$$\mathfrak{o}_q \otimes \mathscr{C}^{(p)} = \bigoplus_{\chi \neq 1} \mathscr{C}^{(p)}(\chi).$$

Let

$$\pi_\chi(\alpha, \beta) = \chi(\alpha + \beta) + \chi(\alpha - \beta) - 2\chi(\alpha) - 2\chi(\beta).$$

We let I_χ be the ideal of $\mathfrak{o}_q$ generated by all elements $\pi_\chi(\alpha, \beta)$. Then on $\mathscr{C}^{(p)}(\chi)$ we have

$$\begin{aligned} \pi(\alpha, \beta)\theta \Big| \mathscr{C}^{(p)}(\chi) &= \pi_\chi(\alpha, \beta) \sum_{\gamma \in C(\pm)} \frac{N}{2} \mathbf{B}_2\left(\left\langle \frac{T\gamma}{N} \right\rangle\right) \bar{\chi}(\gamma) \\ &= \pi_\chi(\alpha, \beta) \frac{N}{4} S_C(\bar{\chi}, Q \circ T). \end{aligned}$$

This gives us the structure of $\mathscr{C}^{(p)}(\chi)$ in terms of the latter sum:

Theorem 1.1. *Let χ be a non-trivial even character of C. Then we have an isomorphism*

$$\mathscr{C}^{(p)}(\chi) \approx \mathfrak{o}_q / N S_C(\bar{\chi}, Q \circ T) I_\chi.$$

In particular, $\mathscr{C}^{(p)}(\chi) = 0$ if and only if $N S_C(\bar{\chi}, Q \circ T) I_\chi = (1)$.

The remaining task is to give the exact order of the ideal

$$N S_C(\bar{\chi}, Q \circ T) I_\chi.$$

As already mentioned, we pass to the most interesting case, $N = p$.

Theorem 1.2. *Let $N = p$ so $X(N) = X(p)$. If $\chi_Z = 1$ then*

$$\mathscr{C}^{(p)}(\chi) = 0.$$

Next we state the result when $\chi_Z \neq 1$. We recall that for any non-trivial character ψ on $\mathbf{Z}(p)^*$, we let

$$B_{2,\psi} = p \sum_{a=1}^{p-1} \psi(a) \mathbf{B}_2\left(\frac{a}{p}\right).$$

We define the **Teichmuller character**

$$\omega : \mathbf{F}_q^* \to \mathfrak{o}_q^*$$

to be the character such that

$$\omega(a) \equiv a \pmod{p}.$$

Then ω has values in the $(q-1)$-th roots of unity. The restriction of ω to $\mathbf{F}_p^*$ is also the Teichmuller character on $\mathbf{F}_p^*$, that is satisfies the same congruence relation.

Any non-trivial *even* character χ on $\mathbf{F}_q^*$ can be written as a power of the Teichmuller character,

$$\chi=\omega^k, \quad 2\leqq k\leqq q-3, \quad k \text{ even}.$$

We define the positive integer $m(k)$ to be the order of the ideal $pS_C(\bar{\chi}, Q\circ T)I_\chi$, so that we have

$$pS_C(\bar{\chi}, Q\circ T)I_\chi=(p^{m(k)}).$$

To determine $m(k)$ in certain cases, we have to introduce more notation. We write

$$k=k_0+k_1 p, \quad 0\leqq k_i\leqq p-1.$$

We note that $k_0+k_1\not\equiv 0 \pmod{p-1}$, otherwise $k\equiv 0 \pmod{p-1}$ and χ_Z is trivial, contrary to assumption. The following table then describes $\mathscr{C}^{(p)}(k)$ and $m(k)$, where $\mathscr{C}^{(p)}(k)=\mathscr{C}^{(p)}(\omega^k)$.

Theorem 1.3. *Assume that χ_Z is not trivial, and again $N=p$.*

Case	k	$\mathscr{C}^{(p)}(k)$	$m(k)$
1	$k=2, 2p, 1+p$	0	$1+\operatorname{ord} B_{2,\bar{\chi}_Z}=0$
2	$k\equiv -2, -2p, -(1+p)$ mod p^2-1	$\neq 0$	$1+\operatorname{ord} B_{2,\bar{\chi}_Z}$
		$k\not\equiv \pm 2, \pm 2p, \pm(1+p) \bmod p^2-1$	
3	$k\equiv 2 \bmod p-1$	0	$1+\operatorname{ord} B_{2,\bar{\chi}_Z}=0$
4	$k\equiv -2 \bmod p-1$	$\mathscr{C}^{(p)}(k)=0\Leftrightarrow p\mid B_{2,\bar{\chi}_Z}$	$\operatorname{ord} B_{2,\bar{\chi}_Z}$
5	$k\not\equiv \pm 2 \bmod p-1$	$\mathscr{C}^{(p)}(k)$ or $\mathscr{C}^{(p)}(-k)\neq 0$	$\operatorname{ord} B_{2,\bar{\chi}_Z}$ if $k_0+k_1<p-1$
6			$1+\operatorname{ord} B_{2,\bar{\chi}_Z}$ if $k_0+k_1>p-1$

The cases have been numbered for ease of reference. Cases 1 and 3 together are those when χ_Z is the square of the Teichmuller character, and $\mathscr{C}^{(p)}(\chi)=0$ in that case. This is analogous to the classical cyclotomic case of p-th roots of unity, when the character is *equal* to the Teichmuller character.

In the table, the orders of the Bernoulli numbers are the orders at p, with respect to the unramified prime in $\mathfrak{o}_q$ from which the Teichmuller character was obtained by reduction mod p.

Part of what we shall prove are the relations

$$1+\operatorname{ord} B_{2,\bar{\chi}_Z}=0$$

under the given congruences (i.e. when $k\equiv 2 \bmod p-1$). These amount to Von Staudt type congruences, which could also be proved by other means than those we use.

Using the same arguments as in [L 2], Chapter XIII, Theorem 3.3, one reduces the Bernoulli numbers with characters to those without characters by means of the following congruences.

Theorem 1.4. *Let* $2 \leqq k \leqq p-2$. *Let* $\omega = \omega_Z$ *be the Teichmuller character on* $\mathbf{Z}(p)^*$. *Then for any integer* $n \geqq 1$ *we have*

$$\frac{1}{n} B_{n, \omega^{k-n}} \equiv \frac{1}{k} B_k \pmod{p}.$$

The proof is the same as in the above reference. When $k = p-1$ then the Bernoulli numbers have poles at p and congruences mod algebraic integers can be derived in a similar way.

In the next section, we give a sequence of lemmas which describe the orders of certain exponential sums. After that, we apply these lemmas to prove the validity of the table. Suppose last that $l \neq p$ is another prime, $l \geqq 7$.

Theorem 1.5. *Assume that* l *does not divide* $p^2 - 1$. *Let* χ *be a non-trivial even character of* $C(p)$.

(i) *If* χ_Z *is trivial, then* $\mathscr{C}^{(l)}(\chi) = 0$.

(ii) *Suppose* χ_Z *is non-trivial. Then* $I_\chi = (1)$ *and*

$$\operatorname{ord}_l pS_C(\bar{\chi}, Q \circ T) I_\chi = \operatorname{ord}_l B_{2, \bar{\chi}_Z}.$$

The isomorphism of Theorem 1.1 yields

$$\mathscr{C}^{(l)}(\chi) \approx \mathfrak{o}_{l^n} / B_{2, \bar{\chi}_Z} \mathfrak{o}_{l^n}.$$

The proof of Theorem 1.5 is a direct consequence of Theorem 2.1, and the fact that the character sums are divisible only by primes dividing p.

Observe that the basic problem raised by Iwasawa [I], p. 176, in the cyclotomic case, whether the ideal class group (-1)-eigenspace is cyclic over the group ring, does not arise here because we know how to get the cusps in a principal homogeneous way from the Cartan group. The analogy of the situations is of course striking, the numbers $B_{2, \bar{\chi}}$ in the modular case playing the role of the numbers $B_{1, \bar{\chi}}$ in the cyclotomic case. Similar algebraic geometric interpretations for the $B_{k, \bar{\chi}}$ in higher dimensional cases remain to be discovered.

The index relation in the present case (and for higher k) analogous to [I] will be studied in a subsequent paper of the series. For instance, we shall find the following expression for the order $h(N)$ of the cuspidal divisor class group on $X(N)$ when $N = p^n$ is a prime power with $p \geqq 5$:

$$h(N) = \frac{6N^2 N^{c(N)}}{c(N)} \prod_{\chi \neq 1} \tfrac{1}{2} B_{2, \chi},$$

where:

$c(N) =$ order of $C(N)/\pm 1 =$ number of cusps;

the characters χ range over the non-trivial characters of $C(N)/\pm 1$;

$$B_{2, \chi} = \sum_{\alpha \in C(N)/\pm 1} \chi(\alpha) \mathbf{B}_2\left(\left\langle \frac{T\alpha}{N} \right\rangle\right).$$

This class number $h(N)$ will be obtained as an index of the ideal of relations in the group ring $\mathbf{Z}[C(N)/\pm 1]$.

Observe that the Bernoulli number $B_{2, \chi}$ is taken *with respect to the Cartan group*. We also denote it by $S(\chi, Q \circ T)$.

When N is composite, one has to look at the ideal generated by Stickelberger elements of all intermediate levels, as in [KL II], [KL III], [KL D], and Sinnott [Si].

Finally, Coates and Sinnott have drawn our attention to their forthcoming paper [C–S], where they prove that a Stickelberger ideal formed also with B_2 annihilates K_2 of the ring of integers of, say, a cyclotomic field. This raises the question whether one can establish a direct connection between such K_2 and the cuspidal divisor class group.

§ 2. p-Adic Orders of Character Sums

In this section we evaluate the order of $pS_C(\bar{\chi}, Q\circ T)$ by means of a sequence of lemmas.

The group C contains $Z=(\mathbf{Z}/p\mathbf{Z})^*$, and if f is a function on C we let f_Z denote its restriction to Z. We may define a sum $S_Z(f_Z, g_Z)$ by a similar sum as $S_C(f,g)$ taken over elements of Z.

We let λ be a non-trivial character of the additive group $\mathbf{F}_q$, restricting to a non-trivial character of $\mathbf{F}_p$.

The next theorem recalls a fact proved in [KL D] for all levels. We reproduce a proof (much simpler) for prime level, following the original suggestion of Tate.

Theorem 2.1. *Assume that χ is non-trivial, and even.*

(i) *If χ_Z is trivial, then*

$$pS_C(\bar{\chi}, Q\circ T)=\frac{p^2-1}{6p}S_C(\bar{\chi}, \lambda\circ T).$$

(ii) *If χ_Z is non-trivial, then*

$$pS_C(\bar{\chi}, Q\circ T)=\frac{1}{p}S_C(\bar{\chi}, \lambda\circ T)S_Z(\chi_Z, \bar{\lambda}_Z)B_{2,\bar{\chi}_Z}.$$

Proof. Let f be any function on $\mathbf{Z}(p)$. Recall the Fourier inversion:

$$f(x)=\sum_{y\in\mathbf{Z}(p)}\hat{f}(y)\lambda(yx)$$

$$\hat{f}(y)=\frac{1}{p}\sum_{x\in\mathbf{Z}(p)}f(x)\lambda(-xy).$$

For purpose of this proof, we write λ instead of λ_Z, because no Fourier transform other than that on $\mathbf{Z}(p)$ will occur. We consider the sum

$$\begin{aligned}S_C(\chi, f\circ T)&=\sum_{\alpha\in C}\chi(\alpha)f(T\alpha).\\&=\sum_{\alpha}\sum_{y}\chi(\alpha)\hat{f}(y)\lambda(T(y\alpha))\\&=\sum_{\alpha}\sum_{y\neq 0}\chi(\alpha)\hat{f}(y)\lambda(T(y\alpha))+\sum_{\alpha}\chi(\alpha)\hat{f}(0).\end{aligned}$$

Since χ is assumed non-trivial, the sum on the right is 0, so changing the order of summation and making a change of variables $\alpha \mapsto y^{-1}\alpha$, we find:

$$= \sum_{y \neq 0} \sum_{\alpha} \chi(\alpha)\bar{\chi}(y)\hat{f}(y)\lambda(T\alpha)$$

$$= \sum_{y \neq 0} \bar{\chi}(y)\hat{f}(y) \sum_{\alpha} \chi(\alpha)\lambda(T\alpha). \tag{1}$$

$$= S_Z(\bar{\chi}_Z, \hat{f}) S_C(\chi, \lambda \circ T).$$

Suppose that χ_Z is trivial. Then

$$\sum_{y \neq 0} \bar{\chi}(y)\hat{f}(y) = \sum_{y \neq 0} \hat{f}(y) = f(0) - \hat{f}(0).$$

In particular, if $f = Q$ we obtain

$$Q(0) - \hat{Q}(0) = \frac{1}{6} - \frac{1}{6p^2} = \frac{p^2 - 1}{6p^2},$$

evaluating $\hat{Q}(0)$ by means of the distribution relation for Bernoulli polynomials, cf. for instance [L 2], Chapter XIII, formula **B6.** This proves (i).

Suppose that χ_Z is non-trivial. Then

$$S_Z(\bar{\chi}_Z, \hat{f}) = \frac{1}{p} \sum_{y \neq 0} \sum_{x} \bar{\chi}(y)\hat{f}(x)\lambda(-xy)$$

$$= \frac{1}{p} \sum_{y \neq 0} \sum_{x \neq 0} \bar{\chi}(y) f(x)\lambda(-xy) + 0$$

$$= \frac{1}{p} \sum_{x \neq 0} \sum_{y \neq 0} \bar{\chi}(y)\chi(x) f(x)\lambda(-y)$$

$$= \frac{1}{p} S_Z(\bar{\chi}, \bar{\lambda}) S_Z(\chi, f). \tag{2}$$

But by definition,

$$B_{2,\chi} = p S_Z(\chi, Q).$$

This proves (ii).

Lemma 1. (i) *If* $k = 2$, $2p$, *or* $1 + p$ *then* $I_\chi = (p)$.

(ii) *For other values of* k, $I_\chi = (1)$.

Proof. In both parts we may reduce the value of the character mod p to get a value for $\pi_\chi(x, y)$ in the finite field, which is

$$(x + y)^k + (x - y)^k - 2x^k - 2y^k, \qquad x, y \in \mathbf{F}_{p^2}.$$

This is a homogeneous polynomial in which the terms of degree k and $k - 1$ in x vanish. Thus the polynomial has degree $\leqq k - 2$ in x, and is equal to

$$k(k-1)x^{k-2}y^2 + \text{lower degree terms in } x.$$

Let $y = 1$ to get a polynomial in one variable

$$(x + 1)^k + (x - 1)^k - 2x^k - 2 = k(k-1)x^{k-2} + \text{lower terms}.$$

If the polynomial is not identically 0, then there is some element $x \in \mathbf{F}_q$, $x \neq 0$, ± 1, at which the polynomial does not take the value 0. Hence we must show that the polynomial is identically 0 only for the three cases $k=2$, $2p$, or $1+p$. Write $k=p^e n$ with n prime to p. Because $k \leqq p^2-3$ we have $e=0$ or 1, and the polynomial can be written in $t=x^{p^e}$ as

$$(t+1)^n+(t-1)^n-2t^n-2=n(n-1)t^{n-2}+\text{lower terms}.$$

Assume that $n \neq 2$. The polynomial is identically 0 only if $n \equiv 1 \pmod p$, so $n=1+mp, m \geqq 1$ and m is odd since k is even. Then $e=0$ (otherwise $n>p^2$). We write the polynomial in the form

$$(t^p+1)^m(t+1)+(t^p-1)^m(t-1)-2t^{pm}t-2.$$

We leave it to the reader to verify that this polynomial is identically 0 only for $m=1$. This concludes the list of cases when I_χ is divisible by p.

Finally we want to show that in Case (i) the ideal is not divisible by p^2. By an inductive argument it can be shown that for any positive integer n not divisible by p, the formal expression

$$(nx)-n^2(x)$$

can be written as linear combination with integer coefficients of "parallelograms"

$$(x+y)+(x-y)-2(x)-2(y),$$

see Kubert [Ku]. Hence the ideal I_χ contains values

$$\chi(nx)-n^2\chi(x).$$

Pick an integer $n=\zeta+pz$, where z is a p-adic unit. Then

$$\zeta^2=\chi(n)$$

is a $(p-1)$-th root of unity. Then I_χ contains

$$\chi(x)(\zeta^2-n^2)$$

which is not divisible by p^2, as desired.

In the next lemma, we use a primitive root w in $\mathbf{F}_p^*$, and we let

$$r=w^{k+2}.$$

Lemma 2. *Assume that χ_Z is non-trivial. Then B_{2,χ_Z} satisfies the following congruences* mod $\mathbf{Z}_p$.

$$B_{2,\chi_Z} \equiv \begin{cases} \dfrac{p-1}{p} & \text{if } k+2 \equiv 0 \pmod{p-1} \\[2mm] \dfrac{1}{p}\dfrac{r^{p-1}-1}{r-1} & \text{if } k+2 \not\equiv 0 \pmod{p-1}. \end{cases}$$

The value in case $k+2 \not\equiv 0 \pmod{p-1}$ is p-integral.

Proof. We have

$$B_{2,\chi_Z} \equiv p \sum_{a=1}^{p-1} \left(\frac{a}{p}\right)^2 \chi_Z(a) \pmod{\mathbf{Z}_p}$$
$$\equiv \frac{1}{p} \sum_{a=1}^{p-1} a^{k+2} \pmod{\mathbf{Z}_p}.$$

When a ranges over $1, \dots, p-1$ the residue classes range over all elements of $\mathbf{F}_p^*$, so that the sum is a geometric series which yields the expression stated in the lemma.

Lemma 3. *Assume that χ_Z is non-trivial. Then $B_{2,\bar{\chi}_Z} I_\chi$, which is $B_{2,\omega^{-k}}$, is equal to the following:*

(i) $pB_{2,\bar{\chi}_Z} \mathfrak{o}_q = (1)$ *if* $k = 2, 2p, 1+p$.

Otherwise, i.e. if $k \neq 2, 2p, 1+p$:

(ii) $B_{2,\bar{\chi}_Z} \mathfrak{o}_q = \frac{1}{p} \mathfrak{o}_q$ *if* $k \equiv 2 \bmod p-1$

(iii) $B_{2,\bar{\chi}_Z} \mathfrak{o}_q =$ *p-integral* *if* $k \not\equiv 2 \bmod p-1$.

Proof. This is immediate from Lemma 1 and Lemma 2, distinguishing all the possible cases.

Next we deal with the divisibility property of the ordinary Gauss sums. These are well known. The references [L 1] Chapter IV, §3, formula **GS 2** and Theorem 9 suffice for our needs here.

We continue to assume that χ_Z is non-trivial. Let

$$S = S_C(\bar{\chi}, \lambda \circ T) S_Z(\chi_Z, \bar{\lambda}).$$

Then **GS 2** shows that

$$S\bar{S} = p^3.$$

The factors other than S in Theorem 2.1 are all elements of the field of $(p-1)$-th roots of unity, and have therefore integral orders at a p-adic valuation because that field is unramified at p. Hence the order of S is in $\mathbf{Z}$. The relation $S\bar{S} = p^3$ shows that

$$\operatorname{ord}_p S = 0, 1, 2, \text{ or } 3.$$

Lemma 4. *The cases* $\operatorname{ord} S = 0$, $\operatorname{ord} \bar{S} = 3$ *(or vice versa) do not occur. In particular* $p^{-1}S$ *is p-integral.*

Proof. By Theorem 9 loc. cit. we know that the Gauss sums $S_C(\bar{\chi}, \lambda \circ T)$ and $S_Z(\chi_Z, \bar{\lambda})$ are divisible by all the primes dividing p in the field of $(q-1)$ and $(p-1)$-th roots of unity respectively. Therefore the order cannot be 0, and it cannot be 3 as one sees by replacing χ with its conjugate. This proves the lemma.

§3. Proof of the Theorems

We put together the lemmas of §2 in order to prove the theorems. We need the exact order given for Gauss sums in terms of the Teichmuller character. We write

$$k = k_0 + k_1 p, \qquad 0 \leqq k_i \leqq p-1$$

and define

$$s_q(k)=k_0+k_1\,.$$

We also have $s_p(k)$. In that case, we have to select

$$k'\equiv -k\ (\mathrm{mod}\ p-1)\quad\text{and}\quad 0\leqq k'<p-1\,.$$

Then

$$s_p(-k)=k'\,.$$

By Theorem 9 loc. cit., we know:

$$\operatorname{ord} S_C(\omega^{-k},\lambda\circ T)=\frac{1}{p-1}(k_0+k_1)$$

$$\operatorname{ord} S_Z(\omega_Z^k,\bar{\lambda})=\frac{1}{p-1}k'\,.$$

Let us start with the case χ_Z trivial. Then $k\equiv 0\ (\mathrm{mod}\ p-1)$, and then $k_0+k_1\equiv 0\ (\mathrm{mod}\ p-1)$. But since $0\leqq k_i\leqq p-1$, we get

$$k_0+k_1=p-1\quad\text{or}\quad 2(p-1)\,.$$

Hence

$$\operatorname{ord} S_C(\bar{\chi},\lambda\circ T)=1\,.$$

By Lemma 1, in the present case we have $I_\chi=(1)$. Using Theorem 1.1 and Theorem 2.1(i) concludes the proof of Theorem 1.2.

Assume that χ_Z is non-trivial. As the technique in all cases is the same, we shall work out in detail the first case and the last case in the table of Theorem 1.3, and leave the others to the reader. By Theorem 1.1 and Theorem 2.1(ii) we have to determine the order of

$$pS_C(\bar{\chi},Q\circ T)I_\chi=\frac{1}{p}S_C(\bar{\chi},\lambda\circ T)S_Z(\chi_Z,\bar{\lambda}_Z)B_{2,\bar{\chi}_Z}I_\chi\,.$$

Let $k=2$, or $2p$, or $1+p$. Then $k_0+k_1=2$. Furthermore,

$$\operatorname{ord} S_C(\omega^{-k})=\operatorname{ord} S_C(\bar{\chi})=\frac{1}{p-1}(k_0+k_1)=\frac{2}{p-1}\,.$$

$$\operatorname{ord} S_Z(\omega^k)=\operatorname{ord} S_Z(\chi)=\frac{1}{p-1}s_p(-2)=\frac{p-3}{p-1}\,.$$

Hence $\operatorname{ord} S=1$. Finally Lemma 2 shows that

$$\operatorname{ord} B_{2,\bar{\chi}}=-1\,,$$

and Lemma 1 shows that $I_\chi=(p)$. This concludes the proof in the present case.

Next, suppose that $k\not\equiv\pm 2\ (\mathrm{mod}\ p-1)$, and $k_0+k_1<p-1$. Then

$$k'=p-1-(k_0+k_1)\,.$$

Hence $\operatorname{ord} S=1$. Lemma 2 shows that $B_{2,\bar{\chi}_Z}$ is p-integral, and Lemma 1 shows that $I_\chi=(1)$. This gives the entry in the table when $k_0+k_1<p-1$.

On the other hand, suppose that $k_0+k_1>p-1$. Then

$$0<k_0+k_1-(p-1)<p-1.$$

Since $k\equiv k_0+k_1(\operatorname{mod} p-1)$ we get

$$-k\equiv k'=2(p-1)-(k_0+k_1).$$

Hence $\operatorname{ord} S=2$. Again, Lemma 2 shows that $B_{2,\bar{\chi}z}$ is p-integral, and Lemma 1 shows that $I_\chi=(1)$. This gives the entry in the table when $k_0+k_1>p-1$, and concludes the proof of the theorem.

§4. The Special Group

In the table of Theorem 1.3 we note that the divisor class group is not a priori zero in Cases 2, 4, 5, 6. In Cases 2 and 6 the order is $1+\operatorname{ord} B_{2,\bar{\chi}z}$. In this section, we describe a special component of order 1 occurring precisely in those cases.

We continue to work on $X(p)$. Notationally, it is more convenient to express modular points in terms of an elliptic curve A, over the complex numbers, assumed in Weierstrass form determined by a lattice, rather than in terms of lattices. The Weierstrass form then determines its associated $\wp$-function. We let $\mathfrak{C}$ denote a cyclic subgroup of order p in A, and let A_p be the subgroup of all points of order p. A pair of points $a, b\in A_p$ is said to be **admissible** for $\mathfrak{C}$ if it satisfies the following conditions.

The cyclic groups (a), (b) generated by a, b are equal. The points a, b, $a\pm b$ are not in $\mathfrak{C}$.

For the moment we let p be any prime $\geqq 5$. Then admissible pairs obviously exist. We define

$$\wp[a,b;\mathfrak{C}]=\wp(a,A/\mathfrak{C})-\wp(b,A/\mathfrak{C}).$$

Write $b=ra$ for some integer r. The association

$$(a,A)\mapsto\wp[a,ra;\mathfrak{C}]=\wp(a,A/\mathfrak{C})-\wp(ra,A/\mathfrak{C})$$

defines a modular form on $X(p)$ whose divisor is the p-th multiple of a divisor on $X(p)$; cf. [KL I], §3, Theorem 3.

Theorem 4.1. *For (a_1,b_1) admissible for $\mathfrak{C}_1$ and (a_2,b_2) admissible for $\mathfrak{C}_2$, the expression*

$$\frac{\wp[a_1,b_1;\mathfrak{C}_1]}{\wp[a_2,b_2;\mathfrak{C}_2]}$$

defines a modular function on $X(p)$. If $p>5$ then the p-th root of this function is not modular of any level.

Proof. This is an easy consequence of [KL IV], and the details are omitted.

From the definition of the Klein forms (which are equal to the sigma function times an exponential factor) and the Siegel functions (having an extra factor η^2), we have the factorization

$$\wp[a,b;\mathfrak{C}] = \frac{\mathfrak{k}(a+b, A/\mathfrak{C})\mathfrak{k}(a-b, A/\mathfrak{C})}{\mathfrak{k}^2(a, A/\mathfrak{C})\mathfrak{k}^2(b, A/\mathfrak{C})}$$

$$= \eta^4(A/\mathfrak{C})\frac{g(a+b, A/\mathfrak{C})g(a-b, A/\mathfrak{C})}{g^2(a, A/\mathfrak{C})g^2(b, A/\mathfrak{C})} \tag{1}$$

$$= \lambda_1 \eta^4(A/\mathfrak{C}) \prod_{c\in\mathfrak{C}} \frac{g(a+b+c, A)g(a-b+c, A)}{g^2(a+c, A)g^2(b+c, A)} \tag{2}$$

for some constant λ_1. This last step is the "distribution relation" of Klein-Ramachandra-Robert, cf. Robert [R] and [KL III], Theorem 1:

Dist 1. $\prod_{c\in\mathfrak{C}} g(a+c, A) = g(a, A/\mathfrak{C})$.

Directly from the q-expansion, one can also verify

Dist 2. $\prod_{\substack{c\in\mathfrak{C}(\pm)\\ c\neq 0}} g(c, A) = \lambda_2 \frac{\eta(A/\mathfrak{C})}{\eta(A)}$

for some constant λ_2. As usual, $\mathfrak{C}(\pm)$ denotes $\mathfrak{C}\pm 1$.

For $p=5$, because of the obvious restriction for an admissible pair, one has the additional relation

$$\prod_{c\in\mathfrak{C}} g(a+b+c, A)g(a-b+c, A) = \lambda_3 \prod_{\substack{c\in\mathfrak{C}(\pm)\\ c\neq 0}} \frac{1}{g(c, A)}.$$

So we obtain:

Theorem 4.2. *For* $p=5$,

$$\left(\frac{\eta(A/\mathfrak{C}_1)}{\eta(A/\mathfrak{C}_2)}\right)^5 = \frac{\wp[a_1, b_1; \mathfrak{C}_1]}{\wp[a_2, b_2; \mathfrak{C}_2]}\lambda_4.$$

The cuspidal divisor class group is trivial when $p=5$, and one expects a priori to have the divisor of the function in Theorem 4.1 expressible as a fifth power, of a modular function of some level. We gave Theorem 4.2 to show explicitly what this function is.

In the sequel we assume that $p \geqq 7$.

We define the **special divisor group** to be the group generated by the divisors of all the functions as in Theorem 4.1, for all choices of $(a_1, b_1; \mathfrak{C}_1)$ and $(a_2, b_2; \mathfrak{C}_2)$, and denote it by $\mathscr{S}$. By [KL I], §3, Theorem 3, we have

$$\mathscr{S} \subset p\mathscr{D}_0.$$

Factoring out the special divisor class group from the full cuspidal divisor class group amounts to considering the factor group

$$\mathscr{C}_{\mathscr{S}} = \mathscr{D}_0/(\mathscr{F} + p^{-1}\mathscr{S}).$$

Again we tensor with $\mathfrak{o}_q$ (where $q=p^2$), and denote by $\mathscr{C}_{\mathscr{S}}^{(p)}(\chi)$ the χ-eigenspace. We want to know when $\mathscr{C}_{\mathscr{S}}^{(p)}(\chi)$ is different from $\mathscr{C}^{(p)}(\chi)$. This again amounts to determining the order of the associated $\mathfrak{o}_q$-ideal, which can differ by at most 1 in the two groups. A priori, we can omit from consideration those cases when $\mathscr{C}^{(p)}(\chi)=0$, so when $\chi_Z=1$, or Cases 1 and 3 of the table in Theorem 1.3.

Theorem 4.2. *Assume that $\chi_Z\neq 1$. Among the Cases 2, 4, 5, 6 of Theorem 1.3, the group $\mathscr{C}_{\mathscr{S}}^{(p)}(\chi)$ differs (necessarily by order 1) from $\mathscr{C}^{(p)}(\chi)$ precisely in Cases 2 and 6.*

The proof will be quite similar to the preceding proofs. In the rest of this section, we describe more closely the defining Stickelberger ideal for the new group $\mathscr{C}_{\mathscr{S}}^{(p)}(\chi)$, and then we analyse its order to finish the proof.

We denote by Γ a cyclic subgroup of $\mathbf{F}_q=\mathbf{F}_{p^2}$. Its elements are of the form

$$r\gamma_0, \quad r=0,\dots,p-1$$

for some single element γ_0.

From formula (2), we see that there exist two cyclic subgroups Γ_1 and Γ_2 such that the divisor of the Weierstrass function of Theorem 4.1 has the following form. We identify cuspidal divisors with elements of the group ring $\mathbf{Q}[C]$.

$$\frac{1}{p}\operatorname{div}\frac{\wp[a_1,b_1;\mathfrak{C}_1]}{\wp[a_2,b_2;\mathfrak{C}_2]}=\frac{1}{2}\sum_{\gamma\in C}\left(Q(T\gamma)-\frac{1}{6}\right)\gamma^{-1}$$
$$+\pi(\alpha_1,\beta_1;\Gamma_1)-s(\Gamma_1)-\pi(\alpha_2,\beta_2;\Gamma_2)+s(\Gamma_2), \tag{3}$$

where we use the notation:

$$\pi(\alpha,\beta;\Gamma)=\sum_{\gamma\in\Gamma}\{(\alpha+\beta+\gamma)+(\alpha-\beta+\gamma)-2(\alpha+\gamma)-2(\beta+\gamma)\}$$

and

$$s(\Gamma)=\sum_{\substack{\gamma\in\Gamma\\ \gamma\neq 0}}(\gamma).$$

The expression $-1/6$ occurs because there is no factor of the Dedekind η-function in the quotient of Weierstrass forms, so the usual term with B_2 has to be corrected by the constant term. It will disappear again when we evaluate with a non-trivial character.

The elements $\alpha,\beta\in C$ correspond to the "vectors" a,b respectively, and form an admissible pair for Γ, namely:

The cyclic groups generated by α and β are equal.
The elements $\alpha,\beta,\alpha\pm\beta$ are not in Γ.

For any character χ on the Cartan group C, we let

$$\pi_\chi(\alpha,\beta;\Gamma)=\sum_{\gamma\in\Gamma}\{\chi(\alpha+\beta+\gamma)+\chi(\alpha-\beta+\gamma)-2\chi(\alpha+\gamma)-2\chi(\beta+\gamma)\}.$$

We assume that χ_Z is a non-trivial even character, in which case

$$\chi(s(\Gamma))=\sum\chi(\gamma)=0.$$

because $\chi(s(\Gamma))=\sum_r\chi_Z(r)\chi(\gamma_0)=0.$

We let $I_{\chi,\mathscr{S}}$ be the ideal generated by all expressions

$$\pi_\chi(\alpha_1, \beta_1; \Gamma_1) - \pi_\chi(\alpha_2, \beta_2; \Gamma_2)$$

with arbitrary groups of order p and admissible pairs for them. In this manner we obtain the analogue of Theorem 1.1.

Theorem 4.3. *Let χ be an even character on C such that χ_Z is non-trivial. Then we have an isomorphism*

$$\mathscr{C}_{\mathscr{S}}^{(p)}(\chi) \approx \mathfrak{o}_q / pS_C(\bar{\chi}, Q \circ T)(I_\chi + p^{-1} I_{\chi,\mathscr{S}}).$$

Our problem is therefore to determine the order of $I_{\chi,\mathscr{S}}$, and in particular to find those cases when $p^{-1} I_{\chi,\mathscr{S}}$ is strictly larger than I_χ. The arguments will simply refine those of §2, Lemma 1, and §3, but will follow the same pattern.

To begin with, we simplify the generators for $I_{\chi,\mathscr{S}}$.

Lemma 5. *The ideal $I_{\chi,\mathscr{S}}$ is generated by all elements*

$$\pi_\chi(\alpha, \beta; \Gamma).$$

In other words, we don't have to take differences of such elements in the definition of the ideal.

Proof. Let $r \in \mathbf{Z}(p)^*$ be such that $\chi_Z(r) \neq 1$. Then

$$\pi_\chi(\alpha, \beta; \Gamma) - \pi_\chi(r\alpha, r\beta; \Gamma) = (1 - \chi(r)) \pi_\chi(\alpha, \beta; \Gamma)$$

has the same p-order as $\pi_\chi(\alpha, \beta; \Gamma)$. The lemma is then obvious.

Lemma 6. *Let $\chi = \omega^k$ with $2 < k < p^2 - 1$. Assume χ_Z non-trivial, and $k \not\equiv 2 \pmod{p-1}$. Then $I_{\chi,\mathscr{S}} = (1)$ if and only if $k_0 + k_1 > p - 1$. If $k_0 + k_1 < p - 1$ then $p | I_{\chi,\mathscr{S}}$.*

Proof. Fix Γ and omit it from the notation. As we are only interested in whether $I_{\chi,\mathscr{S}} = (1)$ or not, we work mod p, so $\pi_\chi(x, y)$ is now interpreted mod p, lying in the finite field. Then

$$\pi_\chi(x, y) = \sum_{z \in \Gamma} \sum_{j=0}^{k} [(x+y)^j z^{k-j} + (x-y)^j z^{k-j} - 2x^j z^{k-j} - 2y^j z^{k-j}].$$

If $j = k$ the sum over z is 0 because of a factor p. The elements of Γ are of the form rz_0 with r in $\mathbf{F}_p$. If

$$k - j \not\equiv 0 \pmod{p-1}$$

then

$$\sum_z z^{k-j} = 0.$$

Interchanging the order of summation shows that

$$\pi_\chi(x, y) = -\sum_{\substack{0 \leqq j < k \\ j \equiv k \pmod{p-1}}} \binom{k}{j} z_0^{k-j} [(x+y)^j + (x-y)^j - 2x^j - 2y^j].$$

The expression on the right is homogeneous of degree k in z_0, x, y. We divide by x^k. Since by assumption of admissibility, the elements x, y generate the same group, putting $x=1$ then implies that y is in the prime field, and so $y^j=y^k$. Hence

$$\pi_\chi(1, y) = -\sum \binom{k}{j} z_0^{k-j}[(1+y)^k+(1-y)^k-2-2y^k],$$

and the sum is taken over j such that

$$0 \leqq j < k \quad \text{and} \quad j \equiv k \ (\mathrm{mod}\ p-1).$$

By Lemma 1, the expression in y is not identically zero as function on the prime field. We are therefore reduced to considering the polynomial

$$P(z) = \sum \binom{k}{j} z^{k-j}$$

summed over the above values of j, and determining when this polynomial is identically zero. Observe that P has degree $\leqq p^2-p$. Since $P(0)=0$, if P is not identically zero, then there exists an element z_0 not in $\mathbf{F}_p$ such that $P(z_0) \neq 0$, and this element generates a group Γ, from which we see that the ideal $I_{\chi, \mathscr{S}}$ is the unit ideal. Thus to prove Lemma 6, it suffices to prove:

P is identically zero if and only if $k_0+k_1 < p-1$.

We obviously have $\mathrm{ord}\, k! = k_1$. Suppose first that $k_0+k_1 < p-1$. Then the possible values for j are of the form

$$j = k_0 + k_1 + r(p-1) \quad \text{with} \quad r \leqq k_1 - 1,$$

because $j<k$. Hence $\mathrm{ord}\, j! = r$, and one verifies at once that $\mathrm{ord}(k-j)! = k_1 - r - 1$. Hence p divides the binomial coefficient, and P is identically zero. On the other hand, suppose that $k_0+k_1 > p-1$. Let $j = k_0 + k_1 - (p-1)$. For this value of j, the binomial coefficient is not divisible by p and hence the polynomial P is not identically zero. This concludes the proof of Lemma 6.

Lemma 7. *Among Cases 2, 4, 5, 6 of the Table in Theorem 1.3, we have $I_{\chi, \mathscr{S}} = (1)$ exactly in Cases 2 and 6. In the other cases, the ideal $I_{\chi, \mathscr{S}}$ is divisible by p.*

Proof. In each case one has to determine k_0 and k_1, and verify that $k_0+k_1 > p-1$ precisely in Cases 2 and 6. The matter is trivial. For example, in Case 2 we have

$$k \equiv -2, -2p, -(1+p) \ (\mathrm{mod}\ p^2-1),$$

so $k = p^2-3$, p^2-2p-1, p^2-p-2. The determination of k_0, k_1 is immediate and shows that their sum is $>p-1$. We leave the rest to the reader.

By Lemma 1, we know that $I_\chi = (1)$ in Cases 2 and 6. Hence

$$I_\chi + p^{-1} I_{\chi, \mathscr{S}} \neq I_\chi$$

precisely in Cases 2 and 6. This concludes the proof of Theorem 4.2.

§5. The Special Group Disappears on $X_1(p)$

We shall prove that the direct image of the special divisor group $\mathcal{S}$ from $X(p)$ to $X_1(p)$ is contained in the p-th multiple of the principal divisors on $X_1(p)$. In particular, this shows:

Theorem 5.1. *The direct image of the special divisor class group*

$$(\mathcal{F}+p^{-1}\mathcal{S})/\mathcal{F}$$

into the cuspidal divisor class group on $X_1(p)$ is trivial.

Proof. We have to analyse the norm of the forms

$$\wp(a, A/\mathfrak{C}) - \wp(b, A/\mathfrak{C}) = \wp[a, b; \mathfrak{C}]$$

from $X(p)$ to $X_1(p)$, and show that they are p-th powers (up to constant factors). We distinguish two cases.

Case 1. Under the identification $A = A_\tau \approx \mathbf{C}/[\tau, 1]$, the group $\mathfrak{C}$ is $\mathfrak{C}_\tau$, generated by $1/p$. Then $\wp[a, b; \mathfrak{C}_\tau]$ is a form on $X_1(p)$, and on $X(p)$ by pull-back from the projection

$$X(p) \to X_1(p).$$

Consequently the norm is the p-th power,

$$\text{Norm of } \wp[a, b; \mathfrak{C}] = \wp[a, b; \mathfrak{C}]^p.$$

Case 2. The group $\mathfrak{C}'$ is not $\mathfrak{C}_\tau$. Then for some constant λ,

$$\text{Norm } \wp[a, b; \mathfrak{C}'] = \lambda(\wp(a, A) - \wp(b, A))^p.$$

Before giving the proof we recall other distribution relations. We let $\mathfrak{D}$ also range over cyclic subgroups of order p in A.

Dist 3. *Let $a \in \mathfrak{C}$. Then*

$$\prod_{\mathfrak{D} \neq \mathfrak{C}} \prod_{d \in \mathfrak{D}} g(a+d, A) \prod_{\substack{c \in \mathfrak{C} \\ c \neq 0}} g(c, A) = \lambda_5 g^p(a, A).$$

This is obvious by separating the terms with $d \in \mathfrak{D}$, $d \neq 0$, and $d = 0$. The product over the former yields a constant, and the product over the latter yields $g^p(a, A)$.

Dist 4. $\prod_{\mathfrak{D}} \eta^4(A/\mathfrak{D}) = \eta^{4(p+1)}(A)$.

This is obvious by looking at the q-expansions.

We now go to the proof of Case 2. We let $\mathfrak{C}=\mathfrak{C}_\tau$. By the periodicity of the Weierstrass function, we can then choose $a, b \in \mathfrak{C}$. Then the desired norm is equal to:

$$\begin{aligned}
&\prod_{\mathfrak{D}\neq\mathfrak{C}} (\wp(a, A/\mathfrak{D})-\wp(b, A/\mathfrak{D}))\\
&\quad=\prod_{\mathfrak{D}\neq\mathfrak{C}} \frac{g(a+b, A/\mathfrak{D})g(a-b, A/\mathfrak{D})}{g^2(a, A/\mathfrak{D})g^2(b, A/\mathfrak{D})}\eta^4(A/\mathfrak{D})\\
&\quad=\prod_{\mathfrak{D}\neq\mathfrak{C}}\prod_{d\in\mathfrak{D}} \frac{g(a+b+d, A/\mathfrak{D})g(a-b+d, A/\mathfrak{D})}{g^2(a+d, A/\mathfrak{D})g^2(b+d, A/\mathfrak{D})}\prod_{\mathfrak{D}\neq\mathfrak{C}}\eta^4(A/\mathfrak{D})\\
&\quad=\lambda_6\frac{g^p(a+b, A)g^p(a-b, A)}{g^{2p}(a, A)g^{2p}(b, A)}\prod_{\substack{c\in\mathfrak{C}\\ c\neq 0}} g^2(c, A)\prod_{\mathfrak{D}\neq\mathfrak{C}}\eta^4(A/\mathfrak{D}) \quad \text{(by \textbf{Dist 3})}\\
&\quad=\lambda_7\frac{g^p(a+b, A)g^p(a-b, A)}{g^{2p}(a, A)g^{2p}(b, A)}\prod_{\mathfrak{D}}\eta^4(A/\mathfrak{D})\eta^4(A)^{-1} \quad \text{(by \textbf{Dist 2})}\\
&\quad=\lambda_7\frac{g^p(a+b, A)g^p(a-b, A)}{g^{2p}(a, A)g^{2p}(b, A)}\frac{\eta^{4(p+1)}(A)}{\eta^4(A)} \quad \text{(by \textbf{Dist 4})}\\
&\quad=\lambda_7\frac{\mathfrak{k}^p(a+b, A)\mathfrak{k}^p(a-b, A)}{\mathfrak{k}^{2p}(a, A)\mathfrak{k}^{2p}(b, A)}\\
&\quad=\lambda_7(\wp(a, A)-\wp(b, A))^p
\end{aligned}$$

thereby proving the theorem.

§6. The Cuspidal Group on $X_1(p)$

The curve $X_1(p)$ has a natural rational map onto $X_0(p)$, and the Galois group G of the covering is cyclic, isomorphic to $\mathbf{Z}(p)^*$.

The curve $X_0(p)$ has two cusps, lying above $j=\infty$, and corresponding to 0 and $i\infty$ in the upper half plane. Their inverse images in $X_1(p)$ form two sets of cusps, each of order $p-1$, and giving rise to two divisor class groups

$$\mathscr{C}_1^0(p) \quad \text{and} \quad \mathscr{C}_1^{i\infty}(p).$$

Of course, we can give a more algebraic description. As before, the cusp $i\infty$ corresponds to the prime at infinity given by the embedding of the modular function field in the power series in $q^{1/p}$, and the cusp 0 corresponds to wP_∞ where w is the involution. The sum of these two groups is contained in $\mathscr{C}_1(p)$, and the complete relationship between this sum and full cuspidal divisor class group $\mathscr{C}_1(p)$ is not known at present (although Mazur has told us partial results). At any rate, the involution on $X_1(p)$ gives an isomorphism between $\mathscr{C}_1^0$ and $\mathscr{C}_1^{i\infty}$. If χ is a character of $G\approx\mathbf{Z}(p)^*$, then the involution switches the eigenspaces, giving an isomorphism

$$\mathscr{C}_1^{0,(p)}(\chi)\approx\mathscr{C}_1^{i\infty,(p)}(\bar{\chi}).$$

(Cf. [L 2], Chapter VI, Lemma 3.) The theory on $X_1(p)$ analogous to the preceding theory on $X(p)$ would deal with a single of these subgroups, say $\mathscr{C}_1^0=\mathscr{C}_1^0(p)$. Klimek

[Kl] has determined the order of $\mathscr{C}_1^0$. By combining the method of quadratic relations of Kubert [Ku], and the characterization of units in the modular function field in [KL IV], one can show that the principal divisor group on $X_1(p)$ having support at the cusps over 0 is generated by the quotients of Weierstrass forms

$$\wp(a, A) - \wp(b, A)$$

where $A = \mathbf{C}/[\tau, 1]$, and a, b belong to the subgroup generated by $1/p$. One may then form the Stickelberger ideal

$$\frac{p}{4} \sum_{c \in \mathbf{Z}(p)^*} \left(\mathbf{B}_2 \left(\left\langle \frac{c}{p} \right\rangle \right) - \frac{1}{6} \right) \sigma_c^{-1} I_p,$$

where I_p is generated by the elements

$$\sigma_{\alpha+\beta} + \sigma_{\alpha-\beta} - 2\sigma_\alpha - 2\sigma_\beta,$$

over $\mathbf{Z}_p[G]$. Then one finds:

Theorem 6.1. *There is an isomorphism*

$$\mathscr{C}_1^{0,(p)}(\chi) \approx \mathbf{Z}_p / B_{2,\bar{\chi}} I_\chi.$$

(i) *If χ is trivial, then $\mathscr{C}_1^{0,(p)}(\chi) = 0$.*

(ii) *If $\chi = \omega^2$, where ω is the Teichmuller character, then again $\mathscr{C}_1^{0,(p)}(\chi) = 0$. In fact, $I_\chi = (p)$ and $B_{2,\bar{\chi}}$ has order -1.*

(iii) *If χ is non-trivial and $\neq \omega^2$ then $I_\chi = (1)$, and $\mathscr{C}_1^{0,(p)}(\chi)$ is cyclic of order* $\operatorname{ord} B_{2,\chi}$.

References

[C-S] Coates, J., Sinnott, W.: Integrality properties of the values of partial zeta functions. Proc. London Math. Soc. 365—384 (1977)

[C-W] Coates, J., Wiles, A.: On the conjecture of Birch and Swinnerton-Dyer. Inv. Math. **39**, (1977)

[I] Iwasawa, K.: A class number formula for cyclotomic fields. Ann. of Math. **76**, 171—179 (1962)

[Iw 2] Iwasawa, K.: On some modules in the theory of cyclotomic fields. J. Math. Soc. Japan **16**, 42—82 (1964)

[Kl] Klimek, S.: Thesis, Berkeley, 1975

[Ku] Kubert, D.: Quadratic relations for generators of units in the modular function field. Math. Ann. **225**, 1—20 (1977)

[KL I] Kubert, D., Lang, S.: Units in the modular function field I. Math. Ann. **218**, 67—96 (1975)

[KL II] Kubert, D., Lang, S.: Units in the modular function field II. Math. Ann. **218**, 175—189 (1975)

[KL III] Kubert, D., Lang, S.: Units in the modular function field III. Math. Ann. **218**, 273—285 (1975)

[KL IV] Kubert, D., Lang, S.: Units in the modular function field. IV. The Siegel function are generators. Math. Ann. **227**, 223—242 (1977)

[KL D] Kubert, D., Lang, S.: Distributions on toroidal groups. Math. Z. **148**, 33—51 (1976)

[L 1] Lang, S.: Algebraic Number Theory. Addison Wesley 1970

[L 2] Lang, S.: Introduction to modular forms. Berlin, Heidelberg, New York: Springer 1977

[Ri] Ribet, K.: A modular construction of unramified extensions of $\mathbf{Q}(\mu_p)$. Inv. Math. **34**, 151—162 (1976)

[Ro] Robert, G.: Unités élliptiques. Bull. Soc. Math. France, Supplement, Decembre 1973, No. 36

[Si] Sinnott, W.: The index of the Stickelberger ideal (to appear)

Received March 3, 1977

Math. Ann. 237, 97—104 (1978) © by Springer-Verlag 1978

Units in the Modular Function Field

V. Iwasawa Theory in the Modular Tower

Daniel S. Kubert*

Department of Mathematics, Cornell University, Ithaca, NY 14853, USA

Serge Lang*

Department of Mathematics, Yale University, New Haven, CT 06520, USA

Iwasawa in a well known series of papers (e.g. [Iw 1], [Iw 2]) has elaborated a substantial theory concerning the projective limit of ideal classes, and the resulting Kummer theory of units, in Γ-extensions, where $\Gamma \approx \mathbf{Z}_p$.

We can carry out a similar theory for the tower of function fields of the modular curves $X(p^n)$, where p is prime, taken to be >3 for simplicity. The role of Γ is here played by a Cartan group (units in the unramified extension of degree 2 over $\mathbf{Z}_p$). Taking into account the results of the [KL] series, all the non-formal work has already been done, and all that remains to do is to put it together with the usual Kummer theory to obtain analogues of some of Iwasawa's results, but with much simpler formulations. This is the purpose of the present short paper, which extracts this Kummer theory.

We may form the usual projective family of group rings, their limit Λ (Iwasawa algebra) which here turns out to be essentially the power series in two variables $\mathbf{Z}_p[[T_1, T_2]]$, and the Galois group G of the extension of the modular function field by p^n-th roots of units. Then G turns out to be a 1-dimensional free module over Λ.

After discussing this modular case, we give a brief description of the cyclotomic case following the same pattern, to emphasize the analogy, assuming the Vandiver conjecture.

Let K_n be the function field of the modular curve $X(p^n)$ with constant field $\mathbf{Q}(\mu_{p^n})$. Let $K_\infty = \bigcup K_n$. The group $\mathrm{GL}_2(\mathbf{Z}_p)$ is represented as a group of automorphisms of K_∞, cf. Shimura [Sh] or [L], Chapter VI, §3. It contains the Cartan group C_p, isomorphic to the group of units in the unramified extension of degree 2 of $\mathbf{Z}_p$, in some fixed representation after a choice of basis over $\mathbf{Z}_p$. See [KL II], §2.

Let U_n be the group generated by the 12-th powers of Siegel functions g_a^{12}, with $a \in \mathbf{Q}^2$ and $a \notin \mathbf{Z}^2$ such that $p^n a \in \mathbf{Z}^2$. Alternatively, we need only take primitive indices a of exact order $p^n \bmod \mathbf{Z}^{(2)}$, cf. [KL II], and especially [KL IV] where it is shown that if h is modular of some level, and h^{p^ν} lies in U_n for some $\nu \geqq 1$, then h lies in U_n modulo constants.

---

* Supported by NSF grants. Kubert is also a Sloane Fellow

0025-5831/78/0237/0097/$01.60

From the formulas giving the operation of $\mathrm{GL}_2(\mathbf{Z}_p)$ in [KL II], formulas **K 1** through **K 6**, one sees that in fact the roots of unity $\mu_{p^{2n}}$ are contained in U_n. The Ramachandra-Robert distribution relations [Ra], [Ro], Proposition 1, see also [KL III], Theorem 3.1 can be used to show easily that $p^{12} \in U_n$, and in fact, the subgroup of U_n consisting of the constants is precisely

$$\{\mu_{p^{2n}}, p^{12}\}.$$

Let:

$$\Gamma = C_p/\pm 1 = C_p(\pm 1), \qquad \Gamma_n = C_p(p^n)/\pm 1 = \Gamma \bmod p^n.$$

$$R_n = \mathbf{Z}(p^n)[\Gamma_n], \qquad \Lambda = \lim R_n.$$

The limit is the projective limit, and defines the **Iwasawa algebra** in the present case. Since C_p contains the units $\equiv 1 (\bmod p)$, which are isomorphic under the exponential map to $\mathbf{Z}_p^{(2)}$, it follows that Λ contains a subring of finite index, isomorphic to

$$\mathbf{Z}_p[[T_1, T_2]],$$

where γ_1, γ_2 are independent generators of the units $\equiv 1 (\bmod p)$, and as usual, $T_i = \gamma_i - 1$.

Let

$$U = \bigcup U_n.$$

In [KL IV], it is proved that U is the group of units in K_∞, except for the fact that we have taken 12-th powers here, which will be irrelevant since the theorem we want will be concerned with Kummer theory in extensions of p-power order. Let

$$\Omega = K_\infty(U^{1/p^\infty}).$$

Then Ω is a Kummer extension of K_∞ and is Galois over $\mathbf{Q}(j)$, and over each field K_n. Let

$$G = \mathrm{Gal}(\Omega/K_\infty).$$

Then G is a Γ-module, and also a Λ-module. The projective structure can be obtained by the cofinal system of finite Kummer extensions

$$\Omega_n = K_\infty(U_n^{1/p^n}), \quad \text{with Galois group} \quad G_n = \mathrm{Gal}(\Omega_n/K_\infty).$$

The field diagram is as follows.

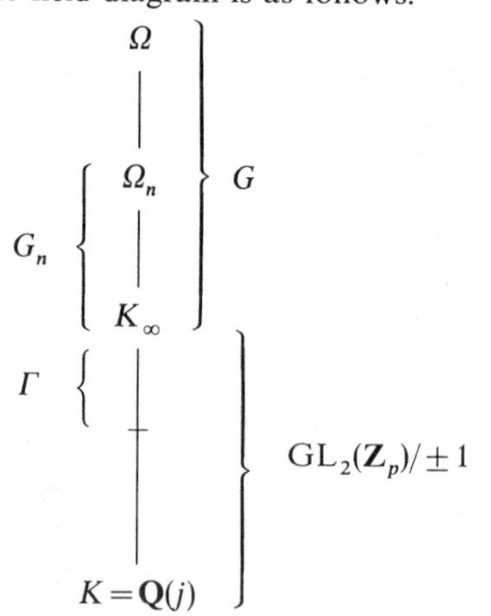

The standard Kummer theory of elementary algebra yields a duality for the finite abelian extension,

$$G_n \times U_n^{1/p^n} K_\infty^* / K_\infty^* \to \mu_{p^n}. \tag{1}$$

We have an isomorphism

$$U_n^{1/p^n} K_\infty^* / K_\infty^* \approx U_n^{1/p^n} / U_n^{1/p^n} \cap K_\infty^*.$$

But

$$U_n^{1/p^n} \cap K_\infty^* = U_n \mu_{(n)}, \tag{2}$$

where $\mu_{(n)}$ is the group of p^{2np^n}-th roots of unity.

Indeed, the Galois group of $\mathbf{Q}(p^{1/p^n}, \mu_{p^n})$ over $\mathbf{Q}$ is a semidirect product of the abelian Galois group of p^n-th roots of unity and a cyclic group of order p^n. The field of roots of unity is maximal subabelian in this Galois extension, and p^{1/p^n} generates an extension of degree p^n over $\mathbf{Q}(\mu^{(p)}) = \mathbf{Q}_\infty$, the tower of roots of unity. Hence it generates an extension of the same degree over K_∞, because $\mathbf{Q}_\infty$ is the constant field of K_∞. By [KL IV], Theorem 1, as already mentioned, we conclude that (2) is true.

Therefore the Kummer theory gives a duality

$$G_n \times U_n^{1/p^n} / U_n \mu_{(n)} \to \mu_{p^n}. \tag{3}$$

Let $\sigma \in G_n$ and $u \in U_n^{1/p^n}$. The Kummer pairing

$$(\sigma, u) \mapsto \langle \sigma, u \rangle$$

is given by the formula

$$\langle \sigma, u \rangle = \sigma u / u = u^{\sigma - 1}.$$

The group Γ of course operates by conjugation on G, and operates on the roots of unity via the determinant. By first principles of functoriality, the symbol satisfies the formula

$$\langle \sigma^\gamma, u^\gamma \rangle = \langle \sigma, u \rangle^\gamma = \langle \sigma, u \rangle^{\det \gamma}, \tag{4}$$

for all $\gamma \in \Gamma$. By σ^γ we mean $\tilde{\gamma} \sigma \tilde{\gamma}^{-1}$, where $\tilde{\gamma}$ is a lifting of γ to an automorphism of Ω. Similarly $u^\gamma = \tilde{\gamma} u$. The value of the symbol is independent of the choice of lifting $\tilde{\gamma}$, because the symbol $\langle \sigma, u \rangle$ is equal to $\langle \sigma, u' \rangle$ if $u^{p^n} = u'^{p^n}$, and because Ω / K_∞ is abelian.

In the Kummer pairing, we want to take projective limits on the Galois groups, and injective limits on the p^n-th roots of units. For this purpose, we use the isomorphism

$$\frac{1}{p^n} \mathbf{Z}_p / \mathbf{Z}_p \approx \mathbf{Z}_p / p^n \mathbf{Z}_p = \mathbf{Z}_p(p^n)$$

obtained by multiplication with p^n. Then we use the isomorphism of the group ring of Γ_n with formal linear combinations of elements in Γ_n with coefficients in

$$\frac{1}{p^n} \mathbf{Z}_p / \mathbf{Z}_p,$$

namely we write

$$\mathbf{Z}(p^n)[\Gamma_n] \approx \frac{1}{p^n} \mathbf{Z}_p / \mathbf{Z}_p[\Gamma_n].$$

With this notation, **we have an isomorphism of Λ-modules**

$$\frac{1}{p^n}\mathbf{Z}_p/\mathbf{Z}_p[\Gamma_n] \to U_n^{1/p^n}/U_n\mu_{p^{2n}} \tag{5}$$

by means of the mapping

$$\gamma \mapsto g_\gamma^{12/p^n},$$

where g_γ is the Siegel function, and the map is well defined since we are allowed to take the p^n-th root mod roots of unity. Since Γ_n operates simply transitively, since the only relation among the Siegel functions is given by the distribution relation that

$$\prod_{\gamma \in C(p^n)} g_\gamma = p \quad \text{modroots of unity},$$

and since the p^n-th roots of p generate an extension disjoint from Ω_n over K_∞, it follows that there cannot be any relation in the homomorphism from the group ring in (5) to the group on the right, and hence that the map is an isomorphism.

The group G_n can also be considered as Λ-module. By duality, **we have an isomorphism**

$$\mathbf{Z}(p^n)[\Gamma_n] \to G_n, \tag{6}$$

and the Kummer duality can be represented by the pairing

$$\mathbf{Z}(p^n)[\Gamma_n] \times \frac{1}{p^n}\mathbf{Z}_p/\mathbf{Z}_p[\Gamma_n] \to \mu_{p^n} \tag{7}$$

given by

$$\left(\sum x(\alpha)\alpha, \sum y(\beta)\beta\right) \mapsto e\left(\sum x(\alpha)y(\alpha)\det\alpha\right),$$

where $e(t)=e^{2\pi i t}$. We have an isomorphism of pairings

$$\begin{array}{ccccc} G_n & \times & U^{1/p^n}/U_n\mu_{p^{2n}} & \longrightarrow & \mu_{p^n} \\ \updownarrow \approx & & \updownarrow \approx & & \updownarrow = \\ \mathbf{Z}(p^n)[\Gamma_n] & \times & \frac{1}{p^n}\mathbf{Z}_p/\mathbf{Z}_p[\Gamma_n] & \longrightarrow & \mu_{p^n}. \end{array} \tag{8}$$

This takes care of the story at finite level.

Passing to the limit, the map $G_{n+1} \to G_n$ corresponds to the natural homomorphism of group algebras. On the other hand we have an *injection*

$$\frac{1}{p^n}\mathbf{Z}_p/\mathbf{Z}_p[\Gamma_n] \to \frac{1}{p^{n+1}}\mathbf{Z}_p/\mathbf{Z}_p[\Gamma_{n+1}]$$

in a natural way. The coefficients are both contained in $\mathbf{Q}_p/\mathbf{Z}_p$, and there is a natural embedding

$$\Gamma_n \to \Gamma_{n+1}$$

which sends an element of Γ_n to the formal sum of elements in Γ_{n+1} lying above it under the canonical surjection $\Gamma_{n+1} \to \Gamma_n$. Passing to the limit may be represented by the arrows:

$$\begin{array}{ccccc} G_{n+1} & \times & \hat{G}_{n+1} & \longrightarrow & \mu_{p^{n+1}} \\ \downarrow & & \uparrow & & \uparrow \\ G_n & \times & \hat{G}_n & \longrightarrow & \mu_{p^n} \end{array} \tag{9}$$

In the limit, we then find a duality of compact and discrete abelian groups,

$$G \times \hat{G} \to \mu^{(p)}, \quad \text{where} \quad \hat{G} = \varinjlim \frac{1}{p^n} \mathbf{Z}_p/\mathbf{Z}_p[\Gamma_n].$$

In any case, it is apparent from the group algebras that

Theorem 1. *G is a 1-dimensional free module over Λ* (11)

The formalism elaborated above gives a clear picture of the situation because we know from the start the structure of the units in the modular function field.

In the classical situation of the tower of p^n-th roots of unity, the picture would be equally clear if a classical conjectures were known. Namely, one has a class number formula for each level

$$h^+ = (E : \mathscr{E})$$

where $\mathscr{E}$ is the group of cyclotomic units. The Vandiver conjecture states that h_n^+ is prime to p, and thus that the factor group $E/\mathscr{E}$ is p-torsion free. Suppose this is true. We may write the cyclotomic p-units:

$$g_c = \zeta^c - 1.$$

Let $K_\infty = \mathbf{Q}(\mu^{(p)})$ and $K_0 = \mathbf{Q}(\mu_p)$. Let E_p be the p-units in K_∞, and let $\mathscr{E}_p$ be the cyclotomic p-units, i.e. p-units generated by the above elements, of all levels. The Kummer extensions $K_\infty(E_p^{1/p^\infty})$ and $K_\infty(\mathscr{E}_p^{1/p^\infty})$ are then equal. The group Γ_n in the present case satisfies

$$\Gamma_n \approx \mathbf{Z}(p^n)^*.$$

One may then go through exactly the same arguments as for the modular case to get the analogous theorem in the cyclotomic case, using the fact that the cyclotomic units satisfy the distribution relations, and no other, due to Bass [Ba]. This is done as follows.

The Iwasawa algebra is

$$\Lambda = \lim \Lambda_n = \lim \mathbf{Z}(p^n)[\Gamma_n].$$

Let $\Gamma_n^+ = \Gamma_n$ mod complex conjugation. Let $V_n = \mathscr{E}_{p,n}/\pm\mu_{p^n}$ be the factor group of p-units at level $\leqq n$ by the roots of unity. The group Γ_n^+ operates simply transitively on the primitive elements of V_n, and the induced homomorphism

$$\mathbf{Z}[\Gamma_n^+] \to V_n \quad \text{such that} \quad \sigma_c \mapsto g_{c/p^n}$$

is an isomorphism, as in Bass [Ba]. In addition, we have:

Lemma 1. *The factor group V_m/V_n for $m \geqq n$ has no torsion.*

Proof. The embedding of V_n into V_m correspond to the embedding of group rings

$$\mathbf{Z}[\Gamma_n^+] \to \mathbf{Z}[\Gamma_m^+]$$

which sends an element σ_c on the element $\sum \sigma_b$, where the sum is taken over $\sigma_b \in \Gamma_m^+(c)$, the set of elements in Γ_m^+ which project on σ_c under the canonical map

$$\Gamma_m^+ \to \Gamma_n^+.$$

If an element

$$\sum_c \sum_{b \in \Gamma_m^+(c)} k(b)\sigma_b, \qquad k(b) \in \mathbf{Z},$$

is a torsion element with respect to $\mathbf{Z}[\Gamma_n^+]$, then all the coefficients $k(b)$ for $b \in \Gamma_m^+(c)$ must be equal to each other, and hence the element already lies in $\mathbf{Z}[\Gamma_n^+]$, as was to be shown.

The lemma plays the analogue in the cyclotomic case of [KL IV], Theorem 1, as used above in the proof of (2).

By following the steps analogous to (5), (6), (7), (8) we end up with a compact-discret duality

$$G \times \varinjlim \Lambda_n^+ \to \mu^{(p)},$$

where

$$\Lambda_n^+ = \mathbf{Z}(p^n)[\Gamma_n^+] \approx V_n(p^n),$$
$$G = \mathrm{Gal}(\Omega/K_\infty) \quad \text{and} \quad \Omega = K_\infty(\mathscr{E}_p^{1/p^\infty}),$$

and the pairing is the usual Kummer pairing.

We see that $G = G^- \approx \Lambda^-$.

Leopoldt [Le] showed at prime level how the Vandiver conjecture that h^+ is prime to p implies that the ideal class group C^- is cyclic over the group ring. We can derive this easily from the present point of view in the whole cyclotomic tower as follows.

Lemma 2. *The maximal unramified p-abelian extension of K_∞ is contained in Ω (under the Vandiver conjecture).*

Proof. Let $K_\infty(\alpha)$ be unramified, with some element α such that α^{p^t} lies in K_∞. We first show that we may select α to be real. Let $C = \lim C_n$ be the projective limit of the ideal class groups, and let

$$G_C = \text{Galois group of maximal unramified } p\text{-abelian extension of } K_\infty.$$

By Vandiver's conjecture, we have $G_C = G_C^-$. Let σ be a generator for $\mathrm{Gal}(K_\infty(\alpha)/K_\infty)$. Then

$$\sigma\alpha = \zeta\alpha \quad \text{for some } p^t\text{-th root of unity } \zeta .$$

Let ϱ be complex conjugation. Then $\varrho\sigma\varrho^{-1} = \varrho\sigma\varrho = \sigma^{-1}$ since $G = G^-$. Hence $\varrho\sigma\varrho\alpha = \zeta^{-1}\alpha$, and therefore

$$\sigma\bar{\alpha} = \zeta\bar{\alpha},$$

so that $\sigma(\bar{\alpha}/\alpha) = \bar{\alpha}/\alpha$. Thus $\bar{\alpha}/\alpha = b$ lies in K_∞. But the norm of b from K_∞ to K_∞^+ is 1 (obvious), so by Hilbert's theorem 90, there exists $\beta \in K_\infty$ such that $\beta/\bar{\beta} = b$. Then $\alpha\beta$ is real, and $K_\infty(\alpha) = K_\infty(\alpha\beta)$. This shows that we may assume α real.

For n sufficiently large, α^{p^t} lies in K_n^+, and $K_n^+(\alpha)$ is unramified over K_n^+ (because we assumed p odd). Hence we have an ideal factorization

$$(\alpha) = \mathfrak{a}^{p^t}$$

for some fractional ideal $\mathfrak{a}$ in K_n^+. The class of this ideal is principal by Vandiver's conjecture. It is then immediate that

$$K_n(\alpha) = K_n(u^{1/p^t})$$

for some unit u, thereby concluding the proof of the lemma.

From the lemma we obtain the tower of fields

$$\underbrace{\Omega \;\text{——}\; \overbrace{\Omega^{\mathrm{ur}} \;\text{——}\; K_\infty}^{G_C}}_{G}$$

Since we saw that $G \approx \Lambda^-$, we obtain G_C in a natural way as a quotient of G, thus showing:

Under the Vandiver conjecture, G_C is cyclic over Λ.

The above arguments are in substance those used in Iwasawa theory (cf. for instance Greenberg's Theorem 5 in [Gr]), but we make more efficient use of the cyclotomic units, in analogy with the modular units, here to get a precise result. The existence of the units conjectured by Stark [St] might be used in a similar fashion to deal with the totally real case.

References

[Ba] Bass, H.: Generators and relations for cyclotomic units. Nagoya Math. J. **27**, 401–407 (1966)

[Gr] Greenberg, R.: On the Iwasawa invariants of totally real number fields. Am. J. Math. **98**, 263–284 (1976)

[Iw 1] Iwasawa, K.: On Γ-extensions of algebraic number fields. Bull. Amer. Math. Soc. **65**, 183–226 (1959)

[Iw 2] Iwasawa, K.: On $\mathbf{Z}_l$-extensions of algebraic number fields. Ann. of Math. **98**, 246–326 (1973)

[KL II] Kubert, D., Lang, S.: Units in the modular function field. Math. Ann. **218**, 67–96 (1975)

[KL III] Kubert, D., Lang, S.: Units in the modular function field: Math. Ann. **218**, 273–285 (1975)

[KL IV] Kubert, D., Lang, S.: The p-Primary Component of the Cuspidal Divisor Class Group on the Modular Curve $X(p)$. Math. Ann. **234**, 25–44 (1978)

[L] Lang, S.: Elliptic functions, Chapter VI, § 3. Addison Wesley 1973
[Le] Leopoldt, H.W.: Zur Arithmetik in abelschen Zahlkörpern. J. reine angew. Math. **209**, 54–71 (1962)
[Ra] Ramachandra, K.: Some applications of Kronecker's limit formula. Ann. of Math. **80**, 104–148 (1964)
[Ro] Robert, G.: Unités élliptiques, Mémoire No. 36. Bull. Soc. Math. France 1973
[Sh] Shimura, G.: Introduction to the arithmetic theory of automorphic functions. Iwanami Shoten and Princeton University Press 1971
[St] Stark, H.: Totally real fields and Hilbert's 12th problem (to appear)

Received May 3, 1977

Math. Ann. 237, 203—212 (1978)

Stickelberger Ideals

Daniel S. Kubert* and Serge Lang*

D. S. Kubert
Mathematics Department, Cornell University, Ithaca, NY 14853, USA

S. Lang
Mathematics Department, Yale University, New Haven, CT 06520, USA

In a series of papers concerned with the cuspidal divisor class group on the modular curves, we have been led to purely algebraic and elementary properties of certain ideals in group rings which it is worth while to extract in a separate publication, independent of the applications. We do this here. We assume that the reader is acquainted with the basic properties of Bernoulli numbers and polynomials as given for instance in [L], Chapter XIII, but we reproduce some fundamental definitions for convenience.

We begin by recalling some integrality properties which appeared both in the work of Mazur and Coates-Sinnott [C-S 1], [C-S 2], and which we complement further, including a characterization of the integralization process for the Bernoulli distribution. We then compute the index of Stickelberger ideals of odd prime power level generalizing to arbitrary k the theorem of Iwasawa for $k = 1$ in [Iw], and the result of [KL 1] for $k = 2$. For $k = 1$ Sinnott [Si] had also worked out the index for composite level, but we do not treat composite levels here.

Finally we give an isomorphism theorem relating Stickelberger factor groups of order k with those order $k-1$, by twisting. Our starting point had been the connection by twisting between the ideal class groups in cyclotomic fields and $K_2\mathfrak{o}$ mentioned by Coates-Sinnott in [C-S 2], see also [Co 1].

We emphasize that the present paper is completely self contained at the elementary level, and thus can be viewed as giving a good introduction to these arithmetic theories.

Contents

* Supported by NSF grants. Kubert is also a Sloan Fellow

0025-5831/78/0237/0203/$02.00

§ 1. Notation

Let k be an integer $\geqq$. Let $N = p^n$ be a prime power with $p \geqq 3$ until § 6. We let:

$$G = \mathbf{Z}(N)^* \quad \text{if } k \text{ is odd}$$

$$G = \mathbf{Z}(N)^*/\pm 1 \quad \text{if } k \text{ is even}.$$

$$R = R_G = \mathbf{Z}[G] \quad \text{and} \quad R_p = \mathbf{Z}_p[G].$$

$\deg : R \to \mathbf{Z}$ is the augmentation homomorphism, such that

$$\deg\left(\sum_{\sigma \in G} m_\sigma \sigma\right) = \sum m_\sigma .$$

This augmentation homomorphism extends to the complex group algebra by linearity.

R_m = ideal of R consisting of those elements whose degree is

$$\equiv 0 \bmod m .$$

If I is an ideal of R, we let $I_m = I \cap R_m$.

$$\text{card } G = |G|.$$

$$s(G) = \sum_{\sigma \in G} \sigma .$$

For any $\xi \in R$ we have

$$\xi s(G) = (\deg \xi) s(G).$$

If J is an ideal of R, we write $d = \deg J$ to mean that d is the smallest integer $\geqq 0$ which generates the $\mathbf{Z}$-ideal of elements $\deg \xi$ with ξ in J.

Let $\mathbf{B}_k(X)$ be the k-th Bernoulli polynomial. We let

$$\theta_k(N) = N^{k-1} \sum_{a \in G} \frac{1}{k} \mathbf{B}_k\left(\left\langle \frac{a}{N} \right\rangle\right) \sigma_a^{-1},$$

$$\theta'_k(N) = N^{k-1} \sum_{a \in G} \frac{1}{k} \left(\mathbf{B}_k\left(\left\langle \frac{a}{N} \right\rangle\right) - \mathbf{B}_k(0)\right) \sigma_a^{-1}$$

$$= \theta_k - \frac{N^{k-1}}{k} B_k s(G),$$

where $B_k = \mathbf{B}_k(0)$ is the k-th Bernoulli number. We have:

$$\deg \theta \neq 0 \quad \text{and} \quad \deg \theta' \neq 0, \quad \text{if } k \text{ is even}.$$

In fact, these degrees can be computed easily. We need only that they are $\neq 0$ when k is even, but the computation is as follows. Suppose k is odd. We use the distribution relation. Summing over all primitive elements, i.e. elements of level p^n yields the value of the distribution summed over all elements of level p^{n-1}. Continuing in this fashion, reduces the computation to level 1. But

$$p^{k-1} \sum_{a \in \mathbf{Z}(p)} \frac{1}{k} \mathbf{B}_k\left(\left\langle \frac{a}{p} \right\rangle\right) = \frac{1}{k} B_k(0) = \frac{1}{k} B_k .$$

The degree of θ arises from the same sum but with the term $a = 0$ omitted. Hence

$$\deg\theta = \frac{1-p^{k-1}}{k} B_k$$

and

$$\deg\theta' = \deg\theta - \frac{N^{k-1}}{k} B_k|G|$$

or

$$\deg\theta' = \left(\frac{1-p^{k-1}}{k} - \phi(p^n)\right) B_k.$$

These formulas would also be valid for k even, except for our convention to take $G = \mathbf{Z}(N)^*/\pm 1$. This requires dividing the formulas by 2 to get $\deg\theta$ and similarly for θ'. The fact that $B_k \neq 0$ when k is even comes from the functional equation of the zeta function and is classical. It is standard that $B_k = 0$ if k is odd, $k \neq 1$.

Next we give the ideals used in integralizing the distributions.

$J^{(k)}(N) =$ ideal of elements $\sum m(b)\sigma_b$ such that

$$\sum m(b)b^k \equiv 0(\operatorname{mod} N),$$

$I^{(k)}(N) =$ ideal of elements $\sigma_c - c^k$ with integers c prime to N.

Since k and N remain fixed, we often write θ and θ' instead of $\theta_k(N)$ and $\theta'_k(N)$. Similarly, we write $J^{(k)}$ and $I^{(k)}$, or J and I. It is obvious that

$$J^{(k)} \supset I^{(k)}.$$

We shall determine the extent to which $J \neq I$ in §2.

We have:

$\deg I^{(k)}(N) = p^t$, where t is the maximum integer such that

$$k \equiv 0 \operatorname{mod} \phi(p^t).$$

This is obvious, because $\deg I^{(k)}(N)$ is generated by the integers $1-c^k$ with c prime to p.

§2. Integrality Properties

Theorem 2.1. (i) *We have*

$$R\theta'_k \cap R = I^{(k)}\theta'_k.$$

In fact, if an element $\xi \in R$ *is such that* $\xi\theta' \in R$, *then* $\xi \in I^{(k)}$.

(ii) *On the other hand, letting* $I_p^{(k)} = \mathbf{Z}_p I^{(k)}$, *we have*

$$R_p\theta_k \cap R_p = I_p^{(k)}\theta_k.$$

If an element $\xi \in R_p$ *is such that* $\xi\theta \in R_p$ *then* $\xi \in I_p^{(k)}$.

Proof. First we prove that

$$I\theta' \subset R, \quad \text{and} \quad I_p\theta \subset R_p.$$

A similar property is due to Mazur and Coates-Sinnott, as mentioned before. Indeed, we have

$$\sigma_c^{-1}(\sigma_c - c^k)\theta_k = N^{k-1} \sum_{a \in G} E_{k,c}^{(N)}(a)\sigma_a^{-1},$$

where

$$E_{k,c}^{(N)}(x) = N^{k-1}\left[\mathbf{B}_k\left(\left\langle\frac{x}{N}\right\rangle\right) - c^k\mathbf{B}_k\left(\left\langle\frac{c^{-1}x}{N}\right\rangle\right)\right].$$

As a function of $x \in \mathbf{Z}/N\mathbf{Z}$, we know from [L], Chapter XIII, formula **E 2**, that $\frac{1}{k}E_{k,c}^{(N)}$ is a distribution whose values are N-integral (replacing N by higher powers N^ν and using the distribution relation). This is also valid for $p = 2$. To get an integrality property at other primes, we use the following lemma.

Lemma 1. *The polynomial* $\frac{1}{k}(\mathbf{B}_k(X) - \mathbf{B}_k(0))$ *maps* $\mathbf{Z}$ *into* $\mathbf{Z}$, *and maps* $\mathbf{Z}_l$ *into* $\mathbf{Z}_l$ *for every prime number* l.

Proof. A standard property of Bernoulli polynomials states that

$$\frac{1}{k}(\mathbf{B}_k(X+1) - \mathbf{B}_k(X)) = X^{k-1}.$$

Hence for any integer m we see recursively that the first assertion of the lemma is true. The second, concerning l-adic integers, follows by continuity. The lemma is also valid for $p = 2$.

We may define $E'_{k,c}$ by using $\mathbf{B}_k(X) - \mathbf{B}_k(0)$ instead of $\mathbf{B}_k(X)$ in the definition of $E_{k,c}$. Then the lemma shows that $I\theta' \subset R$.

For convenience we let

$$\mathbf{B}'_k(X) = \mathbf{B}_k(X) - \mathbf{B}_k(0).$$

Lemma 2. (i) *Let* $\xi \in R$ *and suppose that* $\xi\theta' \in \mathbf{Z}_p[G] = R_p$. *Then* $\xi \in J$.
(ii) *Let* $\xi \in R_p$ *and suppose that* $\xi\theta \in R_p$. *Then* $\xi \in J_p = \mathbf{Z}_p J$.

Proof. Write $\xi = \sum z(b)\sigma_b$ with integral coefficients $z(b)$. Then

$$\xi\theta' = N^{k-1}\sum_c \sum_b z(b)\frac{1}{k}\mathbf{B}'_k\left(\left\langle\frac{bc}{N}\right\rangle\right)\sigma_c^{-1},$$

and therefore

$$\frac{N^{k-1}}{k}\sum_b z(b)\mathbf{B}'_k\left(\left\langle\frac{b}{N}\right\rangle\right) \quad \text{is } p\text{-integral}.$$

But an elementary formula for Bernoulli polynomials, obtained directly from the definition, gives for an integer b,

$$\frac{N^{k-1}}{k}\mathbf{B}_k\left(\frac{b}{N}\right) = \sum_{i=0}^{k} \frac{N^{k-1}}{k}\binom{k}{i} B_i \left(\frac{b}{N}\right)^{k-i}.$$

Comparing the leading term modulo all the lower order terms, and taking into account that $B_1 = -1/2$ is p-integral (here we use $p \neq 2$), and the Kummer theorem that B_i is p-integral for $i < p-1$, we find

$$\frac{\sum m(b)b^k}{kN} \equiv 0 \bmod \frac{1}{k}\mathbf{Z}_p.$$

Multiplying both sides by kN proves the lemma.

Lemma 3. *Let p^s be the smallest power of p such that $p^s\theta'_k$ is p-integral.*

Then

$$s = n + \operatorname{ord}_p k.$$

We have $I^{(k)} \cap \mathbf{Z} = (p^s)$.

Proof. The argument uses the same expression for the Bernoulli polynomial as in the previous lemma. We see that

$$p^s \sum \frac{N^{k-1}}{k}\binom{k}{i} B_i \left(\frac{1}{N}\right)^{k-i} \quad \text{is} \quad p\text{-integral}.$$

The leading term is p^s/kN. The Bernoulli numbers B_i are p-integral for $i < p-1$ by Kummer, and for $i \geqq p-1$ the power N^{k-1} in front integralizes $(1/N)^{k-i}$ by Von Staudt ($B_{r(p-1)}$ has a pole of order 1 at p). It follows that

$$\frac{p^s}{kN} \quad \text{is} \quad p\text{-integral},$$

whence s has the stated value. Since we have already seen that $I\theta \subset R$, it follows that the p-contribution of $I \cap \mathbf{Z}$ is exactly p^s. It is clear that $I \cap \mathbf{Z}$ is equal to (p^s), because we can always select

$$c \equiv 1 \bmod N \qquad \text{and} \qquad c \equiv 0 \bmod l$$

for any prime $l \neq p$ to see that $I \cap \mathbf{Z}$ contains elements prime to l. This proves the lemma.

Lemma 4. *We have $J = I + \mathbf{Z}N$ and $(J : I) = p^{s-n} = p^{\operatorname{ord} k}$.*

Proof. It is clear that $N \in J$. Conversely, write an element of J in the form

$$\sum m(c)(\sigma_c - c^k) + \sum m(c)c^k.$$

The first term is in I, and the second term is an integral multiple of N. This proves the lemma.

We may now conclude the proof of the theorem. We prove (i). Suppose $\xi \in R$ and $\xi\theta' \in R$. By Lemma 2, $\xi \in J$. By Lemma 4, we know that

$$\xi \equiv zN \bmod I \qquad \text{for some } z \in \mathbf{Z}.$$

We know that $I\theta' \subset R$. Hence $zN\theta' \in R$. By Lemma 3, it follows that p^s divides zN, so $\xi \in I$, and the theorem (i) is proved. The part (ii) is proved the same way.

§3. General Comments on Indices

Let V be a finite dimensional vector space over the rationals, and let A, B be lattices in V, that is free $\mathbf{Z}$-modules of the same rank as the dimension of V. Let C be a lattice containing both of them. We define the index

$$(A : B) = \frac{(C : B)}{(C : A)}.$$

It is an easy exercise to prove that this index is independent of the choice of C, and satisfies the usual multiplicativity property

$$(A : D)(D : B) = (A : B).$$

Furthermore, if E is a lattice contained in both A and B then

$$(A : B) = \frac{(A : E)}{(B : E)}.$$

We leave the proofs to the reader.

Suppose that A is not a lattice, but is an algebra over $\mathbf{Z}$. Let θ be an element of $\mathbf{Q}A = V$ and let m be a positive integer such that $m\theta \in A$. Assume that θ is invertible in $\mathbf{Q}A$. Then

$$(A : A\theta) = \pm \det_{\mathbf{Q}A} \theta,$$

where the determinant is taken for the linear transformation of $\mathbf{Q}A$ equal to multiplication by θ. This is easily seen, because

$$(A : A\theta) = (A : Am\theta)(Am\theta : A\theta)$$

and

$$(Am\theta : A\theta) = (A\theta : Am\theta)^{-1}.$$

Since $m\theta$ lies in A, the index $(A : Am\theta)$ is given by the absolute value of the determinant of $m\theta$, which is $m^r \det \theta$, where r is the rank of A. This power m^r then cancels the other index.

Note that the determinant can be computed in the extension of scalars by the complex numbers. In particular, if A is a semisimple algebra, and is commutative, then

$$\det \theta = \prod \chi(\theta),$$

where χ ranges over all the characters of the algebra, counted with their multiplicities. In the applications, the algebra is essentially a group ring, so the multiplicities are 1, and the characters come from characters of the group.

This will applied to the case when $\theta = \theta^{(k)}$. We recall the definition of generalized Bernoulli numbers according to Leopoldt:

$$B_{k,\chi} = N^{k-1} \sum_{a \in G} \chi(a) B_k\left(\left\langle \frac{a}{N} \right\rangle\right).$$

Thus

$$\chi(\theta) = \frac{1}{k} B_{k,\chi}.$$

Note that the Bernoulli number is defined with respect to G, so that for k even, we are summing over $\mathbf{Z}(N)^*/\pm 1$. This convention is the most useful for present applications in §4 and §5. (We revert to the other convention in §6.) For even k, it gives half the other values.

The classical theorem about the non-vanishing of $B_{k,\chi}$ when k and χ have the same parity gives the desired invertibility of the Stickelberger element θ_k in the corresponding part of the group algebra over $\mathbf{Q}$.

§4. The Index for k Even

We let $s = n + \mathrm{ord}_p k$, and t is defined at the end of §1. We regard $R_0 \cap R\theta$ (for k even) as the **Stickelberger ideal**. We shall prove:

$$\boxed{(R_0 : R_0 \cap R\theta) = Np^{\mathrm{ord}\,k - t} \prod_{\chi \neq 1} \pm \frac{1}{k} B_{k,\chi}}\ .$$

First observe that since $\deg\theta$ and $\deg\theta' \neq 0$ we have

$$R_0 \cap R\theta = R_0 \cap R\theta'.$$

By Theorem 2.1, we conclude that

$$R\theta' \cap R = I\theta', \qquad \text{and hence} \qquad R\theta' \cap R_0 = I_0\theta'.$$

But $R_0 + I\theta' = R_d$ where

$$d = \deg I\theta' = (\deg I)(\deg\theta').$$

In §1 we computed $\deg I = p^t$. The factor $\deg\theta'$ will cancel ultimately. In any case, we have:

$$\begin{aligned}
(R_0 : R_0 \cap R\theta) &= (R_0 : R_0 \cap R\theta') \\
&= (R_0 : I_0\theta') \\
&= (R_d : I\theta') \\
&= \frac{(R : I\theta')}{(R : R_d)} \\
&= \frac{1}{d}(R : R\theta')(R\theta' : I\theta') \\
&= \frac{1}{d}\prod \chi(\theta')(R : I).
\end{aligned}$$

The product is taken over all characters χ of G. We separate this product into a factor with the trivial character, giving $\deg\theta'$, canceling that same factor in d, and the product over the non-trivial characters. For χ non-trivial, we have $\chi(\theta) = \chi(\theta')$.

In the final step we also wrote $(R\theta' : I\theta') = (R : I)$. This is because θ' is invertible in the group algebra over $\mathbf{Q}$. Hence the map $\xi \mapsto \xi\theta'$ induces an isomorphism on $R \otimes Q$.

We are therefore reduced to proving a final lemma.

Lemma. $(R : I) = p^s$ *where* $s = n + \mathrm{ord}_p k$.

Proof. We have $(R : I) = (R : J)(J : I)$. Any element ξ in R can be written in the form

$$\xi = \sum m(c)\sigma_c = \sum m(c)(\sigma_c - c^k) + \sum m(c)c^k.$$

From this it is clear that $(R : J) = N$, and the index $(J : I)$ is $p^{\mathrm{ord} k}$ by Lemma 4, thus including the proof.

Remark. The sign in the product is taken to make the product positive. One knows the correct sign from the factorization of the zeta function and L-series evaluated at $1-k$, but this is irrelevant to our purposes here.

§5. The Index for k odd

Assume k is odd. Note that $\theta = \theta'$. Let

$$\varepsilon^- = \tfrac{1}{2}(1-\sigma_{-1})$$

be the idempotent which projects on the (-1)-eigenspace. It is immediate from the definition that θ is odd, that is,

$$\varepsilon^-\theta = \theta.$$

The **Stickelberger ideal** in this case is $R\theta \cap R = I\theta$, and is odd. We shall prove:

$$\boxed{(R^- : R\theta \cap R) = Np^{\mathrm{ord} k} \prod_{\chi \text{ odd}} \pm \frac{1}{2k} B_{k,\chi}}\ .$$

The rest of the section is devoted to the proof.

Lemma 1. *We have* $R^- = 2\varepsilon^- R$ *and* $(\varepsilon^- R : R^-) = 2^{\phi(N)/2}$.

Proof. The inclusion $(1-\sigma_{-1})R \subset R^-$ is clear. Conversely, let P be a set of representatives in $\mathbf{Z}(N)^*$ for $\mathbf{Z}(N)^*/\pm 1$. Let

$$\alpha = \sum z(c)\sigma_c^{-1} \in R^-,$$

with coefficients $z(c) \in \mathbf{Z}$. Thus $\sigma_{-1}\alpha = -\alpha$. Then $z(-c) = -z(c)$. If we let

$$\beta = \sum_{c \in P} z(c)\sigma_c^{-1},$$

then $\alpha = (1-\sigma_{-1})\beta$, thereby proving the lemma, because $\varepsilon^- R$ is a free abelian group of rank $\phi(N)/2$.

We then proceed as in the even case. First we write

$$(R^- : I\theta) = \frac{(\varepsilon^- R : \varepsilon^- I\theta)}{(\varepsilon^- R : R^-)},$$

and then

$$\begin{aligned}(\varepsilon^- R : \varepsilon^- I\theta) &= (\varepsilon^- R : \varepsilon^- R\theta)(\varepsilon^- R\theta : \varepsilon^- I\theta) \\ &= \prod_{\chi \text{ odd}} \chi(\theta)(\varepsilon^- R : \varepsilon^- I)\end{aligned}$$

because θ is invertible in $\varepsilon^{-}\mathbf{Q}[G]$. Furthermore,

$$\begin{aligned}(\varepsilon^{-}R : \varepsilon^{-}I) &= (\varepsilon^{-}R : R^{-})(R^{-} : 2\varepsilon^{-}I)(2\varepsilon^{-}I : \varepsilon^{-}I)\\ &= (R^{-} : 2\varepsilon^{-}I)\end{aligned}$$

because $(2\varepsilon^{-}I : \varepsilon^{-}I) = 2^{-\phi(N)/2}$ since $\varepsilon^{-}I$ is free of rank $\phi(N)/2$. Finally,

Lemma 2. $(R^{-} : 2\varepsilon^{-}I) = p^{s}$ *where* $s = n+\operatorname{ord}_p k$.

Proof. The group $2\varepsilon^{-}I$ is generated by elements of the form

$$(\sigma_c-\sigma_{-c})-c^{k}(\sigma_1-\sigma_{-1}).$$

An element $\xi \in R^{-}$ lies in $\mathbf{Z}(\sigma_1-\sigma_{-1}) \bmod I$. Hence the same argument as in the past case (since $p \neq 2$) gives the desired index.

§6. Twistings and Stickelberger Ideals

The Stickelberger elements θ_k should be really be indexed by the groups to which they correspond. We now want to compare factor groups of the group ring by various Stickelberger ideals, twisted in various ways. Consequently, it is not useful any more to have G different in the even or odd case. For this section, we let $N = p^n$ still, and *we allow* $p = 2$. We let

$$G_n = \mathbf{Z}(p^n)^{*}.$$

We define

$$\theta_{k,c}(p^n) = \sigma_c^{-1}(\sigma_c-c^k)\theta_k(p^n) \in \mathbf{Z}(p^n)[G_n].$$

This makes sense since we know from §1 that $\theta_{k,c}(p^n)$ is p-integral.

Let V be a $\mathbf{Z}(p^n)[G_n]$-module. We define its **twist** to be the tensor product with the roots of unity,

$$V(1) = V \otimes \mu_N.$$

Then σ in G operates diagonally,

$$\sigma(v \otimes \gamma) = \sigma v \otimes \sigma\gamma, \quad \text{and} \quad \sigma_a(v \otimes \gamma) = a(\sigma_a v \otimes \gamma).$$

We let γ be a basis for μ_N over $\mathbf{Z}(N)$. Note that the element a on the right makes sense as an element of $\mathbf{Z}(N)$ since $V \otimes \mu_N$ is a module over $\mathbf{Z}(N)$.

From the definitions we then get the formula

$$\theta_{k,c}(v \otimes \gamma) = \theta_{k-1,c}v \otimes \gamma,$$

resulting from formula **E 2** of [L], Chapter XIII, namely

$$\frac{1}{k}E_{k,c}(a) \equiv a^{k-1}E_{1,c}(a) \bmod N.$$

The distribution relation allows us in **E 2** to replace N by high powers of N at a higher level, and then return to level N to get this congruence.

In particular, if $\theta_{k-1,c}$ annihilates V, then $\theta_{k,c}$ annihilates $V(1)$. The argument simply extracts in a general context the argument given by Coates-Sinnott [C-S 2] in connection with the ideal class groups in cyclotomic fields, see their Theorem 2.1.

Take V to be $\mathbf{Z}(p^n)[G_n]$ itself, so that V is generated by a single element $\sigma_1 \otimes \gamma$. The map

$$\xi \mapsto \xi(\sigma_1 \otimes \gamma)$$

gives an isomorphism

$$\mathbf{Z}(p^n)[G_n] \to \mathbf{Z}(p^n)[G_n] \otimes \mu_{p^n}.$$

Let $\mathcal{S}_k(p^n)$ = ideal of $\mathbf{Z}(p^n)[G_n]$ generated by the elements $\theta_{k,c}(p^n)$. Then the isomorphism induces a bijection

$$\mathcal{S}_k(p^n) \to \mathcal{S}_{k-1}(p^n) \otimes \mu_{p^n}.$$

Hence we get an isomorphism

$$\Lambda_n/\mathcal{S}_k(p^n) \to \Lambda_n \otimes \mu_{p^n}/\mathcal{S}_{k-1}(p^n) \otimes \mu_{p^n},$$

where $\Lambda_n = \mathbf{Z}(p^n)[G_n]$ is the group ring.

We may then pass to the projective limit. The limit of Λ_n is the Iwasawa algebra. We let $\mathcal{S}_k$ be the ideal generated by the elements $\theta_{k,c}$ [projective limit of $\theta_{k,c}(p^n)$]. We obtain an isomorphism with the twist,

$$\Lambda/\mathcal{S}_k \to \Lambda(1)/\mathcal{S}_{k-1}(1).$$

This isomorphism permutes the eigenspaces for the action of μ_{p-1}, and this can be interpreted in terms of congruence relations between Bernoulli-Leopoldt numbers (with characters) in the obvious manner.

Bibliography

[Co 1] J. Coates, *On K_2 and some classical conjectures in algebraic number theory*, Ann. of Math. 95 (1972) pp. 99–116

[Co 2] J. Coates, *K-theory and Iwasawa's analogue of the Jacobian*, Algebraic K-theory II, Springer Lecture Notes 342 (1973) pp. 502–520

[C-L] J. Coates and S. Lichtenbaum, *On l-adic zeta functions*, Ann. of Math. 98 (1973) pp. 498–550

[C-S 1] J. Coates and W. Sinnott, *On p-adic L-functions over real quadratic fields*, Inv. Math. 25 (1974) pp. 253–279

[C-S 2] J. Coates and W. Sinnott, *An analogue of Stickelberger's theorem for higher K-groups*, Invent. Math. 24 (1974) pp. 149–161

[C-S 3] J. Coates and W. Sinnott, *Integrality properties of the values of partial zeta functions*, Proc. London Math. Soc. (1977) pp.

[Iw] K. Iwasawa, *A class number formula for cyclotomic fields*, Ann. of Math. 76 (1962) pp. 171–179

[K-L 1] D. Kubert and S. Lang, *The index of Stickelberger ideals of order 2 and cuspidal class numbers*, Math. Ann. 237 (1978) 213–232

[K-L 2] D. Kubert and S. Lang, *Distributions on toroidal groups*, Math. Zeit. (1976) pp. 33–51

[L] S. Lang, *Introduction to modular forms*, Springer Verlag, 1976

[Si] W. Sinnott, *The index of the Stickelberger ideal*, (to appear)

Received May 3, 1977

Math. Ann. 237, 213—232 (1978)

The Index of Stickelberger Ideals of Order 2 and Cuspidal Class Numbers

Daniel S. Kubert* and Serge Lang*

D. S. Kubert
Mathematics Department, Cornell University, Ithaca, NY 14853, USA

S. Lang
Mathematics Department, Yale University, New Haven, CT 06520, USA

The cuspidal divisor class group on the modular curve $X(N)$ can be represented as a quotient of a group ring by an ideal, called a Stickelberger ideal. Cf. the papers [KL] mentioned in the bibliography and Theorem 2.1. We shall recall the theorem here when N is a prime power, $p \geqq 5$. We also give the proof for a similar theorem on $X_1(N)$, relative to a certain subgroup of the full cuspidal divisor class group on $X_1(N)$ (see Theorem 3.3), since we treated in fully only $X(N)$ in previous papers.

Independently of such representations, we may study the quotients $\mathbf{Z}[G]/\mathcal{S}$ (for appropriate group G and corresponding Stickelberger ideal $\mathcal{S}$) essentially in a purely algebraic context. For instance, one may want to determine the index of the Stickelberger ideal. For Stickelberger ideals of order 1 (formed with the first Bernoulli numbers, and relevant for the ideal class groups of cyclotomic fields) this was done by Iwasawa in the prime power case [Iw], and generalized by Sinnott in the composite case [Si]. The main purpose of this paper is to determine the index for Stickelberger ideals of order 2, formed with the second Bernoulli numbers, and to identify this index with a divisor class number. Two groups play a role:

The group $G = \mathbf{Z}(N)^*/\pm 1$ for the curve $X_1(N)$.

The group $G = C(N)/\pm 1$ for the curve $X(N)$, where $C(N)$ is a "Cartan" group.

The index of the Stickelberger ideal will be determined in each case when N is a prime power, $N = p^n$ and $p \neq 2, 3$. This gives us the order of the cuspidal divisor class group on $X(N)$ in Theorem 2.2. For $X_1(N)$, this leads to the order of the cuspidal divisor class group with support in the cusps lying above 0 on $X_0(p)$, see Theorem 3.4. The method clearly works also when $p = 2, 3$, but the theory of quadratic relations in [Ku] shows that one has to treat these separately. The order for $X_1(p)$ when $N = p$ is prime was previously determined by Klimek [Kl].

It is convenient in some of the proofs on $X(N)$ in §2 to still appeal to the theory and language of modular forms, even though the statement is purely group-theoretic, concerning only objects in the group ring. Occasional possibly tricky arguments in group rings can be replaced by conceptual geometric arguments with

* Supported by NSF grants. Kubert is also a Sloan Fellow

0025-5831/78/0237/0213/$04.00

modular forms. It is of course possible to eliminate all references to modular forms, and in treating the Stickelberger ideals of higher order, it is not only possible, it is necessary. The proofs in §1 concerning the more classical case of $\mathbf{Z}(N)^*/\pm 1$ are completely group theoretic, and generalize in the same manner, as shown in [KL 7], to Stickelberger ideals of arbitrary order k.

The Stickelberger ideal in §1 also occurs in the context of K-theory, specifically $K_2\mathfrak{o}$, where $\mathfrak{o}$ is the ring of integers in the real cyclotomic fields. Coates [Co 1] shows that from work of Tate, $K_2\mathfrak{o}$ is the first twist of the ideal class group. Coates and Sinnott [C-S 1], [C-S 2] show that the Stickelberger ideal annihilates this twist. The Birch-Tate-Lichtenbaum conjecture predicts the order of $K_2\mathfrak{o}$. Our computation of the index coincides with this order, thus giving evidence to conjecture that there is an isomorphism of $K_2\mathfrak{o}$ with $R_0/\mathscr{S}$ in this case, for the p-primary parts.

In that connection it is possible to develop some general statements concerning the twistings of group rings modulo Stickelberger ideals. This is done in [KL 7] also.

Contents

Notation

In each case we let

$$R_G = R = \mathbf{Z}[G].$$

We let

$$\deg \colon R \to \mathbf{Z}$$

be the augmentation homomorphism, such that

$$\deg\Big(\sum_{\sigma \in G} m_\sigma \sigma\Big) = \sum m_\sigma .$$

Let m be an integer. We let:

R_m = ideal of R consisting of those elements whose degree is $\equiv 0 \bmod m$.

In particular, R_0 is the augmentation ideal, consiting of the elements of degree 0. If I is an ideal of R, we let

$$I_m = I \cap R_m .$$

We let:

$$\operatorname{card} G = |G|$$

$$s(G) = \sum_{\sigma \in G} \sigma .$$

For any $\xi \in R$ we have

$$\xi s(G) = (\deg \xi) s(G).$$

If J is an ideal of R, we write $d = \deg J$ to mean that d is the smallest integer $\geqq 0$ which generates the $\mathbf{Z}$-ideal of elements $\deg \xi$ with ξ in J.

If $a \in \mathbf{R}$, we let $\langle a \rangle$ be the smallest real number $\geqq 0$ in the residue class of a $\bmod \mathbf{Z}$.

§1. The Stickelberger Ideal for $G \approx \mathbf{Z}(N)^*/\pm 1$

We let G be as in the title of the section, and let $N = p^n$ where p is prime $\neq 2, 3$. We write elements of G also in the form

$$\sigma_b, \quad b \in \mathbf{Z}(N)^*/\pm 1, \quad \text{where} \quad \mathbf{Z}(N) = \mathbf{Z}/N\mathbf{Z},$$

to suggest the applications to Galois groups. We let:

$I = I_G = I^{(2)}$ be the ideal generated by the elements $\sigma_c - c^2$ with c prime to N, $c \in \mathbf{Z}$.

$$\mathbf{B}_2(X) = X^2 - X + \tfrac{1}{6} = \text{second Bernoulli polynomial.}$$

$$\theta = \theta^{(2)} = \theta_G = N \sum_{b \in G} \tfrac{1}{2}\mathbf{B}_2\left(\left\langle \frac{b}{N} \right\rangle\right)\sigma_b^{-1},$$

$$\theta' = N \sum_{b \in G} \tfrac{1}{2}\left(\mathbf{B}_2\left(\left\langle \frac{b}{N} \right\rangle\right) - \tfrac{1}{6}\right)\sigma_b^{-1} = \theta - \frac{N}{12}s(G).$$

The next lemma is a very special case of Lemma 12 in [C-S 3].

Lemma 1. *Let Q be the ideal in R generated by all elements satisfying the quadratic relations* $\bmod N$, *that is*

$$\sum m(c)\sigma_c \quad \textit{such that} \quad \sum m(c)c^2 \equiv 0 \bmod N.$$

Then $Q = I$, and N itself lies in I.

Proof. It is clear that $N\sigma_1$ lies in Q. As for I, we consider the two elements

$$\sigma_{1+N} - (1+N)^2 \quad \text{and} \quad \sigma_{1-N} - (1-N)^2.$$

Their difference is equal to $4N$. On the other hand, the first of these two elements is also equal to the odd number $N^2 + 2N$, so N lies in I also.

Now we write

$$\sum m(c)\sigma_c = \sum m(c)(\sigma_c - c^2) + \sum m(c)c^2.$$

It is then clear that $Q \subset I$, and the reverse inclusion is immediate, thus proving the lemma.

The next lemma shows there is no ambiguity in the possible definition of the Stickelberger ideal.

Lemma 2. *$R\theta' \cap R = I\theta'$. In fact, if $\xi \in R$ is such that $\xi\theta' \in R$ then $\xi \in Q$. Hence $R_0\theta' \cap R = I_0\theta'$.*

Proof. First note that

$$\left\langle \frac{a}{N} \right\rangle \left(\left\langle \frac{a}{N} \right\rangle - 1 \right) \equiv 0 \bmod 2\mathbf{Z}_2.$$

The element θ' has only a possible N in its denominator, because of the quadratic term. Write

$$\xi = \sum z(b)\sigma_b.$$

Assuming that $\xi\theta' \in R$, and multiplying ξ with θ', we conclude that

$$\sum z(b)b^2 \equiv 0 \bmod N,$$

so $\xi \in Q$. The converse is obvious.

For our purposes, we define the **Stickelberger ideal**

$$\mathcal{S} = I_0\theta' = I_0\theta.$$

Theorem 1.1. *We have* $(R_0 : I_0\theta') = N \prod_{\chi \neq 1} \frac{1}{2}B_{2,\chi}$.

The proof will result from computing a sequence of other indices.

Note: Here and elsewhere, the right hand side is to be interpreted only up to ± 1, taking whichever sign makes it positive. Using other arguments, one could in fact determine the precise sign, but it is not necessary to go into this here.

Since $I\theta' \cap R_0 = I_0\theta'$ (for instance by Lemma 2), we have an injection

$$0 \to R_0/I_0\theta' \to R/I\theta'.$$

The image consists of $(R_0 + I\theta')/I\theta'$, and $R_0 + I\theta' = R_d$ where

$$d = \deg I\theta'.$$

Since $\deg I = 1$ (because $p \neq 2, 3$), we also get $d = \pm\deg\theta'$. It turns out that d will cancel later, so we don't need to know it explicitly, but an easy computation shows that

$$2\deg\theta = \frac{1-p}{12} \quad \text{and} \quad 2\deg\theta' = \frac{1-p}{12} - \frac{N}{12}|G|.$$

The degree of θ is computed using the distribution relation of the Bernoulli polynomials. Be that as it may, we have

$$(R_0 : I_0\theta') = (R_d : I\theta').$$

Furthermore, we have inclusions

$$R \supset R_d \supset I\theta',$$

and obviously $(R : R_d) = d$. On the other hand, we have formally

$$(R : I\theta') = (R : R\theta')(R\theta' : I\theta').$$

Of course, $R\theta'$ need not be contained in R, but the index $(R : R\theta')$ can be easily interpreted as follows. The element $N\theta'$ lies in R. Hence

$$(R : R\theta') = \frac{(R : RN\theta')}{(R\theta' : RN\theta')}.$$

Lemma 3. $(R : R\theta') = (\deg\theta') \prod_{\chi \neq 1} \chi(\theta)$.

Proof. The index $(R : RN\theta')$ is equal to the absolute value of the determinant of $N\theta'$ acting on R by multiplication. This determinant can be computed on the vector space $\mathbf{C}[G]$, which splits into 1-dimensional eigenspaces corresponding to the characters. The trivial character gives the degree. The formula is then obvious since

$$(R\theta' : RN\theta') = N^{|G|},$$

so the extra power of N cancels.

Lemma 4. $(R : I) = N$.

Proof. This is easy, and one can show in fact that

$$R/I \approx \mathbf{Z}/N\mathbf{Z}.$$

We leave the details to the reader. A harder case with a more complicated Cartan group will be treated in full in Lemma 5 of §2.

Putting all the indices together yields the formula as stated in the theorem.

Let $\mathfrak{o}$ be the ring of integers in the real cyclotomic field

$$K = \mathbf{Q}(\mu_N)^+.$$

The Birch-Tate Lichtenbaum conjecture predicts that the order of $K_2\mathfrak{o}$ is given by

$$w_2(N)\zeta_K(-1),$$

where $w_2(N)$ is the order of the group of roots of unity in the composite of all quadratic extensions of K. As we took $N = p^n$ with $p \geqq 5$, it follows that

$$w_2(N) = 12N.$$

On the other hand, decomposing the zeta function into L-series, we find

$$\begin{aligned}\zeta_K(-1) &= \zeta_Q(-1) \prod_{\chi \neq 1} L(1-2, \chi) \\ &= \frac{1}{12} \prod_{\chi \neq 1} -\tfrac{1}{2}B_{2,\chi}.\end{aligned}$$

(Here we determine the sign by the analysis.) Thus the computation of Theorem 1.1 yields the same value.

In the next section we deal immediately with $X(N)$, computing the index of the Stickelberger ideal in that case, and identifying it with the order of the full cuspidal divisor class group. It is easier to treat $X(N)$ than $X_1(N)$, so we postpone the latter to the last two sections.

§2. The Stickelberger Ideal for the Cartan Group $C(N)/\pm 1$

Let C_p be the group of units in the unramified extension of $\mathbf{Z}_p$, of degree 2, and let

$$C(p^n) = C_p \bmod p^n.$$

We let $G \approx C(N)/\pm 1$ with $N = p^n$, and write elements of G in the form

$$\sigma_\gamma, \quad \gamma \in C(N)/\pm 1$$

again to suggest the Galois group situation, acting on the modular function as described in [KL 2], [KL 4], [KL 6]. We let:
I = ideal generated by "parallelograms"

$$\pi(\alpha, \beta) = (\alpha+\beta)+(\alpha-\beta)-2(\alpha)-2(\beta).$$

Note that the degree of a parallelogram is -2. We let

$$\theta = \theta_G = N \sum_{\gamma \in G} \tfrac{1}{2}\mathbf{B}_2\left(\left\langle \frac{T\gamma}{N} \right\rangle\right)\sigma_\gamma^{-1}$$

$$\theta' = \theta'_G = N \sum_{\gamma \in G} \tfrac{1}{2}\left(\mathbf{B}_2\left(\left\langle \frac{T\gamma}{N} \right\rangle\right) - \tfrac{1}{6}\right)\sigma_\gamma^{-1}.$$

The projection

$$T\colon C(N)/\pm 1 \to \mathbf{Z}(N)/\pm 1$$

is defined as follows. We fix a representation of $C(N)$ in $\mathrm{GL}_2(N)$, writing an element of $C(N)$ in the form

$$\gamma = \begin{pmatrix} a & b \\ c & d \end{pmatrix}$$

and we let $T\gamma = a$. The distribution relation for Bernoulli polynomials shows that

$$\deg\theta = 0 \quad \text{and} \quad \deg\theta' = -\frac{N}{12}|G|.$$

So $\theta \in R_0$ but θ' is not in R_0.

The next lemma identifies two possible definitions of the **Stickelberger ideal.**

Lemma 1. $R\theta \cap R = I_{12}\theta$.

Proof. It is convenient to phrase the proof in the language of modular forms. The module $R\theta$ is isomorphic to the group of divisors of products

$$g = \prod_{\alpha \in G} g_\alpha^{m(\alpha)},$$

where g_α are the Siegel functions, and $m(\alpha)$ are integers. Intersecting the group of such divisors with R corresponds to restricting g to have integral order at each cusp on $X(N)$. By a theorem of Fricke-Wohlfhart [Wo], see also Leutbecher [Le], Satz 1, this is equivalent to requiring that g has level N, i.e. that g is on $\Gamma(N)$. The last theorem of Kubert's quadratic relations [Ku] (which is wrongly numbered, and should be renumbered Theorem 8) then implies that the divisor (g) lies in $I_{12}\theta$.

This proves the inclusion

$$R\theta \cap R \subset I_{12}\theta .$$

The reverse inclusion is obvious.

We define the **Stickelberger ideal**

$$\mathcal{S}_G = \mathcal{S} = R\theta \cap R = I_{12}\theta .$$

Theorem 2.1. *Let $\mathcal{C}(N)$ be the cuspidal divisor class group on $X(N)$ for $N = p^n$, p prime $\neq 2, 3$. There is a natural isomorphism*

$$\mathcal{C}(N) \approx R_0/I_{12}\theta = R_0/\mathcal{S} .$$

Proof. This is proved by combining [Ku], §4 and §5 and [KL 4], Theorem 1.

The purely group theoretic result giving the index of the Stickelberger ideal will therefore also yield the order of the cuspidal divisor class group on $X(N)$.

Theorem 2.2. *Let χ be a non-trivial character of G, and*

$$B_{2,\chi} = N \sum_{\gamma \in G} \mathbf{B}_2\left(\left\langle \frac{T\gamma}{N} \right\rangle\right) \chi(\gamma) .$$

Then

$$(R_0 : \mathcal{S}) = \frac{6N^3}{|G|} \prod_{\chi \neq 1} \tfrac{1}{2} B_{2,\chi} .$$

Proof. We go through a sequence of lemmas giving indices of various ideals in each other.

Let Δ be the usual modular form of weight 12 on $\mathrm{SL}_2(\mathbf{Z})$. On $X(N)$ its divisor is

$$(\Delta) = Ns(G) .$$

Indeed, Δ has a zero at each cusp of order N since $X(N)$ is ramified of order N over $j = \infty$. It is suggestive to keep this interpretation in mind when we deal with the element $Ns(G)$ in the group ring.

Lemma 2. $R_0 \cap (I\theta' + RNs(G)) = I_{12}\theta.$

Proof. Suppose $\xi \in I$ and

$$\deg(\xi\theta' + \nu Ns(G)) = 0$$

for some integer ν. Since $\deg\theta' = -N|G|/12$, we get

$$\tfrac{1}{12}\deg\xi = \nu , \qquad \text{and so} \qquad \deg\xi \equiv 0 \bmod 12 .$$

The inclusion $\subset$ is proved. The reverse inclusion is obvious.

From the lemma, we obtain an injection

$$0 \to R_0/I_{12}\theta \to R/(I\theta' + RNs(G)) .$$

We have $R_0+(I\theta'+RNs(G)) = R_d$ for some positive integer d, and

$$R_0/I_{12}\theta \approx [R_0+I\theta'+RNs(G)]/[I\theta'+RNs(G)]$$
$$\approx R_d/(I\theta'+RNs(G)),$$

where $d=$g.c.d.$(\deg(I\theta'), \deg(RNs(G)))$. By the degree of an ideal, we mean the degree of a positive generator for the degrees of all elements in the ideal. It is immediate that

$$d = \frac{N}{6}|G|,$$

because I has degree 2.

We have an inclusion of ideals:

$$R \supset R_d \supset I\theta'+RNs(G) \supset I\theta'.$$

We want the middle index for Theorem 3.2. It is obvious that

$$(R : R_d) = d.$$

We shall compute successively the indices

$$(R : I\theta') \quad \text{and} \quad (I\theta'+RNs(G) : I\theta')$$

to conclude the proof.

Lemma 3. $(I\theta'+RNs(G) : I\theta') = \frac{1}{12}|G|$.

Proof. By Noether isomorphism the left hand side is equal to

$$(RNs(G) : RNs(G) \cap I\theta').$$

We shift to modular forms. The module $I\theta'$ is the module of divisors of products of Klein forms satisfying the quadratic relations, and its intersection with $RNs(G)$ corresponds to the condition that such a product is equal to a power of Δ, that is

$$\prod_\alpha \mathfrak{k}_\alpha^{m(\alpha)} = \Delta^\nu$$

for some integer ν. Comparing weights, we must have

$$\nu = -\tfrac{1}{12}\sum_\alpha m(\alpha).$$

Since the Siegel function g_α satisfies $g_\alpha = \mathfrak{k}_\alpha\eta^2$, it follows that

$$\prod g_\alpha^{m(\alpha)} = \text{constant}.$$

From the independence of the Siegel functions for prime power level [KL 5], it follows that the exponents $m(\alpha)$ have to be constant, independent of α, say equal to an integer m. Then

$$\prod_\alpha \mathfrak{k}_\alpha^m = \Delta^{-m|G|/12}.$$

The lemma follows at once, because the family $\{m(\alpha)\}$ with $m(\alpha) = m$ satisfies the quadratic relations.

Next we have

$$(R : I\theta') = (R : R\theta')(R\theta' : I\theta').$$

The element θ' is invertible in the rational group ring $\mathbf{Q}[G]$ (cf. [KL 2], [KL 3], Theorem 2.3, and [KL 5]). Furthermore, since the Bernoulli polynomial is quadratic, the element $N\theta'$ lies in R. Hence

$$(R : RN\theta') = \pm\det_{\mathbf{C}[G]}N\theta',$$

where the determinant of multiplication by $N\theta'$ can be computed in the group algebra over the complex numbers, where the group algebra splits into its simple components, which are 1-dimensional, and correspond to the characters of G. Since

$$(R : R\theta') = \frac{(R : RN\theta')}{(R\theta' : RN\theta')},$$

and $(R\theta' : RN\theta') = N^{|G|}$, we find:

Lemma 4. $(R : R\theta') = \prod_{\chi} \chi(\theta') = (\deg\theta') \prod_{\chi \neq 1} \chi(\theta)$.

Proof. We have separated the product over all characters into one term with the trivial character, which gives $\deg\theta'$, and the other product, where θ' can be replaced by θ, since the sum of a non-trivial character over all group elements is 0. So the lemma is clear.

Lemma 5. $(R : I) = (R\theta' : I\theta') = N^3$.

Proof. We actually prove an isomorphism

$$R/I \approx \mathbf{Z}(N)^3.$$

By quadratic relations [Ku], §4, Theorem 7, we may first identify the elements of the Cartan group with primitive pairs (r_1, r_2) of integers $\bmod N$, and secondly we may identify the ideal I with the linear combinations of pairs satisfying the quadratic relations $\bmod N$. We then claim that the elements $(0,1)$, $(1,0)$, $(1,1)$ form a basis for R/I, in other words given a primitive pair (r_1, r_2) there exist unique integers x, y, z such that

$$(r_1, r_2) + x(1,0) + y(0,1) + z(1,1)$$

satisfies the quadratic relations $\bmod N$. This means:

$$r_1^2 + x + z \equiv 0 \bmod N,$$

$$r_2^2 + y + z \equiv 0 \bmod N,$$

$$r_1 r_2 + z \equiv 0 \bmod N,$$

or equivalently

$$z \equiv -r_1r_2$$
$$x \equiv -r_1r_2 - r_1^2$$
$$y \equiv -r_1r_2 - r_2^2.$$

This shows x, y, z exist. It is also clear that they are unique mod N, thus proving $(R : I) = N^3$.

The element θ' is invertible in the rational group ring, and so the map

$$x \mapsto x\theta'$$

is an isomorphism on R, so we have $(R : I) = (R\theta' : I\theta')$. This concludes the proof of the lemma.

The theorem is proved by putting all the lemmas together, with the value $d = \frac{N}{6}|G|$.

§3. Application to the Modular Curve $X_1(N)$

We use the standard way of describing cusps on $X(N)$ and $X_1(N)$. First for $X(N)$, the cusps are in bijection with column vectors

$$\pm\begin{bmatrix} x \\ y \end{bmatrix},$$

where $x, y \in \mathbf{Z}(N) = \mathbf{Z}/N\mathbf{Z}$, and $(x, y, N) = 1$. The cusps on $X_1(N)$ are orbit classes under the action of

$$\begin{pmatrix} 1 & 1 \\ 0 & 1 \end{pmatrix},$$

so under the equivalence

$$\pm\begin{bmatrix} x \\ y \end{bmatrix} \sim \pm\begin{bmatrix} x+ry \\ y \end{bmatrix} \quad \text{with} \quad r \in \mathbf{Z}.$$

We can take a normalized representative in such a class with

$$0 \leqq y < N \quad \text{and} \quad 0 \leqq x < (N, y).$$

We then distinguish three kinds of cusps.

C1. The cusps $\pm\begin{bmatrix} 0 \\ y \end{bmatrix}$. In the upper half plane representation these are the cusps lying above 0, and are characterized by $(N, y) = 1$.

C2. The cusps $\pm\begin{bmatrix} x \\ 0 \end{bmatrix}$. These are characterized by $(x, N) = 1$.

C3. All others.

The cusps **C1** and **C2** are interchanged by the involution w_N. The cusps **C1** are the rational cusps in Shimura's model.

The divisor group of degree 0 on $X_1(N)$ whose support lies among the cusps of first type will be denoted by $\mathscr{D}_1^0(N)$, and the corresponding divisor class group will be denoted by

$$\mathscr{C}_1^0(N) = \mathscr{D}_1^0(N)/\mathrm{div}\,\mathscr{F}_1^0(N)$$

where $\mathscr{F}_1^0(N)$ is the group of functions on $X_1(N)$ whose divisors have support in the cusps of first kind.

We let again $G = \mathbf{Z}(N)^*/\pm 1$. The cusps of first kind form a principal homogeneous set over G, so the divisor group $\mathscr{D}_1^0(N)$ can be identified with the augmentation ideal in the group ring $\mathbf{Z}[G]$. We are concerned with determining the subgroup $\mathrm{div}\,\mathscr{F}_1^0(N)$, and we shall describe the extent to which it differs from the Stickelberger ideal, see Theorem 3.3.

We wish to give a characterization of $\mathscr{F}_1^0(N)$. Usually we index the Siegel functions g_a with

$$a = (a_1, a_2) \in \mathbf{Q}^2/\mathbf{Z}^2 \bmod \pm 1, \quad a \neq 0.$$

Here we shall deal here only with a such that $a_1 = 0$. Consequently we shall abbreviate the notation, and write

$$g_a = g_{(0,a)} \quad \text{for} \quad a \in \mathbf{Q}/\mathbf{Z}, \quad a \neq 0, \quad Na = 0.$$

We let

$$T_N = (\mathbf{Q}/\mathbf{Z})_N/\pm 1,$$

where $(\mathbf{Q}/\mathbf{Z})_N$ denotes the subgroup of $\mathbf{Q}/\mathbf{Z}$ consisting of the elements of order N. We let T_N^* be the subset of T_N consisting of primitive elements (those with exact period N). Let:

$$\mathscr{F}_1'(N) = \text{group of products of siegel functions } g = \prod g_a^{m(a)}$$

with $a \in T_N$, $a \neq 0$, and integer exponents $m(a)$, satisfying the relations:

D 1. $\sum m(a) = 0.$

D 2. The quadratic relation mod N, that is:

$$\sum m(a)(Na)^2 \equiv 0 \bmod N.$$

Since $g_a = \mathfrak{k}_a \eta^2$ where $\mathfrak{k}_a$ is the Klein form, we see from condition **D 1** that we could have written $\mathfrak{k}_a$ instead of g_a in the product. By [Ku], the quadratic relations are necessary to make g have level N, and **D 1**, **D 2** are sufficient. (Actually, the quadratic relations and the condition 12 divides $\sum m(a)$ are necessary and sufficient.)

The divisor group of $\mathscr{F}_1'(N)$ is easily determined since we know the order of the Klein form. Indeed, from the q-expansion:

$$\begin{aligned} \mathrm{ord}_{\left[\begin{smallmatrix} x \\ y \end{smallmatrix}\right]} \mathfrak{k}_a &= \frac{1}{2}\left(\mathbf{B}_2(\langle ay \rangle) - \frac{1}{6}\right) \\ &= \frac{1}{2}\langle ay \rangle(\langle ay \rangle - 1). \end{aligned}$$

The order is taken with respect to the parameter $q = e^{2\pi i\tau}$, even though there may be ramification over q. So the order is 0 for any cusp of type **C 2**:

$$\operatorname{ord}_{\left[\begin{smallmatrix} x \\ 0 \end{smallmatrix}\right]} \mathfrak{k}_a = 0.$$

Let $\mathscr{F}_1''$ = subgroup of functions $f \in \mathscr{F}_1'$ such that:

$$\operatorname{ord}_{\left[\begin{smallmatrix} x \\ y \end{smallmatrix}\right]} f = \operatorname{ord}_{\left[\begin{smallmatrix} x' \\ y \end{smallmatrix}\right]} f \quad \text{for any pair } x, x'$$

$$\operatorname{ord}_{\left[\begin{smallmatrix} x \\ 0 \end{smallmatrix}\right]} f = 0.$$

Then $\mathscr{F}_1'' \subset \mathscr{F}_1'$.

Theorem 3.1. *We have* $\mathscr{F}_1' = \mathscr{F}_1''$.

Proof. It is clear that $\mathscr{F}_1''$ has no cotorsion in the group $\mathscr{F}_1(N)$ of units on $X_1(N)$, i.e. $\mathscr{F}_1(N)/\mathscr{F}_1''(N)$ has no torsion. We contend that $\mathscr{F}_1'$ also has no cotorsion. To see this, suppose that

$$g = \prod g_a^{m(a)}$$

is in $\mathscr{F}_1'$ and $g = h^l$, where h is modular and l is prime. Let $m(a) \neq 0$ in the product for some a of maximal level. By Lemma 4.4 of Theorem 3 in [KL 4] we must have

$$m(a) \equiv 0 \bmod l.$$

Indeed, we can use an element $b = \left(\frac{1}{p}, a_2\right)$ so b has the same level as $a_2 = a$. Then $m(b) = 0$ and $m(a) \equiv m(b) \bmod l$. In this way we can absorb in h those factors of the product with a of maximal level, and continue by induction to see that h also has a product expression satisfying **D 1**. If h is on $X_1(N)$ then **D 2** is also satisfied, as desired.

To prove the theorem, it now suffices to prove that $\mathscr{F}_1'$ and $\mathscr{F}_1''$ have the same rank. The functions g_a with $a \in T_N$, $a \neq 0$, are independent, modulo constants. (The argument in Theorem 3.1 yields a proof, selecting infinitely many l for instance.) Hence

$$\operatorname{rank} \mathscr{F}_1'(N) = \frac{N-1}{2} - 1$$

because there are $(N-1)/2$ elements $\pm a$ with $a \neq 0$, and we subtract 1 for the relation $\sum m(a) = 0$.

On the other hand, the group $\mathscr{F}_1''$ is also defined by linear conditions on the divisors, whose orders depend only on y. There are $(N-1)/2$ elements $\pm y$ (other than 0), and we again subtract 1 to account for the fact that the sum of the orders of zeros of a function is 0. This proves Theorem 2.1.

It is clear that

$$\mathscr{F}_1^0 \subset \mathscr{F}_1'.$$

We wish to describe the position of $\mathscr{F}_1^0$ in $\mathscr{F}_1'$. Note that $\frac{1}{p}\mathbf{Z}/\mathbf{Z}$ operates on T_N by addition. The condition

D 3. For every orbit of $\frac{1}{p}\mathbf{Z}/\mathbf{Z}$, $\quad \sum_{a \in \text{orbit}} m(a) = 0$

defines a subgroup of $\mathscr{F}_1'$, which is contained in $\mathscr{F}_1^0$. Note that **D 3** implies **D 1**. Also, if $N = p$, then the sum in **D 3** is to be interpreted as taken over T_p^*.

Theorem 3.2. *The group of elements in $\mathscr{F}_1'$ satisfying* **D 3** *is equal to $\mathscr{F}_1^0$.*

Proof. Denote by $\mathscr{F}_1'(\mathbf{D}\,3)$ the group of elements in $\mathscr{F}_1'$ satisfying **D 3**. The same argument as in Theorem 3.1 shows that $\mathscr{F}_1^0$ and $\mathscr{F}_1'(\mathbf{D}\,3)$ have no cotorsion. Again we are reduced to showing that they have the same rank. The rank of $\mathscr{F}_1^0$ is

$$\operatorname{rank}\mathscr{F}_1^0 = \frac{\phi(N)}{2} - 1.$$

The rank of $\mathscr{F}_1'(\mathbf{D}\,3)$ is determined by a linear condition for each orbit, and the number of orbits of $\frac{1}{p}\mathbf{Z}/\mathbf{Z}$ in T_N is equal to

$$\frac{p^{n-1}-1}{2} + 1.$$

On the other hand, the linear condition **D 1** is implied by **D 3**. Hence

$$\operatorname{rank}\mathscr{F}_1'(\mathbf{D}\,3) = \frac{N-1}{2} - \frac{p^{n-1}-1}{2} - 1 = \frac{\phi(N)}{2} - 1,$$

as was to be shown.

Let pr_0 be the projection operator, which projects a divisor on the subgroup of divisors having support in the cusps of first type **C 1**. Then

$$\mathrm{pr}_0(g_a) = N \sum_{b \in G} \tfrac{1}{2}\mathbf{B}_2(\langle ab \rangle)\sigma_b^{-1}.$$

If $g \in \mathscr{F}_1^0$ and

$$g = \prod g_a^{m(a)},$$

where the family $\{m(a)\}$ satisfies **D 1**, **D 2**, **D 3**, then

$$\begin{aligned}(g) = \mathrm{pr}_0(g) &= N\sum_a m(a) \sum_b \tfrac{1}{2}\mathbf{B}_2(\langle ab \rangle)\sigma_b^{-1} \\ &= N\sum_b \sum_a m(a)\tfrac{1}{2}\mathbf{B}_2(\langle ab \rangle)\sigma_b^{-1}.\end{aligned}$$

But by the distribution relation, lifting non-primitive elements a to primitive ones, we have

$$\sum_a m(a)\mathbf{B}_2(\langle ab \rangle) = \sum_{r=0}^{n-1} p^r \sum_z m(p^r z)\mathbf{B}_2(\langle zb \rangle),$$

where the sum over z is taken for

$$z \in T_N^*,$$

i.e. for primitive elements z. Let

$$Z_N = \mathbf{Z}(N)/\pm 1 \quad \text{and} \quad Z_N^* = \mathbf{Z}(N)^*/\pm 1.$$

Changing variables, letting $b \mapsto c^{-1}b$ where $c = Nz \in Z_N^*$, we find

$$(g) = \sum_{c \in Z_N^*} \sum_{r=0}^{n-1} p^r m(p^r c/N)\sigma_c \cdot \theta,$$

where θ is the Stickelberger element discussed in §1.

Let J_1 = ideal in $R = \mathbf{Z}[G]$ generated by all elements

$$\sum_{c \in Z_N} \sum_{r=0}^{n-1} p^r m(p^r c/N)\sigma_c,$$

where m ranges over all functions satisfying **D 1**, **D 2**, **D 3**. Then we have shown:

Theorem 3.3. *The group of divisors of $\mathscr{F}_1^0(N)$ is given by*

$$\operatorname{div}\mathscr{F}_1^0(N) = J_1\theta,$$

and thus we have a natural isomorphism

$$\mathscr{C}_1^0(N) \approx R_0/J_1\theta.$$

We shall now analyse the ideal J_1 further combinatorially. This does not involve modular forms, only combinatorial juggling with ideals in the group ring. The ideal J_1 is easily seen to be contained in Q_0, cf. §1, Lemma 1, and also Lemma 1 of §4, so

$$J_1\theta \subset Q_0\theta = Q_0\theta' = \mathscr{S}_2,$$

in other words $J_1\theta$ is contained in the Stickelberger ideal, and $R_0/\mathscr{S}$ appears as a natural factor group of $\mathscr{C}_0^1(N)$. We shall determine the index

$$(R_0 : J_1\theta).$$

Since we already know $(R_0 : \mathscr{S})$, it suffices to determine $(\mathscr{S} : J_1\theta)$ (carried out in the next section). We shall find:

Theorem 3.4. $(\mathscr{S} : J_1\theta) = p^{p^{n-1}}/Np^{2n-2}$, *and therefore*

$$\text{order of } \mathscr{C}_1^0(N) = \frac{p^{p^{n-1}}}{p^{2n-2}} \prod_{\chi \neq 1} \tfrac{1}{2}B_{2,\chi}.$$

If $n = 1$, $N = p$ then $\mathscr{S} = J_1\theta$ and

$$\mathscr{C}_1^0(p) \approx R_0/\mathscr{S}.$$

To analyse $\mathscr{S}/J_1\theta$ geometrically, it is necessary to enter into considerations of special divisor classes, analogous to those already discussed in [KL 6], which can be exhibited explicitly and have to be factored out to get a divisor class group precisely isomorphic to $R_0/\mathscr{S}$.

Note that the result for $n = 1, N = p$ that $\mathscr{S} = J_1\theta$ and

$$\mathscr{C}_1^0(p) \approx R_0/\mathscr{S},$$

is compatible with Theorems 5.1 and 6.1 of [KL 6], where there was no special group at level p.

Since θ is invertible in R, the map multiplication by θ is injective on R. Consequently

$$(\mathscr{S} : J_1\theta) = (Q_0\theta : J_1\theta) = (Q_0 : J_1).$$

In the next section, we analyse the factor group

$$Q_0/J_1 \approx \mathscr{S}/J_1\theta.$$

We conclude this section by pointing out that Theorem 3.4 gives an explicit upper and lower bound for the cuspidal class number. For simplicity take $N=p$ prime. Using the functional equation for the L-series at $s=2$, $1-s=-1$, we find:

$$h_1^0(p) = pp^{-(p-3)/4}\left(\frac{p}{2\pi}\right)^{p-3}\prod_{\chi \neq 1} L(2,\chi).$$

The dominant term is therefore $p^{3p/4}$, and other terms behave at most like a constant to the p-th power. Furthermore, we have the bounds

$$c_1^{(p-3)/2} \leqq \prod_{\chi \neq 1} |L(2,\chi)| \leqq c_2^{(p-3)/2}$$

where

$$c_1 = \prod (1+1/l^2)^{-1} \quad \text{and} \quad c_2 = \prod (1-1/l^2)^{-1}.$$

The products are taken over all primes l, and are absolutely convergent. The situation is here easier to handle than the analogous situation for the cyclotomic case, where one has to look at $L(1,\chi)$, as was done by Masley-Montgomery. J. Reine angew. Math. 286 (1976), pp. 248–256.

§4. Computation of a Class Number

Recall that Q, defined in §1, is the ideal of elements in the group ring R satisfying the quadratic relations.

Let J_Q be the ideal in R generated by the elements

$$\sum_{c \in Z_N^*} \sum_{r=0}^{n-1} p^r m\left(p^r \frac{c}{N}\right)\sigma_c$$

with families $\{m(a)\}$ satisfying the quadratic relations **D 2**. Then the coefficient $\{m'(c)\}$ of σ_c is given by the expression

$$m'(c) = \sum_{r=0}^{n-1} p^r m\left(p^r \frac{c}{N}\right).$$

Lemma 1. $J_Q \subset Q$.

Proof. We have

$$\sum_{a \in T_N} m(a)(Na)^2 = \sum_{r=0}^{n-1} \sum_{a \in T_{p^{n-r}}} m(a)(Na)^2.$$

Each $a \in T_{p^{n-r}}$ can be written $a = p^r c/N$ with $c \in Z_N^*$, and there are p^r such possible values of c. Hence for $a \in T_{p^{n-r}}$ we get:

$$m(a)(Na)^2 = \frac{1}{p^r} \sum_{c \in Z_N^*} m\left(p^r \frac{c}{N}\right)\left(Np^r \frac{c}{N}\right)^2$$

$$= \sum_{c \in Z_N^*} p^r m\left(p^r \frac{c}{N}\right) c^2 .$$

Summing over r shows that $\{m(a)\}$ satisfies **D 2** if and only if $\{m'(c)\}$ satisfies the quadratic relations, and proves the lemma.

Let $J_{T_p} = J_{T_p}(N)$ be the ideal in R_0 consisting of elements

$$\sum_{a \in T_N^*} \sum_{r=0}^{n-1} p^r m(p^r a) \sigma_{Na}$$

such that the family $\{m(a)\}$ satisfies the orbit condition **D 3**.

Lemma 2. *We have an injection*

$$Q_0/J_1 \to R_0/J_{T_p},$$

and the index of the image is $N/p = p^{n-1}$.

Proof. The natural map induced by inclusion is injective by Lemma 1. We have to analyse its image. Let

$$\psi \colon R_0 \to \mathbf{Z}/N\mathbf{Z}$$

be the map

$$\sum m'\left(\frac{b}{N}\right)\sigma_b \mapsto \sum m'\left(\frac{b}{N}\right) b^2 .$$

The kernel of ψ is precisely Q_0. It will suffice to show that the image $\psi(J_{T_p})$ is equal to $p^{n-1}\mathbf{Z}/N\mathbf{Z}$. Consider first an orbit of maximal level, and a function m' on this orbit such that

$$\sum_{x \in \text{orbit}} m'(x) = 0,$$

and m' is 0 outside this orbit. Write $x = b/N$, and select some element $x_0 = b_0/N$ in the orbit. Then the image under ψ of

$$\sum m'\left(\frac{b}{N}\right)\sigma_b$$

is

$$\sum_{b \neq b_0} m'\left(\frac{b}{N}\right)(b^2 - b_0^2) .$$

We have $b \equiv b_0 \bmod p^{n-1}$. The function m' can be chosen arbitrarily on $b \neq b_0$, so we can choose it so that one term in the sum is divisible exactly by p^{n-1} and the other terms are 0. It is then clear that the image of elements in J_{T_p} formed with functions m which are 0 outside cosets of maximal level is precisely $p^{n-1}\mathbf{Z}/N\mathbf{Z}$. A similar argument applied to elements in J_{T_p} formed with functions m which are 0 outside cosets of intermediate level shows that the image of such elements is contained in $p^{n-1}\mathbf{Z}/N\mathbf{Z}$. This proves Lemma 2.

It is to a certain extent inconvenient to go back and forth between T_N and Z_N (isomorphic under multiplication with N). Hence we now eliminate the group ring notation, and use functional notation, viewing elements of the group ring as functions on T_N^*. Thus we let:

$F_0(T_N^*)$ = additive group of $\mathbf{Z}$-valued functions on T_N^*, of degree 0, i.e. satisfying

$$\sum_{x \in T_N^*} f(x) = 0.$$

Let $M|N$, $M = p^{n-s}$. Let $\pi_s\colon T_N \to T_M$ be multiplication by p^s. Let φ be a function on T_M^*. We define its **lifting**

$$\pi_s^*\varphi(y) = p^s\varphi(p^s y) \quad \text{for} \quad y \in T_N^*.$$

Thus π_s^* is a homomorphism from $F_0(T_M^*)$ into $F_0(T_N^*)$.

With this functional notation, we see that the ideal J_{T_p} is generated by functions of the following type. We select an orbit C of $\frac{1}{p}\mathbf{Z}/\mathbf{Z}$ in $T_{p^{n-s}}^*$, and we let $m = m_{C,s}$ be any function such that

$$m(x) = 0 \quad \text{if} \quad x \notin C \quad \text{and} \quad \sum_{a \in C} m(a) = 0.$$

We take the lifted function $\pi_s^* m$. Such lifted functions generate J_{T_p}.

We shall characterize J_{T_p} by another condition which allows us to perform an inductive procedure, ultimately leading to the desired computation of the index. Let:

$F_0^{(1)}(T_N^*)$ = subgroup of functions f such that for every r with $1 \leqq r < n$ and every orbit K of $\frac{1}{p^r}\mathbf{Z}/\mathbf{Z}$ in T_N^* we have

$$\sum_{x \in K} f(x) \equiv 0 \bmod p^{2r},$$

and also

$$\sum_{x \in T_N^*} f(x) = 0.$$

Lemma 3. $J_{T_p} = F_0^{(1)}(T_N^*)$.

Proof. We first prove the inclusion $\subset$. Let $m = m_{C,s}$ be one of the functions described above, such that the lifted function

$$\pi_s^* m$$

is one of the generators of J_{T_p}. Let K be an orbit of $\frac{1}{p^r}\mathbf{Z}/\mathbf{Z}$ in T_N^*. We have to verify that for $r < n$,

$$\sum_{z \in K} p^s m(p^s z) \equiv 0 \bmod p^{2r}.$$

Let $K' = \pi_s^{-1}(C)$. Three cases arise:

Case 1. $K \cap K'$ is empty. The congruence condition is clear.

Case 2. $K \supset K'$. The condition is again clear, since

$$\sum_{x \in C} m_{C,s}(x) = 0.$$

Case 3. $K \subset K'$ and $K \neq K'$. Then $r < s+1$, so $r \leqq s$ and $m(p^s z)$ is constant for $z \in K$, so we get the congruence

$$\sum_{z \in K} p^s m(p^s z) \equiv 0 \bmod p^{r+s},$$

whence $\bmod p^{2r}$ as desired.

Conversely, let $f \in F_0^{(1)}(T_N^*)$. We follow an inductive descending procedure, subtracting an appropriate function from J_{T_p} to reduce the level. For each orbit C of $\frac{1}{p}\mathbf{Z}/\mathbf{Z}$ we define

$$g_C(x) = f(x) - \frac{1}{p}\sum_{y \in C} f(y) \quad \text{if} \quad x \in C,$$

$$g_C(x) = 0 \quad \text{if} \quad x \notin C.$$

Let

$$g(x) = \sum_C g_C(x).$$

Then it is immediately verified that $g \in J_{T_p}$ and we can define a function φ by

$$(f-g)(x) = p \sum_{pz = x} \varphi(px),$$

since

$$\sum_{y \in C} f(y) \equiv 0 \bmod p^2.$$

Then $\varphi \in F_0^{(1)}(T_{N/p}^*)$. Since

$$\pi_1^* J_{T_p}(N/p) \subset J_{T_p}(N),$$

the converse inclusion follows by induction.

Let $F^{(1)}(T_N^*)$ be defined by the same congruence relations as $F_0^{(1)}(T_N^*)$, but omitting the degree 0 condition, i.e. only by the congruences for every r with $1 \leqq r < n$ and every orbit K of $\frac{1}{p^r}\mathbf{Z}/\mathbf{Z}$ in T_N^*,

$$\sum_{x \in K} f(x) \equiv 0 \bmod p^{2r}.$$

It is immediately verified that the augmentation homomorphism applied to $F^{(1)}(T_N^*)$ yields

$$\deg F^{(1)}(T_N^*) = (p^{2n-2}).$$

We have an injection

$$R_0/F_0^{(1)} \to R/F^{(1)},$$

from which we already see that Theorem 3.4 follows when $n = 1$, i.e. $R_0/F_0^{(1)}$ is trivial, and Lemmas 2, 3 conclude the proof.

The degree stated above shows that the image of the injection is

$$p^{2n-2}R/F^{(1)}.$$

Suppose that $n \geqq 2$. Then from the value of augmentation we find:

Lemma 4. $(R/F^{(1)} : R_0/F_0^{(1)}) = p^{2n-2}$.

We define a filtration

$$F^{(1)} \subset F^{(2)} \subset \dots \subset F^{(n)} = R$$

by letting:

$F^{(s)}$ = group of functions f on T_N^* such that for every r with $0 \leqq r \leqq n-s$ and every orbit K of $\frac{1}{p^r}\mathbf{Z}/\mathbf{Z}$ we have

$$\sum_{x \in K} f(x) \equiv 0 \bmod p^{2r}.$$

We shall determine the factor group $F^{(s)} \bmod F^{(s-1)}$. We let:

$$\tau(s) = \text{number of orbits of } \frac{1}{p^{n-s}}\mathbf{Z}/\mathbf{Z} \text{ in } T_N^* = \frac{p-1}{2}p^{s-1}.$$

Lemma 5. $(F^{(s)} : F^{(s-1)}) = p^{2\tau(s-1)}$, *for* $s \geqq 2$.

Proof. To each function $f \in F^{(s)}$ we associate a vector

$$f \mapsto \left(\dots, \sum_{x \in K} f(x), \dots\right) \bmod p^{2(n-s+1)}$$

whose components are indexed by orbits K of $\frac{1}{p^{n-s+1}}\mathbf{Z}/\mathbf{Z}$ in T_N^*. This gives rise to a homomorphism

$$\begin{aligned}\psi_s\colon F^{(s)} &\to (p^{2(n-s)}\mathbf{Z}/p^{2(n-s+1)}\mathbf{Z})^{\tau(s-1)} \\ &\cong \mathbf{Z}(p^2)^{\tau(s-1)}\end{aligned}$$

into the product of $\mathbf{Z}(p^2)$ taken $\tau(s-1)$ times. The kernel is clearly $F^{(s-1)}$, and it is easy to verify that the map is surjective. Indeed, we just consider functions divisible by $p^{2(n-s)}$, which thus belong to $F^{(s)}$, and otherwise arbitrary. This proves the lemma.

Taking the product over s in Lemma 5 yields the index

$$(F^{(n)} : F^{(1)}),$$

and $F^{(n)} = R$. Combining this with Lemma 4 yields the value as in Theorem 3.4.

Bibliography

[Co 1] J. Coates, *On K_2 and some classical conjectures in algebraic number theory*, Ann. of Math. 95 (1972) pp. 99–116

[Co 2] J. Coates, *K-theory and Iwasawa's analogue of the Jacobian*, Algebraic K-theory II, Springer Lecture Notes 342 (1973) pp. 502–520

[C-L] J. Coates and S. Lichtenbaum, *On l-adic zeta functions*, Ann. of Math. 98 (1973) pp. 498–550

[C-S] J. Coates and W. Sinnott, *On p-adic L-functions over real quadratic fields*, Inv. Math. 25 (1974) pp. 253–279

[C-S 2] J. Coates and W. Sinnott, *An analogue of Stickelberger's theorem for higher K-groups*, Invent. Math. 24 (1974) pp. 149–161

[C-S 3] J. Coates and W. Sinnott, *Integrality properties of the values of partial zeta functions*, Proc. London Math. Soc. (1977) pp. 365–384

[Iw] K. Iwasawa, *A class number formula for cyclotomic fields*, Ann. of Math. 76 (1962) pp. 171–179

[Kl] P. Klimek, Thesis, Berkeley, 1975

[Ku] D. Kubert, *Quadratic relations for generators of units in the modular function field*, Math. Ann. 225 (1977) pp. 1–20

[KL 1] D. Kubert and S. Lang, *Units in the modular function field I*, Math. Ann. 218 (1975) pp. 67–96

[KL 2] D. Kubert and S. Lang, Idem II, *A full set of units*, pp. 175–189

[KL 3] D. Kubert and S. Lang, Idem III, *Distribution relations*, pp. 273–285

[KL 4] D. Kubert and S. Lang, Idem IV, *The Siegel functions are generators*, Math. Ann. 227 (1977) pp. 223–242

[KL 5] D. Kubert and S. Lang, *Distributions on toroidal groups*, Math. Zeit. 148 (1976) pp. 33–51

[KL 6] D. Kubert and S. Lang, *The p-primary component of the cuspidal divisor class group on the modular curve $X(p)$*, Math. Ann. 234 (1978) pp. 25–44

[KL 7] D. Kubert and S. Lang, *Stickelberger ideals*, Math. Ann. 237 (1978) 203–212

[L] S. Lang, Introduction to modular forms, Springer Verlag, 1976

[Le] A. Leutbecher, *Über Automorphiefaktoren und die Dedekindschen Summen*, Glasgow Math. J. 11 (1970) pp. 41–57

[Si] W. Sinnott, *On the Stickelberger ideal and the circular units of a cyclotomic field*, to appear, Annals of Math.

[Wo] K. Wohlfahrt, *Über Dedekindsche Summen und Untergruppen der Modulgruppe*, Hamburg Abh. 23 (1959) pp. 5–10

Received July 12, 1977

Séminaire Delange-Pisot-Poitou
(Théorie des nombres)
19e année, 1977/78, n° 40, 6 p. 22 mai 1978

Relations de Distributions et Exemples Classiques

Serge Lang

A la suite des travaux d'Iwasawa sur les limites projectives de classes d'idéaux et d'unités cyclotomiques, Mazur a extrait la notion de "distribution" sur un système projectif. Cette structure algébrique élémentaire est comme la prose de Mr Jourdain: Nous en avons tous vu, sans lui avoir donné ce nom. Le but de cet exposé est de donner la définition, et de l'illustrer avec un tas d'exemples classiques. Nous énoncerons aussi un théorème fondamental de Kubert donnant la structure des distributions universelles sur $\mathbf{Q}/\mathbf{Z}$ (et plus généralement sur $\mathbf{Q}^k/\mathbf{Z}^k$), qui sont les plus fréquentes.

Soient $\{X_n\}$ une suite d'ensembles, et $\pi_{n+1}\colon X_{n+1} \to X_n$ des applications surjectives, de sorte qu'on peut considérer la limite projective

$$X \to \cdots \to X_{n+1} \to X_n \to \cdots \to X_1.$$

Soit A un groupe abélien. Pour chaque n, supposons donnée une fonction

$$\varphi_n\colon X_n \to A.$$

On dit que la famille $\{\varphi_n\}$ est une *distribution* si, pour chaque n et $x \in X_n$, on a la relation

$$\varphi_n(x) = \sum_{\pi_{n+1} y = x} \varphi_{n+1}(y).$$

La somme est prise sur tous les éléments $y \in X_{n+1}$ ci-dessus de x.

Soit f une fonction sur X_m pour un entier m donné, à valeur dans un anneau d'opérateurs sur A. Alors, on peut considérer f comme étant définie sur X_n, pour tout $n \geq m$ en composant avec la projection naturelle de X_n sur X_m. On conclut alors de la relation de distribution que

$$\sum_{x \in X_n} f(x)\varphi_n(x) = \sum_{x \in X_m} f(x)\varphi_m(x).$$

Soit X la limite projective des X_n. Une fonction f, comme ci-dessus, est appelée localement constante. On peut alors définir son intégrale

$$\int f\, d\varphi = \sum_{x \in X_n} f(x)\varphi_n(x)$$

d'où le nom de distribution.

On a indexé le système projectif par les entiers positifs. On peut, bien entendu, prendre d'autres indices. Dans les applications que nous avons en vue, on prendra les entiers ordonnés par divisibilité. L'exemple le plus standard est alors le système $\{\mathbf{Z}/N\mathbf{Z}\}$, isomorphe au système $\{((1/N)\mathbf{Z})/\mathbf{Z}\}$, avec le diagramme commutatif pour $M \mid N$:

$$\begin{array}{ccc} & ((1/N)\mathbf{Z})/\mathbf{Z} \xrightarrow{N} \mathbf{Z}/N\mathbf{Z} & \\ \text{mult. par } N/M \downarrow & & \downarrow r_M = \text{réduction mod } M \\ & ((1/M)\mathbf{Z})/\mathbf{Z} \xrightarrow{M} \mathbf{Z}/M\mathbf{Z} & \end{array}$$

On notera que le système projectif $\{(1/N)\mathbf{Z}/\mathbf{Z}\}$ a aussi une structure injective, car tous les $(1/N)\mathbf{Z}/\mathbf{Z}$ sont contenus dans $\mathbf{Q}/\mathbf{Z}$. Une fonction φ sur $\mathbf{Q}/\mathbf{Z}$ sera appelée une *distribution ordinaire* si sa restriction à chaque $(1/N)\mathbf{Z}/\mathbf{Z}$ satisfait à la relation de distribution, à savoir si on a, pour chaque entier positif N, la relation

$$\sum_{i=0}^{N-1} \varphi\left(x + \frac{i}{N}\right) = \varphi(Nx), \quad \text{pour tout } x \in \mathbf{Q}/\mathbf{Z}.$$

On dira que c'est une distribution de *poids d* si

$$N^d \sum_{i=0}^{N-1} \varphi\left(x + \frac{i}{N}\right) = \varphi(Nx).$$

Il arrive, dans la pratique, que ces relations de distributions sont satisfaites, sauf quand $Nx = 0$ dans $\mathbf{Q}/\mathbf{Z}$, auquel cas on dira alors que la distribution est *trouée*.

Si une distribution prend ses valeurs dans un groupe tel que la multiplication par 2 est inversible, alors on peut former sa partie paire et sa partie impaire.

L'image holomorphe d'une distribution est une distribution, et l'on peut donc parler de la distribution universelle ordinaire, par exemple.

Exemple 1. Distribution de Bernoulli. Si $t \in \mathbf{Q}/\mathbf{Z}$, on note $\langle t \rangle$ le nombre tel que $0 \leq \langle t \rangle < 1$ et $\langle t \rangle \equiv t \bmod \mathbf{Z}$. Pour chaque entier positif k, il existe un polynôme monique B_k, de degré k, tel que le système de fonctions

$$x \mapsto N^{k-1} B_k\left(\left\langle \frac{x}{N} \right\rangle\right), \quad \text{pour } x \in \mathbf{Z}/N\mathbf{Z},$$

soit une distribution, et ce polynôme est le polynôme de Bernoulli, défini classiquement par la série

$$\frac{t \exp tX}{(\exp t) - 1} = \sum B_k(X) \frac{t^k}{k!}.$$

Exemple 2. Distribution de Bernoulli-Fourier. Pour θ réel, on a

$$B_k(\langle\theta\rangle) = -\frac{k!}{(2\pi i)^k}\sum_{n\neq 0}\frac{\exp 2\pi i n\theta}{n^k}.$$

On peut donc définir la fonction de distribution sur $\mathbf{R}/\mathbf{Z}$, par la série de Fourier, et la relation de distribution de poids $k-1$ est satisfaite sur $\mathbf{R}/\mathbf{Z}$. On peut aussi ne prendre qu'une partie de la série de Fourier, par exemple la partie holomorphe,

$$f_k(z) = \sum_{n=1}^{\infty}\frac{z^n}{n^k},$$

avec $z = \exp 2\pi i\theta$, de sorte que $\{N^{k-1} f_k\}$ définit une distribution. La partie réelle, pour k pair, et la partie imaginaire, pour k impair, sont des images holomorphes de celle-ci, et donnent lieu à la distribution de Bernoulli.

Exemple 3. Fonctions zêta partielles. Soit

$$\zeta(s, u) = \sum_{n=0}^{\infty}\frac{1}{(n+u)^s}$$

la fonction de Hurwitz, avec $0 < u \leq 1$. On pose $\{t\}$ égal au nombre réel dans la classe de t mod $\mathbf{Z}$ tel que $0 < \{t\} \leq 1$. Alors, il est immédiat que la fonction

$$x \mapsto N^{-s}\zeta\left(s, \left\{\frac{x}{N}\right\}\right), \quad \text{pour } x \in \mathbf{Z}/N\mathbf{Z},$$

définit une distribution trouée.

Exemple 4. La distribution gamma. Posons $G(z) = (1/\sqrt{2\pi})\Gamma(z)$. Une relation classique de la fonction gamma donne

$$\prod_{i=0}^{N-1} G\left(x + \frac{i}{N}\right) = N^{\frac{1}{2}-Nx}G(Nx),$$

si x est un nombre complexe tel que $Nx \not\equiv 0$ mod $\mathbf{Z}$. Si on considère que les valeurs de $G(x)$ sont prises dans $\mathbf{C}^*/\mathbf{Q}_a^*$ (groupe facteur de $\mathbf{C}^*$ par les nombres algébriques $\neq 0$), alors on voit que G définit une distribution trouée, sur $\mathbf{Q}/\mathbf{Z}$, car le facteur $N^{\frac{1}{2}-Nx}$ est alors algébrique. De plus, c'est une distribution impaire, i.e.

$$G(-x) = G(x)^{-1}.$$

Rohrlich a conjecturé que G est la distribution impaire trouée universelle, c'est-à-dire qu'il n'y a pas d'autres relations que les relations de distribution, et la parité. Ceci est une conjecture appartenant à la théorie des nombres transcendants. Elle mène à la question (d'indépendance algébrique) si les relations de distribution et la parité (ainsi que l'équation fonctionnelle) engendrent un idéal de définition sur les nombres algébriques, pour toutes les relations algébriques des valeurs de la fonction gamma prises pour x rationnel $\neq 0$.

D'une manière génerale, on s'attend à ce que toutes les distributions sur $\mathbf{Q}/\mathbf{Z}$, qu'on rencontre de façon "naturelle" soient universelles. On verra plus loin qu'il suffit de déterminer leur rang pour le démontrer.

Exemple 5. Unités cyclotomiques. L'application

$$x \mapsto (\exp 2\pi i x) - 1, \text{ avec } x \in \mathbf{Q}/\mathbf{Z},\ x \neq 0,$$

définit une distribution paire ordinaire trouée à valeurs dans $\mathbf{C}^*/\boldsymbol{\mu}$, où $\boldsymbol{\mu}$ est le groupe des racines de l'unité. C'est immédiat, à partir de la relation

$$X^N - 1 = \prod(\zeta X - 1);$$

le produit étant pris sur toutes les racines N-ièmes de l'unité.

Exemple 6. Unités modulaires. Les fonctions de Siegel g_a, indexées par $a \in \mathbf{Q}^2/\mathbf{Z}^2$, $a \neq 0$, définissent une distribution ordinaire trouée sur $\mathbf{Q}^2/\mathbf{Z}^2$, si on considère ces fonctions modulo les racines de l'unité (relations de Klein, Ramachandra, Robert).

Il existe maints autres exemples semblables à ceux-ci, dans la théorie des unités locales, et dans la K-théorie. Ceux-ci sont en train de s'élaborer en une théorie générale, encore à son début, et nous ne désirons pas entrer dans des considérations plus techniques pour le présent exposé.

Exemple 7. Distribution de Lobačevskij. Milnor a été conduit aux relations de distributions en étudiant les volumes de tétraèdres en géométrie hyperbolique. Je lui dois l'exposé de cet exemple. Considérons la fonction de Lobačevskij

$$\lambda(x) = -\int_0^x \log|2\sin t|\,dt.$$

Essentiellement, c'est l'intégrale

$$\int_0^x \log|(\exp 2\pi i t) - 1|\,dt,$$

et l'on vérifie immédiatement que λ définit une distribution impaire trouée de poids 1 sur $\mathbf{Q}/\mathbf{Z}$.

D'autre part, considérons l'espace hyperbolique à coordonnées (x_1, x_2, y), avec $x_1, x_2 \in \mathbf{R}$ et $y > 0$. Etant donnés 4 points distincts dans le plan (x_1, x_2), c'est-à-dire des points à l'infini, on peut former le tétraèdre dont les arêtes sont des demi-cercles joignant ces points dans le demi-espace supérieur. C'est un théorème classique que les angles (dits diédraux) cotoyant les arêtes opposées sont égaux. On a donc trois paires d'angles α, β, γ. C'est un théorème classique que le volume V du tetraèdre T est donné par la formule

$$V = \iiint_T dx_1 dx_2 dy/y^3 = \lambda(\alpha) + \lambda(\beta) + \lambda(\gamma).$$

Milnor conjecture que la relation de parité et les relations de distribution forment une base des relations pour la fonction de Lobačevskij sur $\mathbf{Q}$.

En vertu d'un théorème de Kubert, que nous allons énoncer, ceci équivaudrait à dire que la fonction de Lobačevskij est la distribution universelle impaire de poids 1 sur $\mathbf{Q}/\mathbf{Z}$.

Théorème de Kubert et distributions universelles. Considérons d'abord les distributions ordinaires de poids zéro. Soit F le groupe abélien libre dont les générateurs sont les éléments de $\mathbf{Q}/\mathbf{Z}$ (on peut appeler les éléments de ce groupe diviseurs). Alors F contient le sous-groupe engendré par les relations de distribution D, et on désire connaître la structure du groupe facteur F/D.

Pour chaque N, on peut aussi considérer le groupe libre F_N, engendré par les éléments de $(1/N)\mathbf{Z}/\mathbf{Z}$, modulo les relations de distribution sur $(1/N)\mathbf{Z}/\mathbf{Z}$, soit F_N/D_N. Kubert montre que F_N/D_N est libre de rang $\phi(N)$ (fonction d'Euler), et construit une base "canonique" comme suit.

Posons $Z_N = (1/N)\mathbf{Z}/\mathbf{Z}$. Soit

$$N = \prod p_i^{n_i}$$

la factorisation de N en facteurs premiers. Ecrivons un élément de Z_N sous la forme

$$\sum a_i/p_i^{n_i} \text{ avec } a_i \in \mathbf{Z}/p_i^{n_i}\mathbf{Z}.$$

Soit T_N l'ensemble de ces éléments tels que:
ou bien a_i est premier à p_i et $a_i \neq 1$;
ou bien $a_i = 0$.

THÉORÈME. *Les éléments de T_N forment une base sur $\mathbf{Z}$ de F_n/D_n.*

On voit donc que, si $M \mid N$ et $(M, N/M) = 1$ la base T_M est contenue dans la base T_N. Le système de bases est compatible pour les niveaux croissants par divisibilité.

Pour la démonstration, ainsi que la généralisation à $\mathbf{Q}^k/\mathbf{Z}^k$, et l'énoncé analogue pour les distributions de poids ≥ 1, voir l'article à paraître de Kubert [Ku 2], voir aussi mon livre récent [La].

Pour les distributions universelles de poids ≥ 1, il se trouve que les éléments primitifs de $(1/N)\mathbf{Z}/\mathbf{Z}$ forment une base pour F_N/D_N sur $\mathbf{Q}$. Milnor avait déjà remarqué, pour $k = 1$, que les éléments primitifs engendrent sur $\mathbf{Q}$ le groupe des diviseurs modulo les relations de distribution de poids 1, et Kubert a démontré le théorème général sur $\mathbf{Q}$ par récurrence (c'est faux en poids 0).

Pour obtenir la distribution universelle paire ou impaire, si on se limite aux distributions à valeurs dans des groupes où la multiplication par 2 est inversible, alors on obtient les théorèmes analogues comme corollaires immédiats.

Pour l'analyse de la 2-torsion, voir le papier de Kubert, qui nécéssite la cohomologie $H(\pm\text{id}, U)$ de $\pm$id dans la distribution universelle.

Puisque la distribution universelle est "libre," chaque fois qu'on veut démontrer qu'une distribution sur $\mathbf{Q}/\mathbf{Z}$ est universelle, il suffit de démontrer qu'elle a le rang approprié, c'est-à-dire $\phi(N)$ sur $(1/N)\mathbf{Z}/\mathbf{Z}$, et la moitié de ça, pour la distribution paire ou impaire.

Pour cela, on peut étendre les scalaires, et se ramener aux composantes correspondant aux caractères de $(\mathbf{Z}/n\mathbf{Z})^*$. Pour la distribution de Bernoulli, on voit alors qu'elle a le bon rang par le résultat classique que $B_{k,\chi} \neq 0$, quand k et χ ont la même parité. La distribution des unités cyclotomiques a le rang maximal par le théorème de Dirichlet, et l'on retrouve de cette façon un théorème de Bass. Les unités modulaires ont le bon rang d'après Kubert-Lang, qui ramènent la question à des nombres de Bernoulli sur un groupe de Cartan, puis aux nombres de Leopoldt-Bernoulli $B_{k,\chi}$ ordinaires. On obtient ainsi des distributions paires. Pour la distribution universelle impaire, voir l'article de Yamamoto [Ya]. On notera que, comme dans Bass, la question de 2-torsion n'est pas traitée de façon appropriée.

Bibliographie

[Ba] Bass (H.). Generators and relations for cyclotomic units, *Nagoya Math. J.*, t. 27, 1966, p. 401–407.

[Ku 1] Kubert (D.). A system of free generators for the universal even ordinary distribution on $\mathbf{Q}^{2k}/\mathbf{Z}^{2k}$, *Math. Annalen*, t. 224, 1976, p. 21–31.

[Ku 2] Kubert (D.). The universal ordinary distribution (à paraître).

[Ku 3] Kubert (D.). Cohomology of ±id in the universal ordinary distribution (à paraître).

[K-L] Kubert (D.) and Lang (S.). Distributions on toroidal groups, *Math. Z.*, t. 148, 1976, p. 33–51.

[La] Lang (S.). *Cyclotomic fields*. Berlin: Springer-Verlag, 1978 (Graduate Texts in Mathematics) (à paraître).

[Ya] Yamamoto (K.). The gap group of multiplicative relationships of gaussian sums, "Symposia mathematica," Vol. 15, p. 427–440. London: Academic Press, 1975 (Istituto nazionale di alta Matematica).

(Texte reçu le 22 mai 1978)

Serge Lang
Mathematics
Yale University, Box 2155 Yale Station
New Haven, Conn. 06520 (Etats Unis)

Math. Ann. 240, 21—26 (1979)

Mathematische Annalen

Cartan-Bernoulli Numbers as Values of L-Series

Daniel S. Kubert* and Serge Lang*

Mathematics Department, Cornell University, Ithaca, NY 14853, and
Mathematics Department, Yale University, New Haven, CT 06520, USA

Bernoulli numbers formed with characters on a Cartan group have played a central role in the study of the cuspidal divisor class group on modular curves [KL2]. On the other hand, general algebraic facts have also emerged which have warranted independent treatment as in [KL1]. The present paper is of the latter type. We show how the Cartan-Bernoulli numbers are values of certain L-series at negative integers, and in the process give a simpler expression and a simpler proof for these values than in [KL1], although we still find a product of certain Gauss sums and classical Bernoulli numbers arising from Dirichlet characters on $\mathbf{Z}/N\mathbf{Z}$.

As in previous papers, we are limiting the treatment here to the unramified Cartan groups, as we have no applications for the other groups, but one should keep in mind the possibilities of generalizations to the ramified case if it ever arises in a natural context, say of algebraic geometry. In any case, the present paper gives another proof of the non-vanishing of the regulator of the modular units formed with the Siegel functions.

§ 1. The Distribution Relation

Let k be an integer $\geqq 1$. For each prime p we let $\mathfrak{o}_p$ be the unramified extension of $\mathbf{Z}_p$ of degree k. Let N be an integer >1. We let

$$\mathfrak{o}_N = \prod_{p|N} \mathfrak{o}_p .$$

We let $\mathfrak{o}(N) = \mathfrak{o}_N / N\mathfrak{o}_N$. Similarly, if $N = \prod p^{n(p)}$ then

$$\mathbf{Z}(N) = \mathbf{Z}/N\mathbf{Z} = \prod_{p|N} \mathbf{Z}(p^{n(p)}) .$$

Let

$$T : \mathfrak{o}_N \to \mathbf{Z}_N$$

* Supported by NSF grants, Kubert is also a Sloan Fellow

0025-5831/79/0240/0021/$01.20

be a $\mathbf{Z}_N$-linear surjective map, for instance, the trace. For each M dividing N, we have a corresponding surjective map

$$T_M : \mathfrak{o}(M) \to \mathbf{Z}(M) .$$

If t is a real number, let $\langle t \rangle$ be the real number in the residue class of t mod $\mathbf{Z}$ such that $0 < \langle t \rangle \leqq 1$. *Note:* This is a different convention from that adopted in our previous papers, adjusted to fit the ordinary formalism of the Hurwitz zeta function

$$\zeta(s, u) = \sum_{n=0}^{\infty} \frac{1}{(n+u)^s} .$$

It is well known and obvious that the ordinary partial zeta functions satisfy the distribution relation of weight $-s$, meaning that if a is a given congruence class mod M, then

$$N^{-s} \sum_{b \equiv a(M)} \zeta\left(s, \left\langle \frac{b}{N} \right\rangle\right) = M^{-s} \zeta\left(s, \left\langle \frac{a}{M} \right\rangle\right) . \tag{1}$$

The sum on the left is taken for b in $\mathbf{Z}(N)$ reducing to a mod M.

Analogously, we have the relation on the projective system $\mathfrak{o}(N)$, namely given a residue class x mod M,

$$N^{-s-(k-1)} \sum_{\substack{y \in \mathfrak{o}(N) \\ y \equiv x(M)}} \zeta\left(s, \left\langle \frac{Ty}{N} \right\rangle\right) = M^{-s-(k-1)} \zeta\left(s, \left\langle \frac{Tx}{M} \right\rangle\right) . \tag{2}$$

To prove this, let $T_M x = a$, and let $b \in \mathbf{Z}(N)$, $b \equiv a \bmod M$. Then a simple argument using elementary divisors and the surjectivity of T_N shows that the number of elements $y \in \mathfrak{o}(N)$ such that $T_N y = b$ is equal to $(N/M)^{k-1}$. Consequently

$$\begin{aligned} \sum_{\substack{y \in \mathfrak{o}(N) \\ y \equiv x(M)}} \zeta\left(s, \left\langle \frac{Ty}{N} \right\rangle\right) &= \sum_{b \equiv a(M)} \zeta\left(s, \left\langle \frac{b}{N} \right\rangle\right) \sum_{T_N y = b} 1 \\ &= (N/M)^{k-1} \sum_{b \equiv a(M)} \zeta\left(s, \left\langle \frac{b}{N} \right\rangle\right) . \end{aligned}$$

Applying (1) concludes the proof of (2).

The multiplicative group $\mathfrak{o}(N)^*$ will be called the *Cartan group*, and we denote it by $C(N)$. Let $\chi = \chi_N$ be a character on $C(N)$. We define the *L-series*

$$L_N(s, \chi_N, T_N) = N^{-s-(k-1)} \sum_{\alpha \in C(N)} \chi(\alpha) \zeta\left(s, \left\langle \frac{T\alpha}{N} \right\rangle\right) .$$

Proposition 1.1. *Let $M|N$ and suppose χ factors through $C(M)$. Then*

$$L_N(s, \chi_N, T_N) = \prod_{\substack{p|N \\ p \nmid M}} \left(1 - \frac{\chi_M(p)}{p^{s+k-1}}\right) L_M(s, \chi_M, T_M) .$$

Proof. We can write

$$L_N(s, \chi_N, T_N) = N^{-s-(k-1)} \sum_{\alpha \in C(M)} \chi(\alpha) \sum_{\substack{\beta \in C(N) \\ \beta \equiv \alpha(M)}} \zeta\left(s, \left\langle \frac{T\beta}{N} \right\rangle\right) . \tag{3}$$

Suppose first that M and N have the same prime factors. Given $\alpha \in C(M)$, any element $\beta \in \mathfrak{o}(N)$ which reduces to $\alpha \bmod M$ also has to be in $C(N)$. The desired relation then follows from (2).

Next, write $N = p^n M$ with M prime to p. Using the first case, we may assume without loss of generality that $n=1$, that is $N = pM$. Given $\alpha \in C(M)$, let $z = z(\alpha)$ be the element of $\mathfrak{o}(N)$ such that in the prime power factorization, $z = \prod z_q$, we have

$$z_q = \alpha_q \quad \text{if} \quad q \neq p \quad \text{and} \quad z_p = 0 .$$

Then

$$\sum_{\substack{\beta \in C(N) \\ \beta \equiv \alpha(M)}} \zeta\left(s, \left\langle \frac{T\beta}{N} \right\rangle\right) = \sum_{\substack{y \in \mathfrak{o}(N) \\ y \equiv \alpha(M)}} \zeta\left(s, \left\langle \frac{Ty}{N} \right\rangle\right) - \zeta\left(s, \left\langle \frac{Tz(\alpha)}{N} \right\rangle\right) .$$

View p^{-1} as an element of $\mathbf{Z}(M)$. Since

$$\frac{Tz(\alpha)}{N} \equiv \frac{T(p^{-1}\alpha)}{M} \bmod \mathbf{Z} ,$$

we see that

$$\zeta\left(s, \left\langle \frac{Tz(\alpha)}{N} \right\rangle\right) = \zeta\left(s, \left\langle \frac{T(p^{-1}\alpha)}{M} \right\rangle\right) .$$

Using the distribution relation (2), and making a change of variables $\alpha \mapsto p^{-1}\alpha$ in relation (3), we get

$$L_N(s, \chi_N, T_N) = (1 - \chi_M(p) p^{-s-(k-1)}) L_M(s, \chi_M, T_M) ,$$

thereby proving the proposition.

§ 2. Values of L-Series at Negative Integers

Let χ be a character on $C(N)$ and let M be its conductor. We define (without subscript)

$$L(s, \chi, T) = L_M(s, \chi_M, T_M) .$$

Proposition 1.1 gives us the reduction of the value of the L-series to the case when χ is primitive.

Let m be an integer $\geqq 1$. We have

$$L_N(1-m, \chi_N, T_N) = N^{m-k} \sum \chi(\alpha) \zeta\left(1-m, \left\langle \frac{T\alpha}{N} \right\rangle\right) ,$$

and by a classical theorem of Hurwitz,

$$\zeta\left(1-m, \left\langle \frac{T\alpha}{N} \right\rangle\right) = -\frac{1}{m} \mathbf{B}_m\left(\left\langle \frac{T\alpha}{N} \right\rangle\right) .$$

where $\mathbf{B}_m$ is the m-th Bernoulli polynomial.

Hence the value of $L(1-m, \chi, T)$ may be viewed as the appropriate Bernoulli number with respect to the Cartan group under consideration, i.e.

$$L(1-m, \chi, T) = -\frac{1}{m} B_{m,\chi} ,$$

where, assuming χ primitive, we define

$$B_{m,\chi} = B_{m,\chi,k,T} = N^{m-k} \sum_{\alpha \in C(N)} \chi(\alpha) \mathbf{B}_m\left(\left\langle \frac{T\alpha}{N} \right\rangle\right) .$$

The dependence on T is almost nil. As $\mathbf{Z}(N)$-module, it is standard that $\mathfrak{o}(N)$ is self dual, so if T'_N is another surjective $\mathbf{Z}(N)$-linear map, then there exists $\beta \in \mathfrak{o}(N)^*$ such that

$$T'_N(x) = T_N(\beta x) .$$

Making a change of variables in the sum, we see that the Bernoulli numbers formed with T'_N instead of T_N differ by a factor of $\bar{\chi}(\beta)$, which is a root of unity, of no consequence for the applications which deal either with the non-vanishing, or with the absolute value (ordinary or p-adic).

Remark. In some applications, it is useful to distinguish cases depending on parity, and to work with $C(N)/\pm 1$ whenever k is even. This is done in earlier papers of the KL series. Consequently, the reader should check each time when using the general Bernoulli numbers which normalization has been taken.

Assume that χ is primitive, so has conductor N. Write

$$\begin{aligned} L(s, \chi, T) &= N^{-s-(k-1)} \sum_{x=1}^{N} f_\chi(x) \zeta\left(s, \frac{x}{N}\right) \\ &= N^{1-k} \zeta(s, f_\chi) , \end{aligned}$$

where

$$f_\chi(x) = \sum_{T\alpha \equiv x(N)} \chi(\alpha) ,$$

and the sum is taken over all α in $C(N)$ such that $T\alpha \equiv x \bmod N$.

The ordinary expression for the partial zeta function formed with a function on $\mathbf{Z}(N)$ is

$$\zeta(s, f) = \frac{1}{2\pi i} \left(\frac{2\pi}{N}\right)^s \Gamma(1-s) [\zeta(1-s, \hat{f}^-) e^{\pi i s/2} - \zeta(1-s, \hat{f}) e^{-\pi i s/2}]$$

and

$$\hat{f}(n) = \sum_{x=1}^{N} f(x) e^{-2\pi i x n/N} .$$

See for instance [L], Chapter XIV, Theorem 2.1. Here we have

$$\begin{aligned} \hat{f}_\chi(n) &= \sum_{x=1}^{N} \sum_{T\alpha = x} \chi(\alpha) e^{-2\pi i x n/N} \\ &= \sum_{\alpha \in C(N)} \chi(\alpha) e^{-2\pi i n T\alpha/N} \\ &= S(\chi, T) \bar{\chi}(-n) , \end{aligned}$$

where $S(\chi, T)$ is the obvious Gauss sum on the Cartan group, formed with the additive character arising from the surjective map

$$T_N : \mathfrak{o}(N) \to \mathbf{Z}(N) .$$

Thus

$$S(\chi, T) = \sum_{\alpha} \chi(\alpha) e^{2\pi i T\alpha/N} .$$

Note that $\hat{f}_\chi(n) = 0$ if $(n, N) \neq 1$. So we find

$$\hat{f}_\chi^-(n) = S(\chi, T)\bar{\chi}(n) ,$$

whence we obtain an expression for the L-function on the Cartan group in terms of an ordinary L-function on $\mathbf{Z}(N)$, namely

Theorem 2.1.

$$L(s, \chi, T) = \frac{N^{1-k}}{2\pi i}\left(\frac{2\pi}{N}\right)^s \Gamma(1-s) S(\chi, T) \cdot [e^{\pi i s/2} - \chi(-1)e^{-\pi i s/2}] L_N(1-s, \bar{\chi}_{\mathbf{Z}})$$

where $\chi_{\mathbf{Z}}$ is the restriction of χ to $\mathbf{Z}(N)$.

We assumed that χ is primitive, but of course $\chi_{\mathbf{Z}}$ need not be primitive. It is easy to give the value of the primitive L-series in terms of the non-primitive one by multiplying with the appropriate factors involving the primes dividing N. We also note that the fudge factors in the analytic expression for the L-function on the Cartan group are essentially the same as the corresponding factors for the ordinary L-function. Consequently, after some obvious cancellations, we find:

Theorem 2.2.

$$L(s, \chi, T) = N^{1-k} \frac{S(\chi, T)}{S_{\mathbf{Z}}(\chi_{\mathbf{Z}})} \prod_{\substack{p \mid N \\ p \nmid c}} \left(1 - \frac{\bar{\chi}(p)}{p^{1-s}}\right) L(s, \chi_{\mathbf{Z}})$$

where $c = c(\chi_{\mathbf{Z}})$ is the conductor of $\chi_{\mathbf{Z}}$, and $S_{\mathbf{Z}}(\chi_{\mathbf{Z}})$ is the standard Gauss sum formed with the standard additive character $x \mapsto e^{2\pi i x/N}$ for $x \in \mathbf{Z}(N)$.

Evaluating the L-function in Theorem 2.2 at negative integers yields the desired values in terms of ordinary Bernoulli numbers.

From the analytic expression of Theorem 2.1, we know at which negative integers does the L-function vanish, and we summarize this as follows:

Corollary 1.

(i) $L(0, \chi, T) = 0$ *if χ is even, $\chi_{\mathbf{Z}}$ non-trivial.*
$L(0, \chi, T) \neq 0$ *if χ is even, $\chi_{\mathbf{Z}}$ is trivial, or if χ is odd.*

(ii) *Let m be an integer ≥ 2. Then $L(1-m, \chi, T) \neq 0$ if and only if χ and m have the same parity.*

Proof. The vanishing or non-vanishing is reduced to that of the ordinary L-function because none of the factors occurring in front of $L(s, \chi_{\mathbf{Z}})$ vanishes at integers $1-m$ with $m \geqq 1$. The assertion of the corollary is then clear from the standard properties of the classical L-function.

As mentioned in the introduction, from this corollary we get what is needed in [KL3] to see that the Siegel functions form a full set of units in the modular function field. We use the map T such that in a matrix representation

$$\alpha = \begin{pmatrix} a & b \\ c & d \end{pmatrix}$$

of the Cartan group, we have $T\alpha = a$. In the theory of modular functions, this is the convenient map which gives the order of a Siegel function at infinity. See [3].

For the record, we also give the explicit value at negative integers.

Corollary 2. *Let m be an integer $\geqq 1$. Then*

$$B_{m,\chi,k,T} = N^{1-k} \frac{S(\chi, T)}{S_{\mathbf{Z}}(\chi_{\mathbf{Z}})} \prod_{\substack{p \mid N \\ p \nmid c}} \left(1 - \frac{\bar{\chi}(p)}{p^m}\right) B_{m,\chi,\mathbf{Z}},$$

except when $\chi_{\mathbf{Z}}$ is trivial, and $m=1$.

Proof. We merely use the classical value of the classical L-series, cf. for instance Theorem 2.3 of Chapter XIV, [L].

References

[KL1] Kubert, D., Lang, S.: Distributions on toroidal groups. Math. Z. **148**, 33–51 (1976)

[KL2] Kubert, D., Lang, S.: The p-primary component of the cuspidal divisor class group on the modular curve $X(p)$. Math. Ann. **234**, 25–44 (1978)

[KL3] Kubert, D., Lang, S.: Units in the modular function field II. A full set of units. Math. Ann. **218**, 175–189 (1975)

[4] Lang, S.: Introduction to modular forms. Berlin, Heidelberg, New York: Springer 1976

Received January 5, 1978

Math. Ann. 240, 191–201 (1979)

Mathematische
Annalen

Independence of Modular Units on Tate Curves

Daniel S. Kubert[1]* and Serge Lang[2]*

1 Mathematics Department, Cornell University, Ithaca, NY 14853, and
2 Mathematics Department, Yale University, New Haven, CT 06520, USA

In a series of papers we have investigated units in the modular function field. In particular, we have shown the independence of the Siegel units which play a fundamental role in the determination of the cuspidal divisor class group. Following the program outlined in the first paper of the series, we shall here be concerned with the independence of these units when specialized to number fields. As mentioned before, three cases arise: elliptic curves with complex multiplication, elliptic curves without complex multiplication but having integral j-invariant, elliptic curves with non-integral j-invariant. The first case is implicit in the work of Ramachandra and Robert, and we shall deal with it elsewhere. The second case is the hardest, and little is known about it. It is clear that cases when the rank of the generic units diminishes under specialization can be constructed at will, for instance by taking two multiplicatively independent units u, v and using a place of the modular function field which sends $u-v$ to 0. No general result is known giving conditions under which the generic units remain independent in this case.

The third case arising from a non-integral j-invariant will be treated here. According to Serre [Se], the Galois group of torsion points is essentially as large as possible, so that in this case, the analysis of independence of the specialized units can be carried out in a manner which is similar to that of the generic situation, although some slightly more delicate points arise when dealing with the intervening Gauss sums.

We state the main theorem in § 1. It gives a sufficient condition under which the units remain independent in terms of the non-degeneracy of the Galois group of torsion points. The proof relies on the non-vanishing of a certain character sum, which is taken care of in § 2. This leads to a question of independent interest, when a Gauss sum is what we call "pure", i.e. equal to a root of unity times a pure power of an integer. It is then convenient to discuss this briefly in the last section § 3, which is stronger than what we need for the immediate applications, but will be used elsewhere.

---

* Supported by NSF grants. Kubert is also a Sloan Fellow

0025-5831/79/0240/0191/$02.20

§ 1. Modular Units on Tate Curves

Let k be a number field. Let A be an elliptic curve defined over k, and let

$$k(A_N/\pm)$$

be the field extension of x-coordinates of N-torsion points. Let $j_A = j(z)$ for some complex number z in the upper half plane. Then:

$$k(A_N/\pm) = \text{field generated by all values } f(z) \text{ of all modular functions } f \text{ of level } N \text{ defined at } z.$$

By definition, the modular functions are taken from the function field having the cyclotomic field $\mathbf{Q}(\mu_N)$ as constant field.

Let $\mathfrak{p}$ be a prime of k lying above the prime number p.

Let V be the multiplicative subgroup of elements $\alpha \in k(A_N/\pm)$, $\alpha \neq 0$, such that $\operatorname{ord}_{\mathfrak{P}} \alpha$ is independent of the prime $\mathfrak{P}|\mathfrak{p}$ in $k(A_N/\pm)$. We observe that V contains the units of $k(A_N/\pm)$, and also contains the cyclotomic numbers $1-\zeta$ where $\zeta \in \mu_N$.

For each prime $l|N$, let $\mathfrak{o}_l$ be the unramified extension of degree 2 of $\mathbf{Z}_l$. Let

$$C(l^n) = (\mathfrak{o}_l/l^n\mathfrak{o}_l)^*$$

be the Cartan group which has been considered all along in the study of modular units (e.g. [3]). If

$$N = \prod l^{n(l)}$$

we let

$$C(N) = \prod C(l^{n(l)}).$$

We often write $C(N)/\pm 1 = C(N)(\pm)$ or $C(N)/\pm$. This Cartan group acts simply transitively on the cusps of the modular curve $X(N)$. Once a basis of A_N over $\mathbf{Z}(N)$ is chosen, this Cartan group has a matrix representation with which we usually identify it. Thus we view $C(N)$ as a subgroup of $\mathrm{GL}_2(N) = \mathrm{GL}_2(\mathbf{Z}/N\mathbf{Z})$. By a theorem of Serre [Se], it is known that the Galois group of $k(A_N)$ over k is "usually" $\mathrm{GL}_2(N)$. For our purposes, we do not need the full GL_2, but will relate the independence of the specialized modular units to the size of the Cartan group in the Galois group. For simplicity, we restrict ourselves to the case when the whole Cartan group is contained in the Galois group, as follows.

As in [3] we let g_a be the Siegel units, with

$$a \in \left(\frac{1}{N}\mathbf{Z}^2/\mathbf{Z}^2\right)\Big/ \pm 1.$$

Then g_a^{12N} is a "geometric unit" in the modular function field and has level N. Let U be the group generated by these Siegel functions. Let $U(z)$ be the group generated by the values

$$g_a^{12N}(z).$$

Then $U(z)$ is a subgroup of $k(A_N/\pm)^*$.

In [KL2] we showed that the rank of U (modulo constants) is equal to

$$|C(N)(\pm)|-1 .$$

Our purpose here is to determine the rank of the specialized group $U(z)$ modulo V when A is a Tate curve.

Theorem 1.1. *Assume:*

(i) $\mathrm{Gal}(k(A_N/\pm)/k)$ *contains* $C(N)(\pm)$.

(ii) *The prime* $\mathfrak{p}$ *divides the denominator of* j_A.

Then

$$\operatorname{rank} U(z)V/V = |C(N)(\pm)|-1 .$$

Proof. Let $\mathfrak{P}$ be a prime of $k(A_N/\pm)$ lying above $\mathfrak{p}$. Let $\mathrm{inj}_{\mathfrak{P}}$ be the injection of $k(A_N/\pm)$ in its completion at $\mathfrak{P}$. The Tate curve has a corresponding element

$$q_A = j_A^{-1}+\dots \quad \text{with} \quad j_A = \frac{1}{q_A}+744+196884q_A+\dots .$$

Cf. for instance [L 2], Chapter XV. In the analytic parametrization of the Tate curve, the N-th torsion points have a basis consisting of some chosen N-the root of q_A and a primitive N-th root of unity ζ. If r is an integer, we write

$$q_1^{r/N} = \zeta^r .$$

If α, α' are two non-zero elements of the $\mathfrak{p}$-adic field, we write

$$\alpha \sim \alpha'$$

to mean that α/α' is a $\mathfrak{P}$-adic unit.

A Siegel unit g_a has the q-expansion at the standard prime at infinity given by:

$$g_a(z) = -q_\tau^{\frac{1}{2}\mathbf{B}_2(a_1)}e^{2\pi i a_2(a_1-1)/2}(1-q_2)\prod(1-q_\tau q_z)(1-q_\tau/q_z),$$

where $z = a_1\tau+a_2$, cf. [L3], Chapter XV, Formula **K3**. All the primes at infinity of the modular function field are conjugate under GL_2. By [KL1], §4, Theorem 9, there exists $\sigma \in \mathrm{GL}_2(N)$ such that

$$\textbf{O1.}\quad \mathrm{inj}_{\mathfrak{P}}\, g_a^{12N}(z) \sim q_A^{\frac{12N}{2}\mathbf{B}_2(\langle(a\sigma)_1\rangle)}(1-q_A^{(a\sigma)_1}q_1^{(a\sigma)_2}) .$$

Lemma. *We can select an* N-th *root* $q^{1/N}$ *and the root of unity* ζ *such that the element* σ *above lies in the Cartan group* $C(N)$.

Proof. As remarked already in [KL 2], we have

$$\mathrm{GL}_2(N) = C(N)G_\infty ,$$

where G_∞ is the isotropy group of $\begin{pmatrix}1\\0\end{pmatrix}$, consisting of the matrices

$$\begin{pmatrix}1 & b\\ 0 & d\end{pmatrix},$$

and every element of $\mathrm{GL}_2(N)$ has a unique decomposition as a product of an element in G_∞ and an element in $C(N)$. Let $\{e_1, e_2\}$ be a basis of $\left(\frac{1}{N}\mathbf{Z}/\mathbf{Z}\right)^2$. Let T be the transformation associated with σ, and let $\{e'_1, e'_2\}$ be a basis for the image. Then

$$\begin{aligned} Te_1 &= ae'_1 + be'_2 , \\ Te_2 &= ce'_1 + de'_2 . \end{aligned}$$

Let $t \in \mathbf{Z}(N)^*$ and $x \in \mathbf{Z}(N)$. Write

$$\begin{aligned} e'_2 &= te''_2 , \\ e'_1 &= e''_1 + xe''_2 . \end{aligned}$$

Then

$$\begin{aligned} Te_1 &= ae''_1 + (bt + ax)e''_2 , \\ Te_2 &= ce_1 + (cx + dt)e''_2 . \end{aligned}$$

Corresponding to the change of basis, the matrix for T is

$$\begin{pmatrix} a & b \\ c & d \end{pmatrix} \begin{pmatrix} 1 & x \\ 0 & t \end{pmatrix} .$$

We can choose t, x such that this matrix lies in $C(N)$. This proves the lemma.

For any element $\alpha \in k(A_N/\pm)$ we have for $\gamma \in C(N)/\pm 1$:

O2. $\operatorname{ord}_{\gamma^{-1}\mathfrak{P}} \alpha = \operatorname{ord}_{\mathfrak{P}} \gamma\alpha .$

Furthermore,

O3. $\operatorname{inj}_{\mathfrak{P}} \gamma g_a^{12N}(z) = \operatorname{inj}_{\mathfrak{P}} g_{a\gamma}^{12N}(z) .$

This and the order formula **O1** give us the order at $\gamma^{-1}\mathfrak{P}$ for the special values of the Siegel functions. We now consider the usual "logarithm map" concentrated at the orbit of $\mathfrak{P}$ under the Cartan group, namely for $\alpha \in k(A_N/\pm)^*$ we let

$$L \colon \alpha \mapsto \sum_{\gamma \in C(N)/\pm} (\operatorname{ord}_{\gamma^{-1}\mathfrak{P}} \alpha) \sigma_\gamma^{-1} .$$

Then by assumption, we see that

$$L \colon V \to \mathbf{Z} \sum_\gamma \sigma_\gamma ,$$

i.e. L maps V into the space in the group algebra corresponding to the trivial character. On the other hand,

$$L \colon g_a^{12N} \mapsto \sum_\gamma r(\gamma) \sigma_\gamma^{-1} ,$$

where

$$r(\gamma) = 12N \tfrac{1}{2} \mathbf{B}_2(\langle (a\gamma\sigma)_1 \rangle) \operatorname{ord}_{\mathfrak{P}} q_A + 12N \operatorname{ord}_{\mathfrak{P}} (1 - q_A^{(a\gamma\sigma)_1} q_1^{(a\gamma\sigma)_2}) .$$

We now analyse the image of the group generated by the Siegel units by means of its position in the various eigenspaces for the characters $\chi \neq 1$ on $C(N)(\pm)$. As in our previous work, in the matrix representation

$$\gamma = \begin{pmatrix} a_1 & a_2 \\ c & d \end{pmatrix}$$

of the Cartan group, we let $T\gamma = a_1$. Let M be the conductor of χ. Let e_χ be the usual idempotent on $C(M)(\pm)$. Then

$$\mathbf{C} \otimes L(U)e_\chi$$

has dimension 0 or 1, and is spanned by

$$\tfrac{1}{2} \sum_{\gamma \in C(M)(\pm)} \mathbf{B}_2\left(\left\langle \frac{T\gamma}{M} \right\rangle\right) \bar{\chi}(\gamma) \operatorname{ord}_{\mathfrak{P}} q_A + \sum_{\substack{T\gamma = 0 \\ \gamma \in C(M)(\pm)}} \operatorname{ord}_{\mathfrak{P}}(1 - \zeta_M)\bar{\chi}(\gamma),$$

where ζ_M is a primitive M-th root of unity; so our assertion is clear from the expression given above for $r(\gamma)$.

We wish to prove that each such eigenspace for a non-trivial character has in fact dimension 1, so we have to prove that each generating element is $\neq 0$.

The first term is a positive multiple of the Cartan-Bernoulli number, which we know is $\neq 0$ since it is expressed as a non-zero multiple of the ordinary Bernoulli-Leopoldt number, see for instance [KL3], Corollary 2 of Theorem 2.2.

The second term

$$\operatorname{ord}_{\mathfrak{P}}(1 - \zeta_M) \sum_{\substack{T\gamma = 0 \\ \gamma \in C(M)(\pm)}} \bar{\chi}(\gamma)$$

is equal to 0 if M is not a prime power, and also is equal to 0 if χ_Z is non-trivial, because if $T\gamma = 0$, then $T(c\gamma) = 0$ for all $c \in \mathbf{Z}(M)(\pm)^*$, and the sum over $T\gamma = 0$ can be decomposed over a sum over cosets of $\mathbf{Z}(M)(\pm)^*$. In case χ_Z is non-trivial, this shows that the generating element is $\neq 0$.

We are left with the case when χ_Z is trivial, and M is a prime power, which we now suppose is the case. Then again [4], Corollary 2 of Theorem 2.2 gives us the value for

$$B_{2,\bar{\chi},T}$$

on the Cartan group, with $m = k = 2$. Since

$$B_{2,\bar{\chi}_Z} = \tfrac{1}{6},$$

we see that

$$B_{2,\bar{\chi},T} = S_C(\bar{\chi}, T)\varrho, \quad \text{where} \quad \varrho = N^{1-2}\left(1 - \frac{1}{p^2}\right)\tfrac{1}{6} > 0.$$

But

$$\sum_{\substack{T\gamma = 0 \\ \gamma \in C(M)(\pm)}} \bar{\chi}(\gamma) = \sum_{T(\gamma) = 0} \bar{\chi}(\gamma).$$

We can therefore apply Proposition 2.2 proved in the next section to conclude the proof of the theorem.

§ 2. The Value of a Gauss Sum

In this section we give the value of the Gauss sum needed to complete the arguments giving the rank of the modular units on a Tate curve. The arguments are self contained.

Let k be an integer $\geqq 1$, p prime. Let $\mathfrak{o}$ be the unramified (ring) extension of $\mathbf{Z}_p$ of degree k, and let

$$\mathfrak{o}(p^n) = \mathfrak{o}/p^n\mathfrak{o} .$$

We call $C = C(p^n) = \mathfrak{o}(p^n)^*$ the **Cartan group.** We let

$$\lambda : \mathfrak{o}(p^n) \to \mu_{p^n}$$

be a character giving rise to the usual self duality. For instance,

$$\lambda(x) = \exp(2\pi i \operatorname{Tr}(x)/p^n)$$

is such a character, where the trace is taken to $\mathbf{Z}_p$.

Let χ be a character on C. We may form the usual **Gauss sum**

$$S_C(\chi, \lambda) = S(\chi, \lambda) = \sum \chi(x)\lambda(x) ,$$

where the sum is taken over $x \in C$, or even $x \in \mathfrak{o}(p^n)$ since χ is defined to be 0 outside C. If χ is primitive, it is standard that the absolute value of the Gauss sum and all its conjugates satisfies

$$|S(\chi, \lambda)| = p^{nk/2} .$$

We shall first prove the proposition which is used in determining the rank of the modular units in the preceding section. After that, we go into a further discussion of the cases in which a Gauss sum is pure.

The group $\mathbf{Z}(p^n)^*$ is contained in the Cartan group. As usual, we let χ_Z denote the restriction of χ to that group. Similarly, we let λ_Z denote the restriction of λ to $\mathbf{Z}(p^n)$.

Suppose that $k = 2$. If λ_Z is trivial, then

$$\operatorname{Ker} \lambda = \mathbf{Z}(p^n)$$

because $\mathfrak{o}(p^n)$ is free of dimension 2 over $\mathbf{Z}(p^n)$. We are of course especially interested in the case $k = 2$ for applications to the modular units.

Proposition 2.1. *Let χ be primitive on the Cartan group. Assume that χ_Z and λ_Z are trivial. Let $k = 2$. Then*

$$S_C(\chi, \lambda) = p^n .$$

Proof. We give the proof when $n > 1$ (it is even easier when $n = 1$). We have

$$S_C(\chi, \lambda) = \sum_{x \in \mathbf{Z}(p^n)^*} 1 + \sum_{x \notin \mathbf{Z}(p^n)} \chi(x)\lambda(x) .$$

The first sum on the right of course gives the Euler function $\phi(p^n)$. As to the second sum, it can be decomposed into a sum over non-trivial cosets of $\mathbf{Z}(p^n)^*$ in the Cartan group. Each coset consists of elements

$$\{xa\}, \quad \text{with} \quad a \in \mathbf{Z}(p^n)^* ,$$

and $\lambda(xa) = \lambda(x)^a$. Furthermore, $\chi(xa) = \chi(x)$, and

$$\sum_a \lambda(x)^a = \operatorname{Tr} \lambda(x),$$

where Tr is the trace from $\mathbf{Q}(\mu_{p^n})$ to $\mathbf{Q}$. Since $\lambda(x) \in \mu_{p^n}$, we see that the trace is $\neq 0$ if and only if $\lambda(x) \in \mu_p$. Let us rewrite the second sum as

$$\sum_{x \notin \mathbf{Z}(p^n)} \chi(x)\lambda(x) = \sum \chi(\bar{x}) \operatorname{Tr} \lambda(\bar{x}),$$

where $\bar{x}$ ranges over representatives of $C/\mathbf{Z}(p^n)^*$ other than the unit coset. Furthermore, since we can limit ourselves to x such that $\lambda(x) \in \mu_p$, since $\operatorname{Ker} \lambda = \mathbf{Z}(p^n)$, and since λ is surjective, it follows that

$$x \equiv a + p^{n-1} y \bmod p^n,$$

with some element $a \in \mathbf{Z}$ and $y \in \mathfrak{o}(p)/\mathbf{Z}(p) - \{0\}$. We may therefore select the representative $\bar{x}$ to be of the form

$$\bar{x} = 1 + p^{n-1} y.$$

For the trace, we find

$$\operatorname{Tr} \lambda(x) = -\frac{\phi(p^n)}{p-1}$$

because the trace of a primitive p-th root of unity to $\mathbf{Q}$ is -1, and we are here taking the trace from the field of p^n-th roots of unity, so -1 has to be multiplied by the appropriate degree. Consequently, our second sum is equal to

$$\sum_y \chi(1+p^{n-1}y)\left(-\frac{\phi(p^n)}{p-1}\right)$$

with $y \in \mathfrak{o}(p^n)/\mathbf{Z}(p^n)$. Since χ is assumed primitive, the sum over y is equal to -1. Hence finally

$$S_C(\chi, \lambda) = \phi(p^n) + \frac{\phi(p^n)}{p-1} = p^n.$$

This proves the proposition.

Proposition 2.2. *Again let $k = 2$. Let t be a real number, $t > 0$. Let χ be a primitive character on $C(p^n)$. Then*

$$S(\chi, \lambda) + t \sum_{\lambda(x) = 0} \chi(x) \neq 0.$$

Proof. If χ_Z is not trivial, then

$$\sum_{\lambda(x) = 0} \chi(x) = 0$$

since the sum can be decomposed over sums on cosets of $\mathbf{Z}(p^n)^*$. The proposition follows because $S(\chi, \lambda) \neq 0$.

Suppose that χ_Z is trivial. Without loss of generality, we may assume that λ_Z is trivial. Indeed, suppose we replace λ by $\lambda \circ \gamma$ for some element $\gamma \in C$. Then the sum

$S(\chi, \lambda)$ changes by a factor $\bar{\chi}(y)$, and so does the other sum over $\lambda(x) = 0$. We can then find some $y \in C$ such that $\lambda \circ y$ is trivial on $\mathbf{Z}(p^n)$, thus reducing the problem to the case when λ_Z is trivial.

Suppose this is the case. Then the desired expression is simply equal to

$$p^n + t(p-1) > 0$$

by Proposition 2.1. This concludes the proof.

§3. Pure Gauss Sums

We return to the general case $k \geqq 1$.

The Gauss sum will be called **pure** if it is equal to a root of unity times $p^{nk/2}$. Let $\mathbf{Q}(\chi, \lambda)$ be the field generated by the values of χ and λ. The following characterization is immediate:

P1. *The Gauss sum is pure if and only if, for every prime ideal $\mathfrak{p}$ of $\mathbf{Q}(\chi, \lambda)$ lying above p, the order* $\operatorname{ord}_{\mathfrak{p}} S(\chi, \lambda)$ *is independent of $\mathfrak{p}$.*

Indeed, if we take the quotient of the Gauss sum by the root $p^{nk/2}$, we obtain a unit all of whose archimedean absolute values are equal to 1, and so a root of unity.

Proposition 3.1. *Let χ be a primitive character of $C(p^n)$. If $n > 1$ then the Gauss sum $S(\chi, \lambda)$ is pure.*

The proof will be based on a certain identity for which we need more notation. We have a direct product decomposition

$$C = C_0 \times C_1,$$

where C_0 is the group of (p^k-1)-th roots of unity, and C_1 consists of those elements $a \equiv 1 \bmod p$. The character χ then has a corresponding decomposition

$$\chi = \chi_0\chi_1,$$

where χ_0 is a character on C_0 and χ_1 is a character on C_1. The desired identity is stated in the next proposition.

Proposition 3.2. *We have:*

$$S_C(\chi, \lambda) \sum_{x \in C_0} \chi_0(x)\chi_1(1-p^{n-1}x) = S_C(\chi_1, \lambda) S_{C_0}(\chi_0, \lambda \circ p^{n-1}).$$

Proof. Starting with the product on the right hand side, we get:

$$\begin{aligned}
&\sum_{y \in C} \chi_1(y)\lambda(y) \sum_{x \in C_0} \chi_0(x)\lambda(p^{n-1}x) \\
&\quad = \sum_{y \in C} \sum_{x \in C_0} \chi_1(y)\chi_0(x)\lambda(y+p^{n-1}x) \\
&\quad = \sum_{y \in C} \sum_{x \in C_0} \chi_1(y-p^{n-1}x)\chi_0(x)\lambda(y) \qquad [\text{using } y \mapsto y+p^{n-1}x] \\
&\quad = \sum_{y, x} \chi_1(y)\chi_1(1-p^{n-1}x)\chi_0(x)\chi_0(y)\lambda(y)
\end{aligned}$$

[by writing $y-p^{n-1}x = y(1-p^{n-1}x/y)$, and letting $x \mapsto xy$]

$$= \sum_{y,\,x} \chi(y)\lambda(y)\chi_0(x)\chi_1(1-p^{n-1}x)\,,$$

which proves the proposition.

We shall now use Proposition 3.2 to prove Proposition 3.1. We note that the function

$$x \mapsto \lambda_1(x) = \chi_1(1-p^{n-1}x)$$

is a non-trivial character on $\mathfrak{o}(p)$, and likewise the function $\lambda \circ p^{n-1}$. Consequently

$$S_{C_0}(\chi_0, \lambda_1) = \sum_{x \in C_0} \chi_0(x)\lambda_1(x) = (\text{root of unity}) \cdot S_{C_0}(\chi_0, \lambda \circ p^{n-1})\,.$$

Therefore by Proposition 3.2, we see that

$$S_C(\chi, \lambda) = (\text{root of unity}) \cdot S_C(\chi_1, \lambda)\,.$$

But $S_C(\chi_1, \lambda)$ lies in the field $\mathbf{Q}(\mu_{p^n})$, and p is totally ramified in that field, so only one prime ideal in that field lies above p. This proves that the Gauss sum is pure, as desired.

By Proposition 3.1, the determination of when a Gauss sum is pure is reduced to the case when $n = 1$, which we now assume. Thus $C = C(p)$. Let ω be the Teichmuller character from $\mathfrak{o}(p)^*$ onto μ_{q-1}, where $q = p^k$. We write a primitive character χ as a power,

$$\chi = \omega^{-a}, \quad \text{where} \quad a = a_0 + a_1 p + \ldots + a_{k-1}p^{k-1},$$
$$\text{and} \quad 0 \leqq a_i \leqq p-1\,.$$

Then as usual we put

$$s(a) = \sum a_i = (p-1)\sum_{i=0}^{k-1} \left\langle \frac{p^i a}{q-1} \right\rangle.$$

Stickelberger's theorem gives us the order of the Gauss sum at a prime p in terms of $s(a)$, cf. for instance [L], Chapter 1, Theorem 2.1. We thus find:

P2. *The Gauss sum $S(\chi, \lambda)$ is pure if and only if $s(ca)$ is constant for $c \in \mathbf{Z}(q-1)^*$.*

Let $\mathbf{B}_1(X) = X - \frac{1}{2}$ be the first Bernoulli polynomial. We consider the Stickelberger element

$$\theta\left(\frac{r}{d}\right) = \sum_{c \in \mathbf{Z}(d)^*} \mathbf{B}_1\left(\left\langle \frac{cr}{d} \right\rangle\right) \sigma_c^{-1}\,,$$

taking d = denominator of $a/(q-1)$.

Proposition 3.3. *The Gauss sum $S(\chi, \lambda)$ is pure if and only if*

$$\sum \theta\left(\left\langle \frac{p^i}{d} \right\rangle\right) = 0\,,$$

where the sum is taken over the powers of p in $\mathbf{Z}(d)^*$.

Proof. Suppose the Gauss sum is pure. By assumption and **P2**, we conclude that for all $c \in \mathbf{Z}(q-1)^*$ the value

$$\sum_{i=0}^{k-1} \left\{ \left\langle \frac{cp^i a}{q-1} \right\rangle - \tfrac{1}{2} \right\}$$

is independent of c. But the function $t \mapsto \mathbf{B}_1(\langle t \rangle)$ is odd. Taking $c = -1$ we see that

$$s(a) - \frac{p-1}{2} k = 0, \quad \text{that is} \quad s(a) = \frac{p-1}{2} k.$$

[This is a necessary condition for $S(\chi, \lambda)$ to be pure. Note that if p is odd, then

$$a \equiv s(a) \equiv 0 \bmod \frac{p-1}{2}.]$$

It now follows that

$$\sum_{i=0}^{k-1} \theta\left(\left\langle \frac{p^i}{d} \right\rangle\right) = 0$$

thus proving one side of the proposition. The other side is essentially trivial.

It is known that the function $t \mapsto \mathbf{B}_1(\langle t \rangle)$ is the universal odd distribution into abelian groups on which multiplication by 2 is invertible, cf. [L], Chapter 2, Theorem 8.3. By Kubert's theorem giving a free basis for the universal distribution, [L], Chapter 2, Theorem 9.2, we conclude:

P3. *The sum* $\sum \theta\left(\left\langle \frac{p^i}{d} \right\rangle\right) = 0$ *if and only if the "divisor"*

$$2 \sum \left(\frac{p^i}{d} \right)$$

lies in the $\mathbf{Z}$*-module generated by the distribution relations and the oddness relation.*

By "divisor" we mean an element of the free abelian group generated by the elements of

$$\frac{1}{d} \mathbf{Z}/\mathbf{Z}.$$

For the distribution relations, see [K].

Remark 1. If -1 lies in the group generated by p in $\mathbf{Z}(d)^*$, then it is clear from **P3** that the Gauss sum is pure.

Remark 2. If $k = 1$ or 2 then

$$s(a) = \frac{p-1}{2} k$$

is necessary and sufficient for the Gauss sum to be pure.

References

[K] Kubert, D.: The universal ordinary distribution (to appear)

[KL 1] Kubert, D., Lang, S.: Units in the modular function field. I. Diophantine applications. Math. Ann. **218**, 67–96 (1975)

[KL 2] Kubert, D., Lang, S.: Units in the modular function field. II. A full set of units. Math. Ann. **218**, 175–189 (1975)

[KL 3] Kubert, D., Lang, S.: Cartan-Bernoulli numbers as values of *L*-series. Math. Ann. **240**, 21–26 (1979)

[L 1] Lang, S.: Cyclotomic fields. Berlin, Heidelberg, New York: Springer 1978

[L 2] Lang, S.: Elliptic functions. Addison Wesley 1973

[L 3] Lang, S.: Introduction to modular forms. Berlin, Heidelberg, New York: Springer 1976

[Se] Serre, J.P.: Propriétés Galoisiennes des points d'ordre fini des courbes élliptiques. Invent Math. **15**, 259–331 (1972)

Received April 10, 1978

Bull. Soc. math. France,
107, 1979, p. 161-178.

MODULAR UNITS INSIDE CYCLOTOMIC UNITS

BY

DANIEL S. KUBERT and SERGE LANG (*)

RÉSUMÉ. — On considère les unités de Siegel-Ramachandra-Robert dans le corps de classes de rayon p sur le corps quadratique imaginaire $\mathbf{Q}(\sqrt{-p})$. Les normes par rapport au corps cyclotomique $\mathbf{Q}(\mu_p)$ sont des unités. On démontre qu'elles sont contenues dans les unités cyclotomiques, et l'on donne une expression explicite des unes en fonction des autres. La démonstration se fait en écrivant les valeurs des séries L en $s = 1$, provenant d'une part de la formule limite de Kronecker, et d'autre part de l'expression usuelle pour les séries L de Dirichlet. On obtient ainsi suffisamment de relations pour résoudre les équations linéaires liant les logarithmes des unités modulaires et les logarithmes des unités cyclotomiques.

ABSTRACT. — We consider the Siegel-Ramachandra-Robert units in the ray class field of conductor p over the imaginary quadratic field $\mathbf{Q}(\sqrt{-p})$. The norms to the cyclotomic field $\mathbf{Q}(\mu_p)$ are units. We prove that they are contained in the cyclotomic units, and give explicit expressions of the former in terms of the latter. This is done by writing the values of L-series at $s = 1$, both from the Kronecker limit formula, and from the usual Drichlet L-series. Enough linear relations are obtained between the logs of the two kinds of units to solve for each in terms of the other.

In a series of papers, we have studied units in the modular function field, and in [KL 1] we already mentioned the possibility of investigating their specializations to number fields. On the other hand, SIEGEL [Si], RAMACHANDRA [Ra], and especially ROBERT [Ro] have investigated certain units in the complex multiplication case, obtained as values of certain theta functions.

In the present paper, we begin the special case of units in the cyclotomic fields of p-th roots of unity, p prime. For ease of exposition, we separate the results in two parts: $p \equiv 1 \bmod 4$ in this part, and $p \equiv -1 \bmod 4$

(*) Texte reçu le 16 mai 1978.
Supported by N.S.F. grants. KUBERT is also a Sloan Fellow.
Daniel S. KUBERT, Mathematics Department, Cornell University, Ithaca, N.Y. 14853 (U.S.A.), and Serge LANG, Mathematics Department, Yale University, New Haven, Conn. 06520 (U.S.A.).

in the next part. We show how the units obtained as values of modular functions (Siegel units) in the cyclotomic field can be expressed in terms of cyclotomic units, by means of an explicit formula. In particular, these modular units are contained in the cyclotomic units.

In paragraph 1, we give general facts and notation. In paragraph 2, we write down a system of linear relations relating the modular units and cyclotomic units by using the decomposition of appropriate L-series. In paragraph 3, we complete these relations for the trivial character. In paragraph 4, we solve for the modular units as power products of the cyclotomic units. This pattern is followed in both parts.

We let

$$K = \mathbf{Q}(\sqrt{-p}) \qquad \text{and} \qquad H = K(\mu_p).$$

We let $\mathfrak{p} = (\sqrt{-p})$ be the prime ideal in the ring of algebraic integers $\mathfrak{o}_K = \mathfrak{o}$. We let $K(1)$ be the Hilbert class field of K, and $K(\mathfrak{p})$ the ray class field of conductor $\mathfrak{p}$. This notation applies to both parts.

Part one : $p \equiv -1 \bmod 4$

1. General facts

We assume $p \equiv -1 \bmod 4$ and $p \geqslant 5$. Then $K \subset \mathbf{Q}(\mu_p) = H$. If $\alpha \in \mathfrak{o}$ and $\alpha \equiv 1 \bmod \mathfrak{p}$, then $\mathbf{N}\alpha \equiv 1 \bmod p$, so H is contained in the ray class field $K(\mathfrak{p})$.

THEOREM 1.1. — $K(1)\,\mathbf{Q}(\mu_p) = K(\mathfrak{p})$.

Proof. — Let $\mathfrak{a}$ be an ideal of K prime to $\mathfrak{p}$. It suffices to show that if $(\mathfrak{a}, K)$ fixes $\mathbf{Q}(\mu_p)$ and $K(1)$, then $(\mathfrak{a}, K)$ fixes $K(\mathfrak{p})$. Since $(\mathfrak{a}, K)$ fixes $K(1)$, it follows that $\mathfrak{a}$ is principal, $\mathfrak{a} = (\alpha)$. Since $(\mathfrak{a}, K)$ fixes H, it follows that $\alpha \equiv \pm 1 \bmod \mathfrak{p}$, so $\mathfrak{a}$ is in the unit class for $K(\mathfrak{p})$, as was to be shown.

We let $\mathrm{Cl}(H/K)$ be the ideal class group isomorphic to $\mathrm{Gal}(H/K)$ under the reciprocity law mapping

$$C \mapsto \sigma_C \quad \text{or} \quad \sigma(C).$$

Observe that all non-trivial characters of $\mathrm{Gal}(H/K) \approx \mathrm{Cl}(H/K)$ are primitive, with conductor $\mathfrak{p}$, because H is totally ramified over K at $\mathfrak{p}$.

The extensions H and $K(1)$ of K are linearly disjoint over K because H is totally ramified at $\mathfrak{p}$. Thre is a natural identification

$$\mathrm{Gal}(H/K) \approx \mathrm{Gal}(K(\mathfrak{p})/K(1))$$

from the diagram

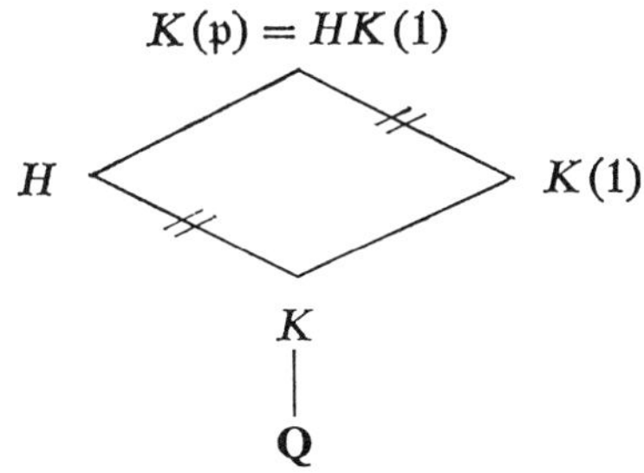

Under the class field theoretic isomorphism between ideal class groups and Galois groups, we have a commutative diagram:

$$\begin{array}{ccc} \mathrm{Cl}(H/K) & \leftrightarrow \mathrm{Gal}(H/K) & \subset \mathrm{Gal}(H/\mathbf{Q}) \\ \updownarrow & \updownarrow & \\ \mathfrak{o}(\mathfrak{p})^{*}/\pm 1 & \longrightarrow \mathbf{Z}(p)^{*2} & \end{array}$$

The arrow on top is the correspondence $C \mapsto \sigma_C$, arising from the above field diagram. The ideal classes of Cl (H/K) are precisely the principal ideal classes, modulo those generated by elements $\equiv 1 \bmod \mathfrak{p}$. This gives rise to the vertical arrow on the left.

The bottom arrow is induced by the norm map. Taking into account the natural isomorphism

$$\mathfrak{o}(\mathfrak{p}) \approx \mathbf{Z}(p),$$

the norm map amounts to the squaring map $x \mapsto x^2$.

The right vertical arrow arises from the usual correspondence between elements of $\mathbf{Z}(p)^*$ and Gal $(H/\mathbf{Q})$:

$$a \mapsto \sigma_a, \quad \text{with} \quad \sigma_a \zeta = \zeta^a.$$

The elements a corresponding to elements of Gal (H/K) are precisely the squares. Consequently an ideal class $C \in \mathrm{Cl}(H/K)$ corresponds uniquely to an element $a \in \mathbf{Z}(p)^{*2}$, and we shall write this correspondence as

$$C_a \leftrightarrow a.$$

Let χ be a non-trivial character of Gal (H/K). The induced character to Gal $(H/\mathbf{Q})$ is the direct sum of two characters χ_1, χ_2, with one of them odd, the other even. Say χ_1 is odd. These are the two characters of Gal $(H/\mathbf{Q})$ restricting to χ on Gal (H/K). If C contains a principal ideal (t), then

$$\chi(C) = \chi((t)) = \chi_1(t^2) = \chi_2(t^2).$$

2. Linear relations for $\chi \neq 1$

We shall apply the above considerations to the L-series. We note that

$$\chi_2 = \chi_1 \chi_K,$$

and χ_K is odd. By classical formulas pertaining to cyclotomic fields (*cf.* for instance [L 1]), we have:

L 1 $$L(1, \chi_1) = \frac{1}{p} \pi i \, S(\chi_1) B_{1, \bar{\chi}_1},$$

L 2 $$L(1, \chi_2) = -\frac{1}{p} S(\chi_2) \sum_{b \in \mathbf{Z}(p)^*} \bar{\chi}_2(b) \log |1 - \zeta^b|.$$

As usual, $S(\psi)$ is the Gauss sum formed with a multiplicative character ψ on $\mathbf{Z}(p)^*$, and the additive character

$$x \mapsto e^{2\pi i x/p} = \zeta^x, \qquad \text{where} \quad \zeta = e^{2\pi i/p}.$$

We have the L-series decomposition (*cf.* [L 3], chapter XII, § 2):

L 3 $$L(\chi, H/K, s) = L(\chi_1, H/\mathbf{Q}, s) L(\chi_2, H/\mathbf{Q}, s),$$

and also

L 4 $$L(\chi, H/K, s) = L(\chi, K(\mathfrak{p})/K, s).$$

The values of the L-series over K at 1 are given in terms of the Siegel functions as follows.

Let $\mathfrak{k}(z, L)$ be the Klein form (*cf.* [KL 2] and [L 2], chapter XV). We define

$$g^{12p}(z, L) = \mathfrak{k}^{12p}(z, L) \Delta(L)^p.$$

Let $\mathrm{Cl}(\mathfrak{p})$ be the ray class group of conductor $\mathfrak{p}$. For $C' \in \mathrm{Cl}(\mathfrak{p})$, we define

$$g_{\mathfrak{p}}(C') = g^{12p}(1, \mathfrak{p}\mathfrak{c}^{-1}),$$

where $\mathfrak{c}$ is any ideal in C'. The value is independent of $\mathfrak{c}$. If $C \in \mathrm{Cl}(H/K)$, *we define*

$$g_H(C) = N_{K(\mathfrak{p})/H} \, g_{\mathfrak{p}}(C'),$$

for any C' lying above C under the canonical homomorphism

$$\mathrm{Cl}(\mathfrak{p}) \to \mathrm{Cl}(H/K).$$

These are the invariants defined by RAMACHANDRA and ROBERT [Ro], paragraphs 2.2 and 2.4 (ROBERT uses the letter φ where we use g). See lso the last section of [KL 1].

From the Kronecker limit formula one obtains the value of the L-series in **L 4** at $s = 1$, as in MEYER [Me], SIEGEL [Si].

L 5 $$L(\chi, H/K, 1) = \frac{-2\pi}{6\, pw_1(\mathfrak{p})\, S_K(\bar{\chi}) \sqrt{d_K}} \sum_{C \in \mathrm{Cl}(H/K)} \bar{\chi}(C) \log |g_H(C)|.$$

We have used the usual notation:

$w_1(\mathfrak{p})$ = number of roots of unity in K which are $\equiv 1 \bmod \mathfrak{p}$. Since we took $p \geqslant 5$, it follows that $w_1(\mathfrak{p}) = 1$;

$\mathbf{d}_K = p$ = absolute value of the discriminant of K;

$S_K(\chi)$ is the Gauss sum relative to K, that is

$$S_K(\chi) = \sum_{x \in \mathfrak{o}(\mathfrak{p})^*} \chi((x))\, e^{2\pi i \operatorname{Tr}(x)/p}.$$

For the proof see also [L 4] (chapter 22, § 2, Theorem 2). In the notation of that chapter, we take $\gamma = 1/p$, and $\mathfrak{d} = (\sqrt{-p})$, so $\mathfrak{d}\mathfrak{p} = (p)$. Furthermore, for any character χ we have

$$\chi(C) = \chi(\bar{C}).$$

Indeed, C contains principal ideals (t), and $t \equiv \bar{t} \bmod \mathfrak{p}$. Finally, as explained in the last section of [KL 1], we have

$$g_{\mathfrak{p}}(C) = \Phi_{\mathfrak{p}}(\bar{C}).$$

These remarks show how the formula in the above reference imply the formula as stated here.

From the values of the L-series at 1, and the factorization **L 3**, we find the following Lemma.

LEMMA 2.1. – *For a non-trivial character* χ *of* Gal $(H/K) \approx$ Cl (H/K):

$$\sum_C \bar{\chi}(C) \log |g_H(C)| = 3i \frac{1}{\sqrt{p}} S(\chi_1) S(\chi_2) S_K(\bar{\chi}) B_{1,\bar{\chi}_1} \sum_{b \in \mathbf{Z}(p)^*} \bar{\chi}_2(b) \log |1 - \zeta^b|.$$

We note that

$$S_K(\bar{\chi}) = \sum_{z \in \mathbf{Z}(p)} \bar{\chi}_1^2(z)\, e^{2\pi i 2z/p}$$

and therefore

$$S_K(\bar{\chi}) = \chi_1^2(2)\, S(\bar{\chi}_1^2).$$

As a special case of the Davenport-Hasse relation (*cf.* for instance [L 1], Chapter 2, § 10), we have

$$S(\chi_1) S(\chi_2) = \bar{\chi}_1^2(2)\, S(\chi_1^2)\, S(\chi_K).$$

Furthermore, the sign of a Gauss sum is always positive (*cf.* for instance [L 3], chapter IV, § 3), and so we have

$$S(\chi_K) = i\sqrt{p}.$$

Therefore the linear relation becomes:

LEMMA 2.2

$$\sum_C \bar{\chi}(C) \log |g_H(C)| = -3\, p\, B_{1,\bar{\chi}_1} \sum_{b \in \mathbf{Z}(p)^*} \bar{\chi}_2(b) \log |1-\zeta^b|.$$

3. The relations for all characters

We now restate the linear relations, including the trivial character.

THEOREM 3.1. — *Let χ be any character of* $\mathrm{Cl}(H/K)$, *trivial or not. Then*

$$\sum_C \chi(C) \log |g_H(C)| = -3\, p\, B_{1,\chi_1} \sum_{b \in \mathbf{Z}(p)^*} \chi_2(b) \log |1-\zeta^b|.$$

If $\chi = 1$, *then* $\bar{\chi}_1 = \chi_1 = \chi_K$.

Proof. — The statement for non-trivial χ has been proved, so we suppose that χ is trivial. Then:

$$\sum_{C \in \mathrm{Cl}(H/K)} \log |g_H(C)|$$
$$= \sum_{C \in \mathrm{Cl}(\mathfrak{p})} \log |g_{\mathfrak{p}}(C)|$$
$$= \sum_{C \in \mathrm{Cl}(1)} \sum_{C' \subset C} \log |g_{\mathfrak{p}}(C')|$$

[where Cl (1) is the group of ordinary ideal classes]

$$= \sum_{C \in \mathrm{Cl}(1)} \frac{p}{2} \log \left| \frac{\Delta(\mathfrak{c}^{-1})}{\Delta(\mathfrak{p}\mathfrak{c}^{-1})} \right| \quad \text{by [Ro] (§ 2.3, Theorem 2 (iii))}$$

[where $\mathfrak{c}$ is any ideal in C], and since $\mathfrak{p} = (\sqrt{-p})$ is principal,

$$= \sum_{C \in \mathrm{Cl}(1)} 3\, p \log p$$
$$= h_K\, 3\, p \log p.$$

But the classical class number formula for K gives

$$h_K = w_K \left(-\frac{1}{2} B_{1,\chi_K} \right) = -B_{1,\chi_K}.$$

The Theorem follows from the obvious value for the sum

$$\sum_b \log |1-\zeta^b| = \log p.$$

4. Modular units as cyclotomic units

For any element u in K^* we consider the *regulator map* ρ given by

$$\rho(u) = \sum_C \log |u^{\sigma(C)}| \sigma_C^{-1}.$$

This map will be applied to p-units. Writing $C = C_a$ with $a \in \mathbf{Z}(p)^{*2}$, the above map can be viewed has having its values in the group algebra $\mathbf{C}[G]$, where

$$G = \mathrm{Gal}(H/K) \approx \mathbf{Z}(p)^{*2}.$$

We shall take $u = g_H(C_1)$, and take into account the fact that

$$g_H(C_1)^{\sigma(C)} = g_H(C).$$

We may write

$$\rho(g_H(C_1)) = \sum_C \log|g_H(C)|\sigma_C^{-1} = \sum_a \log|g_H(C_a)|\sigma_a^{-1}.$$

THEOREM 4.1. – *We have*

$$\rho(g_H(C_1)) = -12\,p\sum_a\sum_b \mathbf{B}_1\left(\left\langle\frac{b^{-1}a}{p}\right\rangle\right)\log|1-\zeta^b|\sigma_a^{-1}.$$

The sums on the right are taken over a, $b \in \mathbf{Z}(p)^{*2}$. *In particular*

$$\log|g_H(C_a)| = -12\,p\sum_b \mathbf{B}_1\left(\left\langle\frac{b^{-1}a}{p}\right\rangle\right)\log|1-\zeta^b|.$$

Proof. – Apply any character χ to $\rho(g_H(C_1))$. From paragraph 3 we find:

$$\sum_a \bar{\chi}(a)\log|g_H(C_a)| = -6\,p\,B_{1,\bar{\chi}_1}\sum_b \bar{\chi}_2(b)\log|1-\zeta^b|.$$

On the other hand applying χ to the right hand side of the formula to be proved, we find:

$$\begin{aligned}
&-12\,p\sum_b\sum_a \mathbf{B}_1\left(\left\langle\frac{b^{-1}a}{p}\right\rangle\right)\log|1-\zeta^b|\bar{\chi}_2(a)\\
&= -12\,p\sum_b\sum_a \mathbf{B}_1\left(\left\langle\frac{a}{p}\right\rangle\right)\log|1-\zeta^b|\bar{\chi}_2(b)\bar{\chi}_2(a)\\
&= -6\,p\,B_{1,\bar{\chi}_1}\sum_b \bar{\chi}_2(b)\log|1-\zeta^b|
\end{aligned}$$

because χ_1 and χ_2 have the same values on squares. Furthermore,

$$2\sum_a \mathbf{B}_1\left(\left\langle\frac{a}{p}\right\rangle\right)\bar{\chi}_1(a) = \sum_{t\in\mathbf{Z}(p)^*} B_1\left(\left\langle\frac{t}{p}\right\rangle\right)\bar{\chi}_1(t) = B_{1,\bar{\chi}_1},$$

because $\mathbf{B}_1$ is an odd function, and -1 is not a square mod p. This proves the Theorem.

We may now translate this result into a multiplicative notation. Let

$$\alpha = \prod_{b\in\mathbf{Z}(p)^{*2}}(1-\zeta^b)^{m(b^{-1}d)}/g_H(C_d).$$

We wish to prove that α is a root of unity. We know that $g_H(C_d)$ is a p-unit because it is obtained as a product of values of Siegel functions whose q-expansions (and those of their conjugates) are p-units in the integral closure of $\mathbf{Z}[j]$ in the modular function field. Furthermore, α has absolute value 1 at all archimedean absolute values by the above calculation. Therefore α must have absolute value 1 at $\mathfrak{p}$ by the product formula. Therefore α has absolute value 1.

Making the change of variables $a \mapsto d$ and $b^{-1} a \mapsto b$ in Theorem 4.1, we have proved the following.

THEOREM 4.2. – *For $d \in \mathbf{Z}(p)^{*2}$ we have*

$$g_H(C_d) = \varepsilon(d) \prod_{b \in \mathbf{Z}(p)^{*2}} (1-\zeta^{b^{-1}d})^{m(b)},$$

where $\varepsilon(d)$ is a root of unity, and

$$m(b) = -12\, p\, \mathbf{B}_1\left(\left\langle \frac{b}{p} \right\rangle\right).$$

The notation is standard: we denote by $\mathbf{B}_1$ the first Bernoulli polynomial,

$$\mathbf{B}_1(X) = X - \frac{1}{2}.$$

For any real number r, we let $\langle r \rangle$ be the unique number satisfying

$$r \equiv \langle r \rangle \bmod \mathbf{Z} \qquad \text{and} \qquad 0 \leqslant \langle r \rangle < 1.$$

Remark. – If we write $\varepsilon(1) = \eta\varepsilon_0$ with $\varepsilon_0 = \pm 1$, and $\eta \in \mu_p$, then

$$\varepsilon(d) = \eta^d \varepsilon_0.$$

The numbers $g_H(C)$ are p-units, and their quotients

$$g_H(C)/g_H(C'),$$

are units (RAMACHANDRA and ROBERT). Let:

$\Phi_{p,H} = \Phi_p$ = group generated by all values $g_H(C)$ and μ_H;

$\Phi_H = \Phi$ = group generated all quotients $g_H(C)/g_H(C')$ and μ_H;

$\Phi_p(w_H)$ = group of p-units of the form;

$$\prod g_H(C)^{n(C)},$$

where the exponents $n(C)$ satisfy the condition

$$\sum_C n(C)\, \mathbf{N}\, \mathfrak{a}(C) \equiv 0 \bmod w_H,$$

where $\mathfrak{a}(C)$ is any ideal prime to w_H in the class C, and $w_H = 2p$ is the number of roots of unity in H. In the present case, we can write these units in the form

$$\prod_d g_H(C_d)^{n(d)} = \prod_d \varepsilon(d)^{n(d)} \prod_b (1-\zeta^b)^{-12pv(b)},$$

where

$$\sum n(d)\, d \equiv 0 \bmod p \qquad \text{and} \qquad \sum n(d) \equiv 0 \bmod 2.$$

The exponent $\nu(b)$ is given by the formula:

$$\nu(b) = \sum_d n(d)\, \mathbf{B}_1\left(\left\langle \frac{b^{-1}d}{p} \right\rangle\right) \in \mathbf{Z}.$$

Since $\varepsilon(d) = \eta^d \varepsilon_0$ (*cf.* remark above), we have

$$\prod_d \varepsilon(d)^{n(d)} = \varepsilon_0^{\Sigma n(d)} \eta^{\Sigma n(d)\, d} = 1,$$

so that

$$\prod_d g_H(C_d)^{n(d)} = \prod_b (1-\zeta^b)^{-12p\nu(b)}.$$

The root of unity factor has gone out.

Using similar notation, we let

$$\Phi_H(w_H) = \text{subgroup of } \Phi_H \text{ satisfying the above condition.}$$

An element of $\Phi_{p,H}$ lies in Φ_H if, and only if, the exponents $n(C)$ satisfy

$$\sum_C n(C) = 0.$$

We have also given a proof in the present instance for Robert's result that the elements of $\Phi(w_H)$ (or $\Phi_p(w_H)$) are $12\,p$-th powers in H. As in ROBERT [Ro], this allows us to take $12\,p$-roots, and we define:

$E_{p,\mathrm{cyc}}$ = group generated by μ_H and by the cyclotomic p-units $\zeta^b - 1$, with $b \in \mathbf{Z}(p)^{*2}$;

$E_{p,\mathrm{mod}}$ = group generated by μ_H and all elements $\alpha \in H$ such that $\alpha^{12p} \in \Phi_p(w_H)$.

We define E_{cyc} and E_{mod} in a similar way, taking the elements of degree 0 to get units instead of p-units. We call the groups $E_{p,\mathrm{mod}}$ or E_{mod} the groups of *modular p-units* or *modular units* respectively. The latter could also be called the group of *Robert units.* For any element $\alpha \neq 0$ of H, we let

$$\rho(\alpha) = \sum_{\sigma \in G} \log|\sigma\alpha|\, \sigma^{-1},$$

where $G \approx \mathbf{Z}(p)^{*2}$ is the Galois group of H over K. Thus ρ is the usual ,'regulator'' map. Then

$$E_{p,\mathrm{cyc}}/E_{p,\mathrm{mod}} \approx \rho(E_{p,\mathrm{cyc}})/\rho(E_{p,\mathrm{mod}}).$$

Let $R = \mathbf{Z}[G]$, and let ξ be the element of $\mathbf{C}\, R$ given by

$$\xi = \sum_a \log|\zeta^a - 1|\, \sigma_a^{-1}.$$

Let the *Stickelberger element* be

$$\Theta = \sum_b \mathbf{B}_1\left(\left\langle \frac{b}{p} \right\rangle\right) \sigma_b^{-1}.$$

Let:

I_H = ideal of elements $\sum n(d)\sigma_d$, $n(d) \in \mathbf{Z}$, d prime to $2p$, satisfying the conditions

$$\sum n(d)d \equiv 0 \bmod p \qquad \text{and} \qquad \sum n(d) \equiv 0 \bmod 2.$$

$\mathscr{S}$ = *Stickelberger ideal* = $I_H \Theta \subset R$ because 2 divides w_H. Then

$$\rho(E_{p,\,\mathrm{cyc}}) = R\xi, \qquad \rho(\Phi_p) = 12\,p\,R\xi\Theta.$$

$$\rho(E_{p,\,\mathrm{mod}}) = \mathscr{S}\xi.$$

Consequently we obtain an isomorphism.

THEOREM 4.3. — $E_{p,\,\mathrm{cyc}}/E_{p,\,\mathrm{mod}} \approx R\xi/\mathscr{S}\xi \approx R\,\mathscr{S}$.

A similar isomorphism is obtained for $E_{\mathrm{cyc}}/E_{\mathrm{mod}}$ by considering the elements of degree 0.

Remark. — For the record, it may be useful to have the expression of the cyclotomic units as *rational* power products of the modular units. In Theorem 4.1, we apply a character $\bar{\chi}$, divide by $-6\,p\,B_{1,\,\chi_1}$, and sum over $\bar{\chi}$. We find:

$$-\frac{1}{6\,p}\sum_C \log g_H(C) \sum_\chi \frac{1}{B_{1,\,\chi_1}}\,\chi(C)$$
$$= \sum_{b\,\in\,(\mathbf{Z}(p)^*)^2} \log|1-\zeta^b| \sum_\chi \chi_2(b)$$
$$= \frac{p-1}{2}\log|1-\zeta|.$$

Therefore, we have the following Theorem.

THEOREM 4.4

$$\log|1-\zeta| = \frac{-1}{3\,p(p-1)}\sum_C m'(C)\log|g_H(C)|,$$

where

$$m'(C) = \sum_\chi \frac{1}{B_{1,\,\chi_1}}\,\chi(C).$$

In multiplicative notation, up to a root of unity, this yields

$$1-\zeta^a = \varepsilon'(a)\prod_C g_H(C)^{-\frac{1}{3\,p(p-1)}m'(C/C_a)}.$$

Part two: $p \equiv 1 \bmod 4$

1. General facts

As before, we have the following Theorem.

THEOREM 1.1. — $K(\mathfrak{p}) = K(1)\,\mathbf{Q}(\mu_p)$.

The proof is similar and can be omitted. We let $K' = \mathbf{Q}(\sqrt{p})$. We let

$$H = K(\mu_p) = \mathbf{Q}(\mu_{4p}).$$

Then

$$[H:K] = [K(\mu_p):K] = p-1,$$

so $\mathrm{Gal}(K(\mu_p)/K) \approx \mathbf{Z}(p)^*$ in a natural way. We have the following diagram of fields.

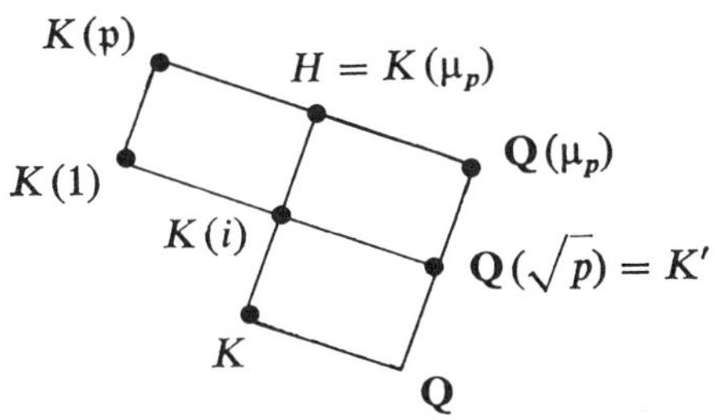

Note that $K(i)$ is unramified over K (because only 2 can ramify, and $K(i)$ is also obtained by adjoining $\sqrt{p}$ to K). Thus $K(i) = K(1) \cap K(\mu_p)$, and we denote

$$K(i) = H_1 = H \cap K(1).$$

We have a diagram similar to the other case.

$$\begin{array}{ccc} \mathrm{Cl}(H/H_1) & \overset{\approx}{\leftrightarrow} & \mathrm{Gal}(H/H_1) \\ \updownarrow & & \updownarrow \\ (\mathfrak{o}/\mathfrak{p})^*/\pm 1 & \xrightarrow[\text{square}]{} & \mathbf{Z}(p)^{*2} \end{array}$$

Let χ be a character on $\mathrm{Gal}(H/K)$. We denote by $\chi_{\mathbf{Q}}$ the corresponding character on the isomorphic group $\mathrm{Gal}(\mathbf{Q}(\mu_p)/\mathbf{Q})$, and also view $\chi_{\mathbf{Q}}$ as a character on $\mathbf{Z}(p)^*$ under the usual isomorphism.

$$a \mapsto \sigma_a.$$

As before, for any class $C \in \mathrm{Cl}(H/K)$, we have

$$\chi(C) = \chi(\bar{C}),$$

because for any ideal $\mathfrak{c}$ in C,

$$\chi(C) = \chi_{\mathbf{Q}}(\mathbf{N}\,\mathfrak{c}) = \chi_{\mathbf{Q}}(\mathbf{N}\,\bar{\mathfrak{c}}) = \chi(\bar{C}).$$

Note that $\chi^2 \neq 1$ *if and only if conductor of* $\chi = \mathfrak{p}$.

There are three characters of order 2 on $\mathbf{Z}(4p)^*$, corresponding to the three subfields of degree 2 over $\mathbf{Q}$, namely

$$\chi_K, \quad \chi_{K'}, \quad \chi_{\mathbf{Q}(i)}, \quad \text{and we have} \quad \chi_{K'} = \chi_K \chi_{\mathbf{Q}(i)}.$$

2. Linear relation for $\chi^2 \neq 1$

We assume $\chi^2 \neq 1$. We apply the Kronecker limit formula as in [L 4], that is

$$L(\chi, H/K, 1) = -\frac{2\pi}{6\,p\,S_K(\bar{\chi})\sqrt{d_K}} \sum_C \bar{\chi}(C) \log |g_H(C)|.$$

In the notation of [L 4], $S_K(\bar{\chi}) = \bar{\chi}(\gamma\,\mathfrak{d}\mathfrak{p})\,T(\bar{\chi}, \gamma)$, where $\mathfrak{d} = (2\sqrt{-p})$, and we can take $\gamma = 1/2p$. Thus

$$S_K(\bar{\chi}) = S(\bar{\chi}_{\mathbf{Q}}^2).$$

Also, $d_K = 4p$. We use the decomposition

$$L(\chi, H/K, 1) = L(\chi_{\mathbf{Q}}, H/\mathbf{Q}, 1)\,L(\chi'_{\mathbf{Q}}, H/\mathbf{Q}, 1),$$

where $\chi'_{\mathbf{Q}}$ is the other character on $\mathbf{Z}(4p)^*$ restricting to χ. *A priori*, we do not know which of $\chi_{\mathbf{Q}}$ or $\chi'_{\mathbf{Q}}$ is even or odd, and we use again χ_1, χ_2 to denote the odd and even characters respectively equal to $\chi_{\mathbf{Q}}$ or $\chi'_{\mathbf{Q}}$. In the specific determination of the values of the L-series, we shall have to distinguish corresponding cases. We have

$$\chi'_{\mathbf{Q}} = \chi_{\mathbf{Q}}\chi_K = \chi_{\mathbf{Q}}\chi_{K'}\chi_{\mathbf{Q}(i)}.$$

We let m_2 = conductor of χ_2. Then

$$L(\chi, H/K, 1) = \frac{1}{i/p}\pi\, i\, S(\chi_1)\, B_{1,\bar{\chi}_1}\left(-\frac{1}{p}\right) S(\chi_2) \times \sum_{b \in \mathbf{Z}(m_2)^*} \bar{\chi}_2(b) \log |1 - \zeta^b|,$$

where $\zeta = ie^{2\pi i/p}$ (resp. $\zeta = {}^{2\pi i/p}$) according as $m_2 = 4p$ or $m_2 = p$. From this we get the following relation.

LEMMA 2.1

$$\sum_C \bar{\chi}(C) \log |g_H(C)| = \frac{3i}{2\sqrt{p}} S(\chi_{\mathbf{Q}})\, S(\chi'_{\mathbf{Q}})\, S(\bar{\chi}_{\mathbf{Q}}^2)\, B_{1,\bar{\chi}_1} \sum_{b \in \mathbf{Z}(m_2)^*} \bar{\chi}_2(b) \log |1 - \zeta^b|.$$

We now simplify the Gauss sums and their products.

LEMMA 2.2. — $S(\chi_{\mathbf{Q}})\, S(\chi'_{\mathbf{Q}}) = \bar{\chi}^2_{\mathbf{Q}}(2)\, S(\chi^2_{\mathbf{Q}})\, S(\chi_K)$.

Proof. — This is a special case of the Davenport-Hasse distribution relation (*cf.* for instance [L 1], chapter 2, § 10).

In the present case, we have from the prime power decomposition (with respect to the primes p and 2),

$$S(\chi_K) = S(\chi_{K'})\, S(\chi_{\mathbf{Q}(i)}) = 2\, i\sqrt{p},$$

because the signe of the Gauss sum is always positive (*see* for instance [L 3], chapter IV, § 3). This yields:

LEMMA 2.3

$$\begin{aligned}&\textstyle\sum_C \bar{\chi}(C) \log |g_H(C)| \\ &\quad = -3\, p\, \bar{\chi}^2_{\mathbf{Q}}(2)\, B_{1,\bar{\chi}_1} \textstyle\sum_{b \in \mathbf{Z}(m_2)^*} \bar{\chi}_{\mathbf{Q}}(b) \log |1-\zeta^b|.\end{aligned}$$

We must then distinguish two cases.

Case 1: $\chi_{\mathbf{Q}} = \chi_2$. — The expression in Lemma 2.3 is then equal to

$$-3\, p\, \bar{\chi}^2_{\mathbf{Q}}(2)\, B_{1,\bar{\chi}_{\mathbf{Q}}\chi_K} \textstyle\sum_{b \in \mathbf{Z}(p)^*} \bar{\chi}_{\mathbf{Q}}(b) \log |1-\zeta^b|.$$

Case 2: $\chi_{\mathbf{Q}} = \chi_1$. — The expression in Lemma 2.3 is then equal to

$$\begin{aligned}&-3\, p\, \bar{\chi}^2_{\mathbf{Q}}(2)\, B_{1,\bar{\chi}_{\mathbf{Q}}} \textstyle\sum_{b \in \mathbf{Z}(4p)^*} \bar{\chi}_{\mathbf{Q}}\chi_K(b) \log |1-\zeta^b| \\ &= -3\, p\, \bar{\chi}^2_{\mathbf{Q}}(2)\, B_{1,\bar{\chi}_{\mathbf{Q}}} \Big[\textstyle\sum_{b \in \mathbf{Z}(4p)^*,\, b \equiv 1 \bmod 4} \bar{\chi}_{\mathbf{Q}}(b) \left(\dfrac{b}{p}\right) \log |1-\zeta_p^b \zeta_4| \\ &\qquad - \textstyle\sum_{b \in \mathbf{Z}(4p)^*,\, b \equiv -1 \bmod 4} \bar{\chi}_{\mathbf{Q}}(b) \left(\dfrac{b}{p}\right) \log |1-\zeta_p^b \zeta_4^{-1}| \Big],\end{aligned}$$

where (b/p) is the quadratic symbol, $\zeta_p = e^{2\pi i/p}$, $\zeta_4 = i$.

3. Linear relations for $\chi^2 = 1$

We distinguish two cases, depending on whether χ is trivial or not.

Case $\chi = 1$. — Then case 1 of Lemma 2.3 also holds here, that is

$$\textstyle\sum_C \log |g_H(C)| = 3\, p (\log p)\, h_K = -3\, p (\log p)\, B_{1,\chi_K}.$$

The proof is the same as in the case $p \equiv -1 \bmod 4$, we did not need any special property of p for this relation.

Case $\chi \neq 1$, **so** $\chi_{\mathbf{Q}} = \chi_{K'}$. — Then

$$\textstyle\sum_C \chi(C) \log |g_H(C)| = 0.$$

Proof. — As in the proof of the other case, we first write the sum as a sum over elements of $Cl(\mathfrak{p})$, and then as a sum

$$\sum_{c \in \mathrm{Cl}(1)} \chi(c) \log \left| \frac{\Delta(\mathfrak{c}^{-1})}{\Delta(\mathfrak{c}^{-1}\mathfrak{p})} \right| = 0,$$

because $\mathfrak{p}$ is principal, so we can use the homogeneity of the delta function, and end up with the sum of the non-trivial character over all elements of Cl (1), thus yielding 0.

4. Modular units as cyclotomic units

Let $\zeta_n = e^{2\pi i/n}$. We wish to give an expression for the modular units

$$g_H(C_1) = \varepsilon \prod_b (1-\zeta_p^b)^{m(b)} (1-\zeta_p^b \zeta_4)^{r'(b)} (1-\zeta_p^b \zeta_4^{-1})^{r''(b)},$$

where $m(b)$, $r'(b)$, $r''(b)$ are rational numbers, and the product is taken for $b \in \mathbf{Z}(p)^*$. Since

$$(1-\zeta_p^b \zeta_4)(1-\zeta_p^b \zeta_4^{-1}) = 1+\zeta_p^{2b} = \frac{1-\zeta_p^{4b}}{1-\zeta_p^{2b}},$$

we may assume that the expression has the form

$$g_H(C_1) = \varepsilon \prod_b (1-\zeta_p^b)^{m(b)} \left(\frac{1-\zeta_p^b \zeta_4}{1-\zeta_p^b \zeta_4^{-1}} \right)^{r(b)},$$

where ε is a root of unity, and

$$b \mapsto m(b) \text{ is an even function,}$$
$$b \mapsto r(b) \text{ is an odd function.}$$

THEOREM 4.1. — *There is an expression as above, with*

$$m(b) = -3p\left(\frac{b}{p}\right)\left[\mathbf{B}_1\left(\left\langle \frac{b^{-1}}{p} + \frac{1}{4} \right\rangle\right) - \mathbf{B}_1\left(\left\langle \frac{b^{-1}}{p} - \frac{1}{4} \right\rangle\right)\right],$$

$$r(b) = -3p\left(\frac{b}{p}\right)\mathbf{B}_1\left(\left\langle \frac{4^{-1}b^{-1}}{p} \right\rangle\right).$$

We have the regulator relation:

$$\sum_C \log |g_H(C)| \sigma_C^{-1}$$
$$= \sum_a \left[\sum_b m(ba^{-1}) \log |1-\zeta_p^b| + \left(\frac{a}{p}\right) r(ba^{-1}) \log \left| \frac{1-\zeta_p^b \zeta_4}{1-\zeta_p^b \zeta_4^{-1}} \right| \right] \sigma_a^{-1}.$$

If $\mathfrak{c}$ is an ideal in the class C, and $a = \mathbf{N}\mathfrak{c}$, we also put

$$C = C_a.$$

Then we may also write the above relation in the form:

$$\log|g_H(C_a)| = \sum_b m(ba^{-1})\log|1-\zeta_p^b| + \left(\frac{a}{p}\right) r(ba^{-1})\log\left|\frac{1-\zeta_p^b\zeta_4}{1-\zeta_p^b\zeta_4^{-1}}\right|.$$

Remark 1. — The expression in brackets [ ] in the definition of $m(b)$ is always 1/2 or $-1/2$. The product is taken over all b in $\mathbf{Z}(p)^*$. Consequently combining the values for $m(b)$ and $m(-b)$, we see that the factor involving $m(b)$ already gives an integral representation in terms of the cyclotomic numbers $1-\zeta_p^b$. A similar remark applies to the factor involving $r(b)$. As before, considering the subgroup generated by the $g_H(C)$ satisfying the Robert congruence conditions on the exponents shows that elements of this subgroup have p-th roots in the cyclotomic units.

Remark 2. — The factor (a/p) in front of $r(ba^{-1})$ in the formula arises from the Galois action on the 4-th roots of unity.

Proof of Theorem 4.1. — We apply an arbitrary character $\chi_{\mathbf{Q}}$, to the expression on the right hand side (RHS) of the regulator relation to be proved, and verify that it gives the desired value from paragraphs 2 and 3

Suppose first that $\chi_{\mathbf{Q}}$ is even. The sum over the terms containing $r(ba^{-1})$ will be 0, because $a \mapsto \chi_{\mathbf{Q}}(a)$ is even, and

$$a \mapsto \left(\frac{a}{p}\right) r(ba^{-1}),$$

is odd. After a change of variables, we thus obtain

$$\begin{aligned}\chi_{\mathbf{Q}}(\mathrm{RHS}) &= \textstyle\sum_a\sum_b \chi_{\mathbf{Q}}(a)\bar{\chi}_{\mathbf{Q}}(b)\, m(a)\log|1-\zeta_p^b| \\ &= \textstyle\sum_a \chi_{\mathbf{Q}}(a)\, m(4a).\sum_b \bar{\chi}_{\mathbf{Q}}(4b)\log|1-\zeta_p^b| \\ &= S_a S_b, \text{ say.}\end{aligned}$$

Furthermore,

$$\begin{aligned}S_a &= \textstyle\sum_a \bar{\chi}_{\mathbf{Q}}(a)\, m(4^{-1}a^{-1}) \\ &= \textstyle\sum_a \bar{\chi}_{\mathbf{Q}}(a)\left(\frac{a}{p}\right)\left[\left\langle\frac{4^{-1}a}{p}+\frac{1}{4}\right\rangle - \left\langle\frac{4^{-1}a}{p}-\frac{1}{4}\right\rangle\right] \\ &= \textstyle\sum_{t\in\mathbf{Z}(4p)^*,\, t\equiv 1 \bmod 4} \bar{\chi}_{\mathbf{Q}}(t)\left(\frac{t}{p}\right)\mathbf{B}_1\left(\left\langle\frac{4}{4p}\right\rangle\right) \\ &\quad -\textstyle\sum_{t\in\mathbf{Z}(4p)^*,\, t\equiv -1 \bmod 4} \bar{\chi}_{\mathbf{Q}}(t)\left(\frac{t}{p}\right)\mathbf{B}_1\left(\left\langle\frac{t}{4p}\right\rangle\right).\end{aligned}$$

Note that

$$S_a = B^{(4p)}_{1, \bar{\chi}_{\mathbf{Q}}\chi_K} = \sum_t \bar{\chi}_{\mathbf{Q}} \chi_K(t) \mathbf{B}_1\left(\left\langle \frac{t}{4p} \right\rangle\right),$$

is the sum defining the Bernoulli-Leopoldt number at level $4p$ with respect to the character $\bar{\chi}_{\mathbf{Q}} \chi_K$.

If $\chi_{\mathbf{Q}} \neq \chi_{K'}$ then $\bar{\chi}_{\mathbf{Q}} \chi_K$ has conductor $4p$, and we get the desired value corresponding to Lemma 2.3.

If $\chi_{\mathbf{Q}} = \chi_{K'}$, then this Bernoulli number is 0 (corresponding to the value found in paragraph). Indeed, the standard reduction for computing Bernoulli numbers from one level to a lower level with one fewer prime factor introduces the factor

$$1 - \bar{\chi}_{\mathbf{Q}}(p) \chi_K(p) = 0,$$

because $\chi_K \chi_{K'}(p) = \chi_{\mathbf{Q}(i)}(p) = 1$ (*cf.* for instance [L 1], the Lemma of chapter 2, § 8). This concludes the proof of the case when $\chi_{\mathbf{Q}}$ is even.

Suppose now that $\chi_{\mathbf{Q}}$ is odd, so $\chi_{\mathbf{Q}} = \chi_1$. Then

$$\chi_2 = \chi_1 \chi_K,$$

and the conductor of χ_2 is $4p$. Then the term with m drops out, for parity reasons again and we get a sum with the terms containing $r(ba^{-1})$. For simplicity, abbreviate

$$Z(b) = \frac{1 - \zeta_p^b \zeta_4}{1 - \zeta_p^b \zeta_4^{-1}}.$$

Then we find:

$$\chi_{\mathbf{Q}}(\text{RHS}) = -3p \sum_b \sum_a \bar{\chi}_{\mathbf{Q}}(a) \left(\frac{a}{p}\right)\left(\frac{ba^{-1}}{p}\right) \mathbf{B}_1\left(\left\langle \frac{4^{-1} b^{-1} a}{p} \right\rangle\right) \log|Z(b)|$$
$$= -3p \bar{\chi}_{\mathbf{Q}}(4) S_a S_b,$$

where

$$S_a = \sum_a \bar{\chi}_{\mathbf{Q}}(a) \mathbf{B}_1\left(\left\langle \frac{a}{p} \right\rangle\right) = B_{1, \bar{\chi}_{\mathbf{Q}}}$$
$$S_b = \sum_b \bar{\chi}_{\mathbf{Q}}(b) \left(\frac{b}{p}\right) \log|Z(b)|.$$

It now suffices to verify that S_b is equal to the sum in brackets in case 2 of Lemma 2.3, namely we must show

$$S_b = \sum_{t \in \mathbf{Z}(4p)^*} \bar{\chi}_2(t) \log|1 - \zeta^t|.$$

To do this, we decompose the right hand side, writing $\zeta = \zeta_p \zeta_4$, and sum over $b \in \mathbf{Z}(p)^*$ corresponding to values

$$t \equiv b \bmod p, \qquad t \equiv 1 \bmod 4$$

and also

$$t \equiv b \bmod p, \qquad t \equiv -1 \bmod 4.$$

The desired equality drops out.

As in the previous case, let

$$\alpha = \prod_b (1-\zeta_p^b)^{m(b)} Z(b)^{r(b)} / g_H(C_1).$$

We want to prove that α is a root of unity. By the above calculations, α has absolute value 1 at all archimedean absolutes values. Moreoever, α is a unit outside of primes deviding p in $\mathbf{Q}(\mu_{4p})$. We have

$$p = (\mathfrak{p}_1 \mathfrak{p}_2)^{(p-1)/2}.$$

We note that the valuation of the numerator of α at $\mathfrak{p}_1$ equals the valuation of the numerator at $\mathfrak{p}_2$ since the contributions come from elements of $\mathbf{Q}(\mu_p)$. The same is true of the denominator, as one can see from the distribution relations (ROBERT-RAMACHANDRA). Thus this is true for α, and then by the product formula, α must be a unit at $\mathfrak{p}_1$ and at $\mathfrak{p}_2$. Therefore α is a root of unity. This proves Theorem 4.1.

Remark. — In the present case, when H contains a non-trivial unramified extension of K, the group generated by the values $g_H(C)$ is not of finite index in the cyclotomic p-units. From ROBERT [Ro], we know that unramified units formed with the delta function must also be taken into account to get a full group of units. One can follow the same method to carry this out. This will be done elsewhere. *Cf.* KERSEY'S thesis for a treatment of the general case, of an arbitrary imaginary quadratic field and nth roots of unity.

BIBLIOGRAPHY

[KL 1] KUBERT (D.) and LANG (S.). — Units in the modular function field, I, *Math. Annalen*, t. 218, 1975, p. 67-96.

[KL 2] KUBERT (D.) and LANG (S.). — Units in the modular function field, II: A full set of units, *Math. Annalen*, t. 218, 1975, p. 175-189.

[L 1] LANG (D.). — *Cyclotomic fields*. — Berlin, Springer-Verlag, 1978.

[L 2] LANG (S.). — *Introduction to modular forms*. — Berlin, Springer-Verlag, 1976 (*Grundlehren der mathematischen Wissenschaften*, 222).

[L 3] LANG (S.). — *Algebraic number theory*. — Reading, Addison-Wesley, 1970 (*Addison-Wesley Series in Mathematics*).

[L 4] LANG (S.). — *Elliptic functions.* — Reading, Addison-Wesley, 1973 (*Addison-Wesley Series in Mathematics*).

[Me] MEYER (C.). — *Die Berechnung der Klassenzahl Abelscher Körper über quadratischen Zahlkorpern.* — Berlin, Akademie-Verlag, 1957 (*Mathematische Lehrbücher und Monographien*, 2).

[Ra] RAMACHANDRA (K.). — Some applications of Kronecker's limit formula, *Annals of Math.*, Series 2, t. 80, 1964, p. 104-148.

[Ro] ROBERT (G.). — Unités elliptiques, *Bull. Soc. math. France*, Mémoire n° 36, 1973, 77 p.

[Si] SIEGEL (C. L.). — *Lectures on advanced analytic number.* — Bombay, Tata Institute of fundamental Research, 1961 (*Tata Institute, Lectures on Mathematics*, 23.

Extrait de *L'Enseignement mathématique*, T. XXVII, fasc. 3-4, 1981

FINITENESS THEOREMS IN GEOMETRIC CLASSFIELD THEORY

by Nicholas M. KATZ [1]) and Serge LANG [1])

(with an appendix by Kenneth A. RIBET)

0. INTRODUCTION

The geometric classfield theory of the 1950's was the principal precursor of the Grothendieck theory of the fundamental group developed in the early 1960's (cf. SGA I, Exp. X, 1.10). The problem was to understand the abelian unramified coverings of a variety X, or, as we would say today, to understand $\pi_1(X)^{ab}$. When X is "over" another variety S, the functoriality of π_1^{ab} gives a natural homomorphism.

$$\pi_1(X)^{ab} \to \pi_1(S)^{ab}$$

whose kernel Ker (X/S) measures the extent to which the abelian coverings of X fail to "come from" abelian coverings of S.

In the language of the 1950's, we can make the problem "explicit" in terms of galois theory. Thus we consider the case when $S = \text{Spec}(K)$, with K a field, and X a smooth and geometrically connected variety over K. Let F denote the function field of X, and denote by E/F the compositum, inside some fixed algebraic closure of F, of all finite abelian extensions E_i/F which are unramified over X in the sense that the normalization of X in E_i is finite etale over X. Then $\pi_1(X)^{ab}$ is "just" the galois group Gal (E/F).

Each finite extension L_i/K of K gives rise to a constant-field extension $F \cdot L_i$ over F which is abelian and unramified over X, so that if we denote by K^{ab} the maximal abelian extension of K, we have a diagram of fields and galois groups

[1]) Supported by NSF grants.

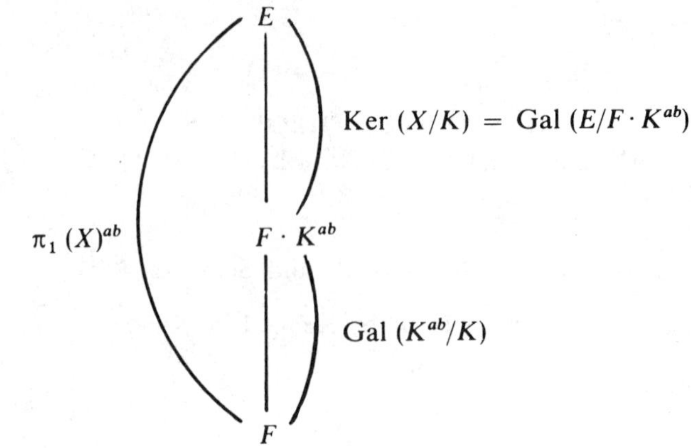

and a corresponding exact sequence

$$\begin{array}{ccccccccc} 0 & \to & \mathrm{Ker}\,(X/K) & \to & \pi_1\,(X)^{ab} & \to & \mathrm{Gal}\,(K^{ab}/K) & \to & 0 \\ & & \| & & \| & & \| & & \\ & & \mathrm{Gal}\,(E/F\cdot K^{ab}) & & \mathrm{Gal}\,(E/F) & & \mathrm{Gal}\,(F\cdot K^{ab}/F) & & \end{array}$$

If we suppose further that X admits a K-rational point x_0, then we can "descend" the extension $E/(F\cdot K^{ab})$ to an extension E_0/F by the following device: we define E_0 to be the union of those finite abelian extensions E_i/F which are unramified over X, and such that the fibre over x_0 of the normalization X_i of X in E_i consists of deg (E_i/F) distinct K-rational points of X_i. Then E_0/F is a geometric extension, i.e. K is algebraically closed in E_0, and E is the compositum $E_0\cdot K^{ab}$. Thus we have a diagram of fields and galois groups

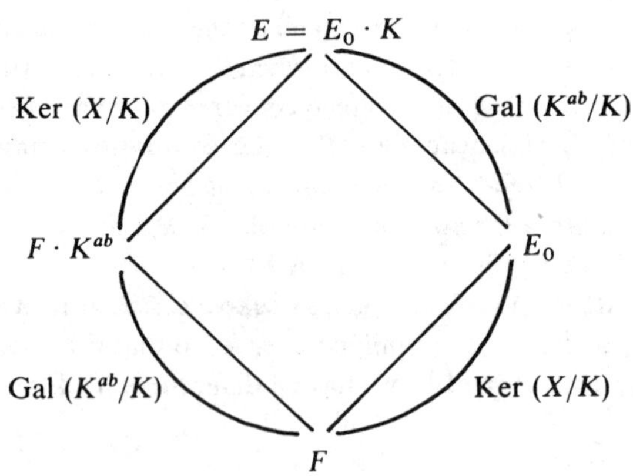

and a corresponding splitting of the above exact sequence

$$\begin{array}{ccccccccc} 0 & \to & \mathrm{Ker}\,(X/K) & \to & \pi_1\,(X)^{ab} & \to & \mathrm{Gal}\,(K^{ab}/K) & \to & 0 \\ & & \| & & \| & & \| & & \\ & & \mathrm{Gal}\,(E_0/F) & & \mathrm{Gal}\,(E/F) & & \mathrm{Gal}\,(E/F_0) & & \end{array}$$

We will see that when K is a number field, i.e. a finite extension of $\mathbf{Q}$, then the group Ker (X/K) is finite, or in other words the extension $E/(F \cdot K^{ab})$ (as well as the extension E_0/F when X has a K-rational point x_0) is *finite*.

Our main result is a finiteness theorem for the kernel group Ker (X/S) for a reasonably wide class of situations X/S which are sufficiently "of absolutely finite type" (cf. Theorems 1 & 2 for precise statements). When in addition we have *a priori* control of $\pi_1\,(S)^{ab}$, [as provided by global classfield theory when S is the (spectrum of) the ring of integers in a number field], or a systematic way of ignoring $\pi_1\,(S)^{ab}$ [e.g. if X/S has a section] we get "absolute" finiteness theorems (cf. Theorems 3, 4, 5 for precise statements). Following Deligne ([2], 1.3) and Grothendieck, we also give an application of our result to the theory of l-adic representations of fundamental groups of varieties over absolutely finitely generated fields (cf. Theorem 6 for a precise statement). In fact, it was this application, already exploited so spectacularly by Deligne in the case of varieties over finite fields, which aroused [resp. rearoused] our interest in the questions discussed here. For an application of these theorems to K-theory, we refer to recent work of Spencer Bloch [1] and A. H. Parshin [12].

The idea behind our proof is to reduce first to the case of an open curve over a field, by using Mike Artin's "good neighborhoods" and an elementary but useful exact homotopy sequence (cf. Preliminaries, Lemma 2). We then reduce to the case of an abelian variety over a field by using the theory of the generalized Jacobian. A standard specialization argument then reduces us to the case of an abelian variety over a finite field. In this case, we use Weil's form ([12], thm. 36) of the Lefschetz trace formula for abelian varieties to reduce our finiteness theorem to the fact that the number of rational points on an abelian variety over a finite field is finite and non-zero!

In explicating our results in the case of an abelian variety over a number field (cf. Section II, Remark 2), we were led to the conjecture that if A is an abelian variety over a number field k, and if $k\,(\mu)$ is the extension of k obtained by adjoining to k all roots of unity, then the torsion subgroup of A in $k\,(\mu)$ is finite. We shall prove this conjecture when A has complex multiplication. In an appendix, Ribet extends this result to a proof of the conjecture in general.

II. Preliminaries

Let S be a connected, locally noetherian scheme, and s a geometric point of S (i.e., s is a point of S with values in an algebraically closed field). The fundamental group $\pi_1(S, s)$ in the sense of SGA I is a profinite group which classifies the finite etale coverings of S. Given two geometric points s_1 and s_2 each choice of "chemin" $c(s_1, s_2)$ from s_1 to s_2 determines an isomorphism

$$c(s_1, s_2): \pi_1(S, s_1) \xrightarrow{\sim} \pi_1(S, s_2)$$

and formation of this isomorphism is compatible with composition of chemins. If we fix s_1 and s_2 but vary the chemin, this isomorphism will (only) change by an inner automorphism of, say, $\pi_1(S, s_2)$.

Therefore the *abelianization* of $\pi_1(S, s)$ (in the category of profinite groups) is canonically independant of the auxiliary choice of base point; we will denote it $\pi_1(S)^{ab}$. This profinite abelian group classifies (*fppf*) torsors over S with (variable) finite abelian structure group, i.e. for any finite abelian group G we have a canonical isomorphism

$$\text{(1.1)} \qquad \operatorname{Hom}_{gp}(\pi_1(S)^{ab}, G) \xrightarrow{\sim} H^1_{et}(S, G).$$

The total space of the G-torsor T/S is connected if and only if its classifying map $\pi_1(S)^{ab} \to G$ is surjective.

Given a morphism $f: X \to S$ between connected locally noetherian schemes, a geometric point x of X and its image $s = f(x)$ in S, there is an induced homomorphism

$$\pi_1(X, x) \to \pi_1(S, s)$$

of fundamental groups. The induced homomorphism

$$\pi_1(X)^{ab} \to \pi_1(S)^{ab}$$

is independent of the choice of geometric point x; indeed for any finite abelian group G the transposed map

$$\operatorname{Hom}\left(\pi_1(S)^{ab}, G\right) \to \operatorname{Hom}\left(\pi_1(X)^{ab}, G\right)$$

is naturally identified with the map "inverse image of G-torsors"

$$f^*: H^1_{et}(S; G) \to H^1_{et}(X; G).$$

We will denote by Ker (X/S) the kernel of the map of π_1^{ab}'s. Thus we have a tautological exact sequence

$$0 \to \operatorname{Ker}(X/S) \to \pi_1(X)^{ab} \to \pi_1(S)^{ab}. \tag{1.2}$$

When X/S has a section

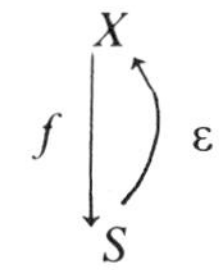

there is a simple interpretation of Ker (X/S); it classifies those torsors on X with finite abelian structure group whose inverse image via ε is trivial on S, i.e. whose restriction to the section, viewed as a subscheme of X, is completely decomposed. There is a natural product decomposition

$$\pi_1(X)^{ab} \simeq \pi_1(S)^{ab} \times \operatorname{Ker}(X/S) \tag{1.3}$$

corresponding to the expression of a G-torsor on X as the "sum" of a G-torsor on X whose restriction to ε is completely decomposed and the inverse image by f of a G-torsor on S. In particular, given, a G-torsor T/X whose restriction to ε is completely decomposed, T is connected if and only if its classifying map Ker $(X/S) \to G$ is surjective. In the absence of a section, there seems to be no simple physical interpretation of Ker (X/S).

There are two elementary functorialities it is convenient to formulate explicitly. Consider a commutative diagram of morphisms of connected, locally noetherian schemes

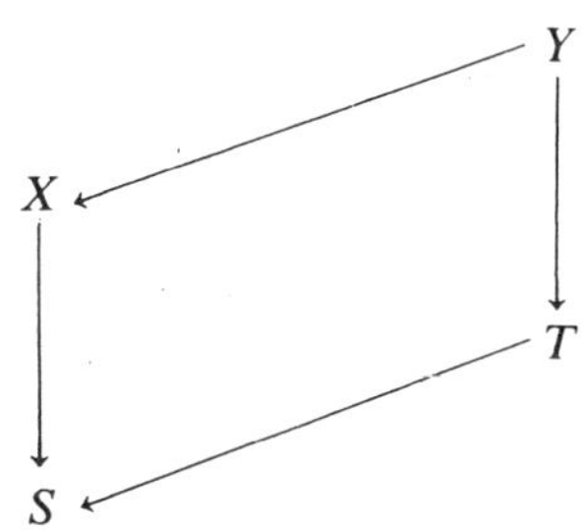

Proceeding down to the left, we have an exact sequence

$$0 \to \operatorname{Ker}(Y/X) \to \operatorname{Ker}(Y/S) \to \operatorname{Ker}(X/S). \tag{1.4}$$

Proceeding across, we have an induced map

$$\operatorname{Ker}(Y/T) \to \operatorname{Ker}(X/S) \tag{1.5}$$

which sits in a commutative diagram

$$\begin{array}{ccccccc} 0 & \longrightarrow & \operatorname{Ker}(Y/T) & \longrightarrow & \pi_1(Y)^{ab} & \longrightarrow & \pi_1(T)^{ab} \\ & & \downarrow & & \downarrow & & \downarrow \\ 0 & \longrightarrow & \operatorname{Ker}(X/S) & \longrightarrow & \pi_1(X)^{ab} & \longrightarrow & \pi_1(S)^{ab}. \end{array}$$

Let X be a geometrically connected noetherian scheme over a field K, (i.e. $X \otimes \overline{K}$ is connected, where $\overline{K}$ denotes an algebraic closure of K). Let $\bar{x}$ be a geometric point of $X \otimes \overline{K}$, x its image in X, and s its image in Spec $(K) = S$. The fundamental exact sequence (SGA I, IX, 6.1)

$$(1.6) \qquad 0 \to \pi_1(X \otimes \overline{K}, \bar{x}) \to \pi_1(X, x) \to \pi_1(S, s) \to 0$$

yields, upon abelianization, an exact sequence

$$(1.7) \qquad \pi_1(X \otimes \overline{K})^{ab} \to \pi_1(X)^{ab} \to \pi(S)^{ab} \to 0$$

The exact sequence (1.6) allows us to define an action "modulo inner automorphism" of $\pi_1(S, s)$ on $\pi_1(X \otimes \overline{K}, \bar{x})$ (given an element $\sigma \in \pi_1(S, s)$, choose $\tilde{\sigma} \in \pi_1(X, x)$ lying over it and conjugate $\pi_1(X \otimes \overline{K}, \bar{x})$ by this $\tilde{\sigma}$). The induced action of $\pi_1(S, s)$ on $\pi_1(X \otimes \overline{K})^{ab}$ is therefore well-defined. (This same action is well-defined, and trivial, on $\pi_1(X)^{ab}$.)

Therefore the map $\pi_1(X \otimes \overline{K})^{ab} \to \pi_1(X)^{ab}$ factors through the coinvariants of the action of $\pi_1(S, s)$ on $\pi_1(X \otimes \overline{K})^{ab}$: we have an exact sequence

$$(1.8) \qquad \left(\pi_1(X \otimes \overline{K})^{ab}\right)_{\pi_1(S,s)} \to \pi_1(X)^{ab} \to \pi_1(S)^{ab} \to 0.$$

If we identify $\pi_1(S, s)$ for $S = \operatorname{Spec}(K)$ with the galois group Gal $(\overline{K}/K)$, (which we may do canonically (only) up to an inner automorphism), then this last exact sequence may be rewritten

$$(1.9) \qquad \left(\pi_1(X \otimes \overline{K})^{ab}\right)_{\operatorname{Gal}(\overline{K}/K)} \to \pi(X)^{ab} \to \operatorname{Gal}(\overline{K}/K)^{ab} \to 0.$$

Consider the special case in which X has a K-rational point x_0; if we choose for $\bar{x}$ the geometric point "x_0 viewed as having values in the overfield $\overline{K}$ of K" then the morphism x_0: Spec $(K) \to X$ which "is" x_0 gives a splitting of the exact sequence (1.6)

$$(1.10) \qquad 0 \to \pi_1(X \otimes \overline{K}, \bar{x}) \to \pi_1(X, x) \underset{\xleftarrow{x_0}}{\to} \pi_1(S, s) \to 0$$

so that we have a semi-direct product decomposition

$$(1.11) \qquad \pi_1(X, x) \simeq \pi_1(X \otimes \overline{K}, \bar{x}) \ltimes \operatorname{Gal}(\overline{K}/K).$$

"Physically", the action of Gal $(\overline{K}/K)$ on $\pi_1(X \otimes \overline{K}, \bar{x})$ is simply induced by the action of Gal $(\overline{K}/K)$ on the coefficients of the defining equations of finite etale coverings of $X \otimes \overline{K}$; this action is well defined on $\pi_1(X \otimes \overline{K}, \bar{x})$ precisely because $\bar{x}$ is a $\overline{K}$-valued point which is fixed by Gal $(\overline{K}/K)$; if $\bar{x}$ were not fixed, an element $\sigma \in$ Gal would "only" define an isomorphism

$$\pi_1(X \otimes \overline{K}, \bar{x}) \xrightarrow{\sim} \pi_1(X \otimes \overline{K}, \sigma(\bar{x})) .$$

The semi-direct product decomposition (1.11) yields, upon abelianization, a product decomposition

$$\pi_1(X)^{ab} \xrightarrow{\sim} ((\pi_1(X \otimes \overline{K})^{ab})_{\mathrm{Gal}(\overline{K}/K)} \times \mathrm{Gal}(\overline{K}/K)^{ab}; \tag{1.12}$$

in other words, the existence of a K-rational point on X assures that the right exact sequence (1.9) is actually a split short exact sequence

$$0 \to ((\pi_1(X \otimes \overline{K})^{ab})_{\mathrm{Gal}(\overline{K}/K)} \to \pi_1(X)^{ab} \rightleftarrows \mathrm{Gal}(\overline{K}/K)^{ab} \to 0 .$$

For ease of later reference, we explicitly formulate the following lemma.

LEMMA 1. *Let X be a geometrically connected noetherian scheme over a field K. Then* Ker (X/K) *is the image of $\pi_1(X \otimes \overline{K})^{ab}$ in $\pi_1(X)^{ab}$. The natural surjective homomorphism*

$$\pi_1(X \otimes \overline{K})^{ab} \to \mathrm{Ker}(X/K)$$

factors through a surjection

$$(\pi_1(X \otimes \overline{K})^{ab})_{\mathrm{Gal}(\overline{K}/K)} \twoheadrightarrow \mathrm{Ker}(X/K) \tag{1.14}$$

which is an isomorphism if X has a K-rational point. Given any algebraic extension L/K, the natural map

$$\mathrm{Ker}(X \underset{K}{\otimes} L/L) \to \mathrm{Ker}(X/K) \tag{1.15}$$

is surjective.

Proof. The only new assertion is the surjectivity of (1.15), and this follows immediately from the surjectivity of the indicated maps in the commutative diagram

$$\begin{array}{ccccccc}
 & & (\pi_1(X \otimes \bar{K})^{ab})_{\mathrm{Gal}(\bar{K}/L)} & \twoheadrightarrow & \mathrm{Ker}(X \underset{K}{\otimes} L/L) & \hookrightarrow & \pi_1(X \underset{K}{\otimes} L)^{ab} \\
 & \nearrow & \downarrow & & \downarrow & & \downarrow \\
\pi_1(X \otimes \bar{K})^{ab} & & & & & & \\
 & \searrow & & & & & \\
 & & (\pi_1(X \otimes \bar{K})^{ab})_{\mathrm{Gal}(\bar{K}/K)} & \twoheadrightarrow & \mathrm{Ker}(X/K) & \hookrightarrow & \pi_1(X)^{ab}.
\end{array}$$

Now consider a normal, connected locally noetherian scheme S with generic point η and function field K. We fix an algebraic closure $\bar{K}$ of K, and denote by $\bar{\eta}$ the corresponding geometric point of S. The fundamental group $\pi_1(S, \bar{\eta})$ is then a quotient of the Galois group Gal $(\bar{K}/K)$; the functor "fibre over η"

$$\{\text{connected finite etale coverings of } S\} \to \{\text{finite separable extensions } L/K\}$$

is fully faithful, with image those finite separable extensions L/K for which the normalization of S in L is finite etale over S.

LEMMA 2. *Let S be normal, connected and locally noetherian, with generic point η and function field K. Let $f: X \to S$ be a smooth surjective morphism of finite type, whose geometric generic fibre $X_{\bar{\eta}}$ is connected. Then*

(1) *X is normal and connected.*

(2) *For any geometric point $\bar{x}$ in $X_{\bar{\eta}}$ with image x in X and s in S, the sequence*

$$\pi_1(X_{\bar{\eta}}, \bar{x}) \to \pi_1(X, x) \to \pi_1(S, s) \to 0$$

is exact.

(3) Ker (X/S) *is the image of* $\pi_1(X_{\bar{\eta}})^{ab}$ *in* $\pi_1(X)^{ab}$.

(4) *The natural map*

$$\mathrm{Ker}(X_\eta/K) \to \mathrm{Ker}(X/S)$$

is surjective.

Proof. (1) Because X is smooth over a normal scheme, it is itself normal (SGAI, Exp II, 3.1). To see that X is connected, we argue as follows. The map f, being flat (because smooth) and of finite type over a locally noetherian scheme, is *open* (SGAI, Exp IV, 6.6). Therefore any nonvoid open set $U \subset X$ meets X_η (because $f(U)$ is open and non-empty in S, so contains η). But X_η is connected (because $X_{\bar{\eta}}$ is !) and therefore the intersection of any two-non-empty open sets in X meets X_η.

(2) Because X is normal and connected, it has a generic point ξ and a function field F, and its function field F is none other than the function field of X_η (itself normal (because smooth over K) and connected). Therefore the natural map

$$\pi_1(X_{\bar{\eta}}, \bar{\xi}) \to \pi_1(X, \bar{\xi})$$

must be surjective, because it sits in the commutative diagram

$$\begin{array}{ccc} & & \pi_1(X_\eta, \bar{\xi}) \\ & \nearrow & \\ \mathrm{Gal}(\bar{F}/F) & & \downarrow \\ & \searrow & \\ & & \pi_1(X, \bar{\xi}) \end{array}$$

Comparing our putative exact sequence with its analogue for X_η/K, we have a commutative diagram

$$\begin{array}{ccccccccc} 0 & \longrightarrow & \pi_1(X_{\bar{\eta}}, \bar{x}) & \longrightarrow & \pi_1(X_\eta, x) & \longrightarrow & \mathrm{Gal}(\bar{K}/K) & \longrightarrow & 0 \\ & & \| & & \downarrow & & \downarrow & & \\ & & \pi_1(X_{\bar{\eta}}, \bar{x}) & \xrightarrow{\alpha} & \pi_1(X, x) & \xrightarrow{\beta} & \pi_1(S, s) & \longrightarrow & 0 \end{array}$$

whose top row is exact. Therefore β is surjective, and $\beta \circ \alpha = 0$. To show the exactness, given the surjectivity of β, we must show (cf. SGA I, Exp V, 6.6) that any connected etale covering Y of X which admits a section over $X_{\bar{\eta}}$ is isomorphic to the inverse image of a connected etale covering of S. Given such Y, its restriction Y_η to X_η is still connected; so the existence of a section over $X_{\bar{\eta}}$ and the exactness of (1.6) imply that Y_η is the normalization of X_η in a constant-field extension $F \cdot L$, where L is a finite separable extension of K. Therefore the function field of Y is $F \cdot L$, whence Y is the normalization of X in $F \cdot L$. Let S' denote the normalization of S in L. Then S' is finite over S. We will show that S' is finite etale over S, and that Y is the inverse image over X of this covering. By (1) applied to $X \times_S S'/S'$, the scheme $X \times_S S'$ is normal and connected, and finite over X. Therefore $X \times_S S'$ is just the normalization of X in its function field, i.e. in $F \cdot L$. Therefore $Y = X \times_S S'$. It remains only to see that S'/S is finite etale. But this follows by *fpqc* descent from that fact that $Y = X \times_S S'$ is finite etale over X.

(3) This follows immediately from the exact sequence established in (2), by abelianization.

(4) This follows immediately from (3), and the commutativity of the diagram of maps induced by the obvious inclusions

$$\begin{array}{ccc} \pi_1(X_\eta)^{ab} & \longrightarrow & \pi_1(X_\eta)^{ab} \\ & \searrow \quad \swarrow & \\ & \pi_1(X)^{ab}. & \end{array}$$

LEMMA 3. *Let X be a smooth geometrically connected variety of finite type over a field K, and let $U \subset X$ be any non-empty open set. Then the natural map*

$$\mathrm{Ker}(U/K) \to \mathrm{Ker}(X/K)$$

is surjective.

Proof. The variety $X \otimes \overline{K}$ is normal and connected, as is the non-empty open $U \otimes \overline{K}$ in it. Therefore the natural map $\pi_1(U \otimes \overline{K}) \to \pi_1(X \otimes \overline{K})$ is surjective (because both source and target are quotients of the galois group of their common function field). The result now follows from the indicated surjectivities in the commutative diagram

$$\begin{array}{ccc} \pi_1(U \otimes \overline{K})^{ab} & \twoheadrightarrow & \mathrm{Ker}(U/K) \\ \downarrow & & \downarrow \\ \pi_1(X \otimes \overline{K})^{ab} & \twoheadrightarrow & \mathrm{Ker}(X/K). \end{array}$$

II. THE MAIN THEOREM

Recall that a field K is said to be absolutely finitely generated if it is a finitely generated extension of its prime field, i.e. of $\mathbf{Q}$ or of $\mathbf{F}_p$.

THEOREM 1. *Let S be a normal, connected, locally noetherian scheme, whose function field K is an absolutely finitely generated field. Let $f: X \to S$ be a smooth surjective morphism of finite type, whose geometric generic fibre is connected. Then the group* $\mathrm{Ker}(X/S)$ *is finite if K has characteristic zero, and it is the product of a finite group with a pro-p group in case K has characteristic p.*

Proof. We will first reduce to the case in which X/S is an elementary fibration in the sense of M. Artin (SGA 4, Exp XI, 3.1), i.e. the complement, in a proper and smooth *curve* C/S with geometrically connected fibres, of a divisor $D \subset C$ which is finite etale over S. By lemma 2, part (4), Ker (X/S) is a quotient of Ker (X_η/K), so we are reduced to the case $S = \text{Spec}(K)$. If L is a finite extension of K, then Ker (X/K) is a quotient of Ker $(X \otimes L/L)$ (by lemma 1), so we may further reduce to the case when X/K has a K-rational point, say x_0. Thanks to M. Artin's theory of good neighborhoods (SGA 4, Exp XI, 3.3), at the expense of once again passing to a finite extension field L of K, we can find a Zariski open neighborhood U of x_0 in $X \underset{K}{\otimes} L$ which sits atop a finite tower

$$
\begin{array}{ll}
U = U_0 & \\
\quad\;\;\downarrow & f_0 \\
\quad\;\; U_1 & \\
\quad\;\;\downarrow & f_2 \\
\quad\;\; U_2 & \\
\quad\;\;\downarrow & \\
\quad\;\;\vdots & \\
\quad\;\;\downarrow & \\
\quad\;\; U_n = \text{Spec}(L) &
\end{array}
\tag{2.1}
$$

in which each morphism f_i is an elementary fibration. By lemma 1 again, it suffices to prove the theorem for $X \otimes L/L$, and for this it suffices, by lemma 3, to prove it for a good neighborhood U/L. By the exact sequence (1.4), it suffices to prove the theorem for each step U_i/U_{i+1} individually.

This completes the reduction to the case of an elementary fibration. By lemma 2, part (4) we may further reduce to the case $S = \text{Spec}(K)$. Again passing to a finite extension L/K, which is allowable by lemma 1, we may assume that our elementary fibration $X/K\,(= (C-D)/K)$ has a K-rational point x_0 and that the divisor D of points at infinity consists of a finite set of distinct K rational points of C. We must show that the prime-to-p-part ($p = \mathit{char}\,(K)$) of the group of Galois coinvariants

$$
(\pi_1 (X \otimes \overline{K})^{ab})_{\text{Gal}(\bar{K}/K)}
$$

is *finite*.

For this, we must recall the explicit description of the prime-to-p part of $\pi_1(X \otimes \overline{K})^{ab}$ as the Tate module of a generalized Jacobian. Let J denote the Jacobian $Pic^0_{C/K}$, and let J_D denote the generalized Jacobian of C/K with respect to the modulus D. Thus J_D is a smooth commutative group-scheme over K which represents the functor on {schemes/K}

(2.2) $W \longrightarrow$ { the group of W-isomorphism classes of pairs $(\mathcal{L}, \varepsilon)$ consisting of an invertible sheaf $\mathcal{L}$ on $C \underset{K}{\times} W$ which is fibre-by-fibre of degree zero, together with a trivialization ε of the restriction of $\mathcal{L}$ to $D \times W$.

"Forgetting ε" defines a natural map $J_D \to J$, which makes J_D an extension of J by a $\#(D) - 1$ dimensional split torus:

(2.3) $$0 \to (\mathbf{G}_m)^{\#(D)}/\mathbf{G}_m \to J_D \to J \to 0\,.$$

Kummer theory (cf. SGA 4, Exp. XVIII, 1.6 for a "modern" account) furnishes a canonical isomorphism between the prime-to-p part of $\pi_1(X \otimes \overline{K})^{ab}$ and the prime-to-p Tate module of J_D; for any finite abelian group G killed by an integer N prime to the characteristic p of K, it gives a canonical isomorphism

(2.4) $$H^1_{et}(X \otimes \overline{K}, G) \xrightarrow{\sim} \operatorname{Hom}\left(J_D(\overline{K})_N, G\right)$$

where $\left(J_D(\overline{K})\right)_N$ is the "abstract" subgroup of points of order N in $J_D(\overline{K})$. In terms of the prime-to-p Tate module

(2.5) $$T_{\text{not } p}\left(J_D(\overline{K})\right) \overset{\text{dfn}}{=} \varprojlim_{p \nmid N} \left(J_D(\overline{K})\right)_N \simeq \prod_{l \neq p} T_l\left(J_D(\overline{K})\right),$$

we can rewrite this

(2.6) $$\operatorname{Hom}\left(\pi_1(X \otimes \overline{K})^{ab}, G\right) \xrightarrow{\sim} \operatorname{Hom}\left(T_{\text{not } p}(J_D(\overline{K}), G\right),$$

whence finally a canonical isomorphism

(2.7) $$\pi_1(X \otimes \overline{K})^{ab} \xrightarrow{\sim} T_{\text{not } p}\left(J_D(\overline{K})\right) \times (\text{a pro-}p\text{-}group).$$

Thus we are reduced to showing the finiteness of the group

$$\left(T_{\text{not } p}\left(J_D(\overline{K})\right)\right)_{\mathrm{Gal}(\overline{K}/K)}\,.$$

The exact sequence (2.3)

$$0 \to (\mathbf{G}_m)^{\#(D)-1} \to J_D \to J \to 0$$

gives an exact sequence of $\bar{K}$-valued points

$$0 \to \mathbf{G}_m(\bar{K})^{\#(D)-1} \to J_D(\bar{K}) \to J(\bar{K}) \to 0$$

Applying the snake lemma to the endomorphism "multiplication by N" of this exact sequence, and passing to the inverse limit over N's prime to p, we get a short exact sequence of prime-to-p Tate modules

$$(2.8)\quad 0 \to T_{\text{not } p}(\mathbf{G}_m(\bar{K}))^{\#(D)-1} \to T_{\text{not } p}(J_D(\bar{K})) \to T_{\text{not } p}(J(\bar{K})) \to 0\,.$$

Because formation of Gal $(\bar{K}/K)$-coinvariants is right-exact, we are reduced to showing separately the finiteness of the groups

$$(T_{\text{not } p}(\mathbf{G}_m(\bar{K})))_{\text{Gal}(\bar{K}/K)}\,, \qquad (T_{\text{not } p}(J(\bar{K})))_{\text{Gal}(\bar{K}/K)}\,.$$

In fact, these groups are finite even if we replace $T_{\text{not } p}$ by the entire Tate module $T = T_p \times T_{\text{not } p}$.

THEOREM 1 (bis). *Let K be an absolutely finitely generated field, and A/K an abelian variety. The groups*

$$(T(\mathbf{G}_m(\bar{K}))_{\text{Gal}(\bar{K}/K)}\,, \qquad T(A(\bar{K}))_{\text{Gal}(\bar{K}/K)}$$

are finite.

Proof. We will reduce to the case when K is finite. Because K is absolutely finitely generated, it is standard that we can find an integrally closed sub-ring R of K, with fraction K, which is finitely generated as a $\mathbf{Z}$-algebra, together with an abelian scheme $\mathbf{A}$ over R whose generic fibre $\mathbf{A} \otimes_R K$ is A. If K has characteristic $p > 0$, we may further suppose that geometric fibres of $\mathbf{A}/R$ have constant p-rank (if $g = \dim \mathbf{A}/R$, simply localize on R until the rank of the g'th iterate of the p-linear Hasse-Witt operation on $H^1(\mathbf{A}, \mathcal{O}_A)$ is constant).

Suppose first that K has characteristic $p > 0$. Then the Gal $(\bar{K}/K)$ representations $T(G_m(\bar{K}))$ and $T(A(\bar{K}))$ are unramified over Spec (R), i.e. they are actually representations of the fundamental group $\pi_1(\text{Spec}(R), \bar{\eta})$, viewed as a quotient of Gal $(\bar{K}/K)$.

Let $\mathfrak{p}$ be a maximal ideal of R, i.e. a closed point of Spec (R), $\mathbf{F}_\mathfrak{p}$ its residue field, $\bar{\mathbf{F}}_\mathfrak{p}$ an algebraic closure of $\mathbf{F}_\mathfrak{p}$, and $\bar{\mathfrak{p}}$ the corresponding geometric point of Spec (R) (namely $R \to R/\mathfrak{p} = \mathbf{F}_\mathfrak{p} \hookrightarrow \bar{\mathbf{F}}_\mathfrak{p}$). Pick a "chemin" from $\mathfrak{p}$ to the geometric generic point $\bar{\eta}$ (which is $R \hookrightarrow K \hookrightarrow \bar{K}$), i.e. letting $\bar{R}$ denote the integral closure of R in $\bar{K}$, pick a homomorphism $\bar{R} \to \bar{\mathbf{F}}_\mathfrak{p}$ which extends $\bar{\mathfrak{p}}$. Then we get isomorphisms of $\hat{\mathbf{Z}}$-modules

$$T(\mathbf{A}(\bar{\mathbf{F}}_\mathfrak{p})) \xleftarrow[\substack{\text{chosen chemin}\\ \bar{R} \to \bar{\mathbf{F}}_\mathfrak{p}}]{\sim} T(\mathbf{A}(\bar{R})) \xrightarrow[\bar{R} \hookrightarrow \bar{K}]{\sim} T(A(\bar{K}))$$

which is Gal $(\bar{\mathbf{F}}_\mathfrak{p}/\mathbf{F}_\mathfrak{p})$ equivariant when we make Gal $(\bar{\mathbf{F}}_\mathfrak{p}/\mathbf{F}_\mathfrak{p})$ operate on $T(A(\bar{K}))$ via the composite

$$\mathrm{Gal}(\bar{\mathbf{F}}_\mathfrak{p}/\mathbf{F}_\mathfrak{p}) = \pi_1(\mathrm{Spec}(\mathbf{F}_\mathfrak{p}); \bar{\mathfrak{p}}) \xrightarrow{\text{“}\mathfrak{p}\text{”}}$$

$$\pi_1(\mathrm{Spec}(R), \bar{\mathfrak{p}}) \xrightarrow[\sim]{\text{“chemin”}} \pi_1(\mathrm{Spec}(R), \bar{\eta})$$

Passing to coinvariants now yields a diagram

$$\begin{array}{ccc} (T(\mathbf{A}(\bar{\mathbf{F}}_\mathfrak{p})))_{\mathrm{Gal}(\bar{\mathbf{F}}_\mathfrak{p}/\mathbf{F}_\mathfrak{p})} & \stackrel{\sim}{\to} & (T(A(\bar{K})))_{\mathrm{Gal}(\bar{\mathbf{F}}_\mathfrak{p}/\mathbf{F}_\mathfrak{p})} \\ & & \downarrow \\ & & (T(A(\bar{K})))_{\pi_1(\mathrm{Spec}(R), \bar{\eta})} \\ & & \| \\ & & T(A(\bar{K}))_{\mathrm{Gal}(\bar{K}/K)}, \end{array}$$

in which the vertical arrow is trivially surjective (because Gal $(\bar{\mathbf{F}}_\mathfrak{p}/\mathbf{F}_\mathfrak{p})$ operates through its image in $\pi_1(\mathrm{Spec}(R), \bar{\eta})$). Similarly for $\mathbf{G}_m$.

When K is of characteristic zero, and A/K has been "spread out" to an abelian scheme $\mathbf{A}/R$, we argue as follows. Fix a closed point $\mathfrak{p}$ of Spec (R). For each prime $l \neq p = \mathrm{char}(\mathbf{F}_\mathfrak{p})$, the l-adic Tate module $T_l(A(\bar{K}))$ is unramified over Spec$(R[1/l])$ and the above specialization argument gives a surjection, for each $l \neq p$,

$$T_l(\mathbf{A}(\bar{\mathbf{F}}_\mathfrak{p}))_{\mathrm{Gal}(\bar{\mathbf{F}}_\mathfrak{p}/\mathbf{F}_\mathfrak{p})} \twoheadrightarrow T_l(A(\bar{K}))_{\mathrm{Gal}(\bar{K}/K)}.$$

Therefore the prime-to-p part of the order of $(T(A(\overline{K})))_{\mathrm{Gal}(\overline{K}/K)}$ divides the order of $(T(\mathbf{A}(\overline{\mathbf{F}}_p)))_{\mathrm{Gal}(\overline{\mathbf{F}}_p/\mathbf{F}_p)}$.

Now choose a second closed point λ of Spec (R), with residue characteristic $l \neq p$. [This is possible because, K being a characteristic zero, Spec (R) necessarily dominates Spec $(\mathbf{Z})$, and hence by Chevalley's theorem all but finitely many primes occur as residue characteristics of closed points of Spec (R)]. Then the p-part (and indeed the prime to-l part) of the order of $(T(A(\overline{K})))_{\mathrm{Gal}(\overline{K}/K)}$ divides the order of $(T(\mathbf{A}(\overline{\mathbf{F}}_\lambda)))_{\mathrm{Gal}(\overline{\mathbf{F}}_\lambda/\mathbf{F}_\lambda)}$. Similarly for $\mathbf{G}_m$.

Thus we have reduced theorem 1 (bis) to the case of finite fields, where it is "classical". Explicitely, the result is

THEOREM 1 (ter). *Let k be a finite field, $q = \# k$, and A an abelian variety over k. Then we have the explicit formulas*

$$\begin{cases} \#(T(A(\overline{k})))_{\mathrm{Gal}(\overline{k}/k)} = \# A(k) \\ \#(T(\mathbf{G}_m(\overline{k})))_{\mathrm{Gal}(\overline{k}/k)} = \# \mathbf{G}_m(k) = q-1\,. \end{cases}$$

Proof. Let $F \in \mathrm{Gal}(\overline{k}/k)$ denote the arithmetic Frobenius automorphism of $\overline{k}/k$ (i.e. $F(x) = x^q$) which is a topological generator of Gal $(\overline{k}/k)$. In any Gal $(\overline{k}/k)$-module T, the coinvariants are simply the cokernel of $1-F$:

$$T/(1-F)\,T \xrightarrow{\sim} (T)_{\mathrm{Gal}(\overline{k}/k)}\,.$$

In the case $T = T(\mathbf{G}_m(\overline{k}))$, T is a free module of rank one over $\prod_{l \neq p} \mathbf{Z}_l$ on which F operates as multiplication by q, whence the asserted result. In the case $T = T(A(\overline{k}))$, we have $T = \prod_l T_l(A(\overline{k}))$, the product extended to all primes l. Each module $T_l(A(\overline{k}))$ is a free $\mathbf{Z}_l$-module of finite rank (2 dim $\mathbf{A}$ for $l \neq p$, the "p-rank" of A for $l = p$). Because $\# A(k)$ is non-zero, it is enough to prove that, for each l, we have an equality of l-adic ordinals:

$$\mathrm{ord}_l\left(\#\left(T_l(A(\overline{k}))/(1-F)\,T_l(A(\overline{k}))\right)\right) = \mathrm{ord}_l\left(\# A(k)\right).$$

By the theory of elementary divisors, we have

$$\mathrm{ord}_l\left(\#\left(T_l/(1-F)\,T_l\right)\right) = \mathrm{ord}_l\left(\det(1-F\,|\,T_l)\right).$$

Now for $l \neq p$, we have Weil's celebrated equality ([16], thm. 36)

$$\det\left(1-F\,|\,T_l(A(\overline{k}))\right) = \# A(k) \qquad (l \neq p)\,.$$

For $l = p$, we have (cf. [13]) only the weaker, but adequate

$$\det\left(1 - F \mid T_p\left(A(\bar{k})\right)\right) = \left(\# A(k)\right) \times (\text{a } p\text{-adic unit}). \qquad \text{QED}$$

Remarks. (1) Given an abelian variety A over any field K, Kummer theory and duality lead to a canonical isomorphism

$$\pi_1(A \otimes \bar{K}) \xrightarrow{\sim} T(A(\bar{K})).$$

Because abelian varities have rational points (e.g. their origins) we have canonically

$$\operatorname{Ker}(A/K) \xrightarrow{\sim} \left(T(A(\bar{K}))\right)_{\operatorname{Gal}(\bar{K}/K)}.$$

From this point of view, Theorem 1 (bis) is simply the abelian variety case of Theorem 1 with the added information that even the p-part is finite.

Now consider the special case when $K = k$ is a *finite* field. Then Theorem 1 (ter) gives us

$$\# \operatorname{Ker}(A/k) = \# A(k).$$

In fact, there is a canonical isomorphism of groups

$$\operatorname{Ker}(A/k) \xrightarrow{\sim} A(k).$$

To see this recall the interpretation of Ker (A/k) as the inverse limit of the galois groups of connected finite etale A-schemes $\mathbf{E}/A$ which are galois over A with abelian galois group, and completely decomposed over the origin (cf. 1.3). The Lang isogeny

$$\begin{array}{ll} A & \\ \downarrow \; 1-F & \qquad (F \text{ the Frobenius endomorphism of } A/k) \\ A & \end{array}$$

is precisely such a covering, with structural group $A(k)$. Therefore we have a surjective homomorphism

$$\operatorname{Ker}(A/k) \twoheadrightarrow A(k)$$

which is the required isomorphism (since source and target have the same cardinality!).

(2) The $\mathbf{G}_m$ case of Theorem 1 (bis) could have been handled directly by remarking that for any field K, the cardinality (as a supernatural number) of the group of coinvariants $\left(T(\mathbf{G}_m(\bar{K}))\right)_{\operatorname{Gal}(\bar{K}/K)}$ is equal to the number of roots of unity in the field K. But how, in fact, do we know that this number is finite for an

absolutely finitely generated field? The proof by specialization is pretty much the simplest one! Another approach, after "fattening" K into its finitely generated sub-ring R, is to prove the stronger assertion, in Mordell-Weil style, that the group $\mathbf{G}_m(R) = R^\times$ of units in such an absolutely finitely ring is a finitely generated abelian group.

(3) In the case of an abelian variety A over an absolutely finitely generated field K, the multiplicative upper bounds we get for $\# \, T(A(\overline{K}))_{\mathrm{Gal}(\overline{K}/K)}$ (essentially $\# \mathbf{A}(k)$ whenever we specialize to a finite field k, with the proviso that we must ignore the p-parts when it's a mixed-characteristic specialization) are *exactly the same bounds* usually used to control the size of the torsion subgroup of $A(K)$. There is a simple galois-theoretic interpretation of the group $(T(A(\overline{K})))_{\mathrm{Gal}(\overline{K}/K)}$, or at least its prime-to-p part, in terms of "*twisted-rational*" *torsion points*, which is perhaps worth pointing out. Thus let $A^\vee$ denote the dual abelian variety to A, p the characteristic of K, $\mathrm{Tors}_{\mathrm{not}\,p} A^\vee(\overline{K})$ the $\mathrm{Gal}(\overline{K}/K)$-module of all torsion points of order prime-to-p on $A^\vee$ and

$$(\mathrm{Tors}_{\mathrm{not}\,p} A^\vee(\overline{K}))(-1)$$

the $\mathrm{Gal}(\overline{K}/K)$-module obtained from this one by tensoring with the *inverse* of the cyclotomic character χ of $\mathrm{Gal}(\overline{K}/K)$. Alternately, we could describe this last module as the $\mathrm{Gal}(\overline{K}/K)$-module

$$\mathrm{Hom}\left(T(\mathbf{G}_m(\overline{K})), \mathrm{Tors}_{\mathrm{not}\,p} A^\vee(\overline{K})\right).$$

The e_N-pairings define a $\mathrm{Gal}(\overline{K}/K)$-equivariant pairing

$$T_{\mathrm{not}\,p}(A(\overline{K})) \times (\mathrm{Tors}_{\mathrm{not}\,p} A^\vee(\overline{K}))(-1) \to \mathbf{Q}/\mathbf{Z}.$$

which makes the compact abelian group $T_{\mathrm{not}\,p}$ and the discrete abelian group $(\mathrm{Tors}_{\mathrm{not}\,p})(-1)$ the Pontryagin duals of each other. Thus we obtain a perfect pairing

$$T_{\mathrm{not}\,p}(A(\overline{K}))_{\mathrm{Gal}(\overline{K}/K)} \times ((\mathrm{Tors}_{\mathrm{not}\,p}(A^\vee(\overline{K}))(-1))^{\mathrm{Gal}(\overline{K}/K)} \to \mathbf{Q}/\mathbf{Z}.$$

The group $((\mathrm{Tors}_{\mathrm{not}\,p} A^\vee(\overline{K}))(-1))^{\mathrm{Gal}(\overline{K}/K)}$ is none other than the group $(\mathrm{Tors}_{\mathrm{not}\,p} A^\vee(\overline{K}))^\chi$ of all prime-to-p ($p = \mathrm{char}(K)$) torsion points in $A^\vee(\overline{K})$ which transform under $\mathrm{Gal}(\overline{K}/K)$ by the cyclotomic character χ. Thus we obtain

SCHOLIE. *Over any field* K *of characteristic zero, the Pontryagin dual of* $\mathrm{Ker}(A/K)$ *is the group* $(\mathrm{Tors}\, A^\vee(\overline{K}))^\chi$.

(4) The same reasoning as in (3) above, if carried "scheme-theoretically", leads to a concrete interpretation of the Pontryagin dual of the entire group $T(A(\bar{K}))_{\mathrm{Gal}(\bar{K}/K)}$ "in terms of" μ-type subgroupschemes" of $A^\vee$;

SCHOLIE. *Over any field* K, *the Pontryagin dual of the compact group* $T(A(\bar{K}))_{\mathrm{Gal}(\bar{K}/K)}$ *is the discrete group*

$$\varinjlim_N \mathrm{Hom}_{K-gp}(\mu_N, A^\vee),$$

where Hom *is taken in the category of* K-*groupschemes, and the transition maps are those induced by* $\mu_{NM} \xrightarrow{\text{"}M\text{"}} \mu_N$.

Still by Theorem 1 (bis), this group is *finite* for an absolutely finitely generated field K.

For any given curve X over, say, $\mathbf{Q}$, it is an interesting problem to compute the maximal μ-type subgroup of its Jacobian. For example, let p be an odd prime, and consider the modular curves $X_0(p)$ and $X_1(p)$. Then $X_1(p)$ is a ramified covering of $X_0(p)$, cyclic of degree $(p-1)/2$, which is completely split over the rational cusp at infinity. Let

$$N = \text{numerator of } (p-1)/12 .$$

The unique intermediate covering of $X_0(p)$ of degree N is unramified; it is called the Shimura covering. According to Mazur [20], the corresponding μ_N inside $J_0(p)$ is the maximal μ-type subgroup of $J_0(p)$ over $\mathbf{Q}$. Therefore we have

$$\mathrm{Ker}(X_0(p)/\mathbf{Q}) \simeq \mathbf{Z}/N\mathbf{Z}$$

with the Shimura covering as the maximal abelian unramified geometric covering of $X_0(p)$ defined over $\mathbf{Q}$ in which the rational cusp at infinity splits completely.

On the other hand, we may extend $X_0(p)$ to a normal scheme $\mathbf{X}_0(p)$ over $\mathbf{Z}$. At the prime p, the covering $X_1(p)$ (and hence also the Shimura covering) becomes *completely* ramified over one of the two components of $\mathbf{X}_0(p) \otimes \mathbf{F}_p$. Therefore

$$\mathrm{Ker}(\mathbf{X}_0(p)/\mathbf{Z}) = 0,$$

so that Spec $(\mathbf{Z})$ being simply connected, we have

$$\pi_1(\mathbf{X}_0(p)^{ab}) = 0 .$$

(5) Consider the case when K is a finitely generated extension of an algebraically closed constant field K_0, and suppose that A/K is an abelian variety over K which has *no fixed part* relative to K_0. Because K_0, and hence K, contains all roots of unity, the cyclotomic of character of Gal $(\overline{K}/K)$ is trivial. Therefore the Pontryagin dual of $T_{\text{not } p}(A(\overline{K}))_{\text{Gal}(\overline{K}/K)}$ is simply the group of K-rational torsion points of prime-to-p order on $A^\vee$. By the *Mordell-Weil theorem* in the function field case (cf. [4], V, thm. 2) the group $A(K)$ of all K-rational points on A is finitely generated so in particular its torsion subgroup is finite. Therefore the group $T_{\text{not } p}(A(\overline{K}))_{\text{Gal}(\overline{K}/K)}$ is also finite in this "geometric" case.

Whether or not the p-part $(T_p(A(\overline{K}))_{\text{Gal}(\overline{K}/K)}$ is also finite under these assumptions is unknown in general. When A/K is a non-constant elliptic curve, this finiteness can be established by considering the ramification properties of the "V-divisible group" of A near a supersingular point on the moduli scheme. However, the general case would seem to require new ideas.

(6) Theorem 1 (bis) implies the finiteness of the group $(\text{Tors } A^\vee(\overline{K}))^\chi$ when K is a finitely generated extension of $\mathbf{Q}$, e.g. a number field. Let $K(\mu)$ be the field obtained by adjoining to K all roots of unity. We clearly have the inclusion

$$(\text{Tors } A^\vee(\overline{K}))^\chi \subset \text{Tors } A^\vee(K(\mu)).$$

This leads to the conjecture:

For any abelian variety A over a number field K, the group Tors $A(K(\mu))$ *of $K(\mu)$-rational torsion points on A is finite.*

When A is an elliptic curve without complex multiplication, this is an immediate consequence of Serre's theorem that the Galois group of the torsion points is open in $\prod GL_2(\mathbf{Z}_p)$.

For an arbitrary abelian variety, Imai [Im] shows that the group of torsion points in $K(\mu_{p^\infty})$ is finite for a fixed prime p. We shall prove below that the conjecture is true when A admits complex multiplication. This was extended to a proof of the conjecture in general by Ribet, cf. the appendix.

First we need a lemma.

LEMMA. *Let k be a number field. There exists a positive integer m such that, if F is any finite extension of k ramified at only one prime number p, and contained in some cyclotomic field, then*

$$F \subset k(\mu_{p^\infty}, \mu_m).$$

Proof. There exists a finite set of primes S such that

$$\mathrm{Gal}\,(k\,(\mu)/k) = G_S \times \prod_{l \notin S} G_l$$

where $G_l \approx \mathbf{Z}_l^*$, and G_S contains a subgroup

$$H_S = \prod_{l \in S} H_l$$

where H_l is open in $\mathbf{Z}_l^*$. Without loss of generality, we may assume that S contains p and all primes which ramify in k. If $l \notin S$, then the inertia group at l contains G_l (embedded as a component of the product). If $l \in S$, then the inertia group at l contains a subgroup H_l' open in H_l. Consequently the subgroup of the Galois group generated by all the inertia groups at primes $l \neq p$ contains

$$\prod_{\substack{l \in S \\ l \neq p}} H_l' \times \prod_{l \notin S} G_l\,.$$

This proves the lemma.

Now let A be an abelian variety defined over a number field k, and with complex multiplication. Suppose that $A_{\mathrm{tor}}\,(k\,(\mu))$ is infinite, so contains points of arbitrarily high order. We consider separately the two cases when there is a point of prime order p rational over $k\,(\mu)$ for arbitrarily large p, or when for some fixed p, there is a point of order p^n with large n.

After extending k by a finite extension if necessary, we may assume without loss of generality that A has good reduction at every prime of k. Let $k' = k\,(\mu_m)$ where m is chosen as in the lemma. Let x be a point on A of order a power of the prime p. Then $k\,(x)$ is ramified only at p, and it follows that

$$k'\,(x) \subset k'\,(\mu_{p^\infty})\,.$$

Let K be the field of complex multiplication, which we may also assume contained in k'. Furthermore, after an isogeny of A if necessary, we may assume that the ring of algebraic integers in K acts on A via an embedding

$$\iota\colon \mathfrak{o}_K \to \mathrm{End}\,(A)\,.$$

Let

$$p\mathfrak{o}_K = \mathfrak{p}_1^{e_1} \cdots \mathfrak{p}_r^{e_r}$$

be the prime ideal decomposition of p in K, and let $\mathfrak{p}_1 = \mathfrak{p}$, say.

Suppose that x has order p, and that p is large, so p is unramified in k'. By projection on the $\mathfrak{p}$-component, we may assume that x is a point of order $\mathfrak{p}$, that is $\iota(\mathfrak{p})\,x = 0$. If $r \geqq 2$, and $\mathfrak{P}'$ is a prime ideal of k' dividing one of $\mathfrak{p}_2, \ldots, \mathfrak{p}_r$, then

$\mathfrak{P}'$ is unramified in $k'(x)$. But since p is unramified in k', then $k'(\mu_{p^\infty})$ is totally ramified above every prime dividing p in k'. Therefore $r = 1$ and p remains prime in k'.

In that case, $k'(x) = k'(A_p)$ and A_p is a cyclic module over $\mathfrak{o}_K$, or also a vector space of dimension 1 over $\mathfrak{o}_K/p\mathfrak{o}_K$. Furthermore, Gal $(k'(A_p)/k')$ can be identified with a subgroup of $(\mathfrak{o}_K/p\mathfrak{o}_K)^*$, which has order $\mathbf{N}\mathfrak{p} - 1$, and in particular is prime to p. By a theorem of Ribet [Ri], we have

$$| \operatorname{Gal}(k'(A_p)/k) | \gg p^2 ,$$

where the sign $\gg$ means that the left hand side is greater than some positive constant times the right hand side. However, the prime-to-p part of Gal $(k(\mu_{p^\infty})/k)$ has order $\ll p$. This contradiction proves the theorem in the present case.

Consider finally the case when there is a point x_n of order p^n with p fixed but n arbitrarily large. Without loss of generality, we may assume that μ_p is contained in k'. We shall prove again that $r = 1$. For some prime $\mathfrak{p} = \mathfrak{p}_1$ dividing p in K, the point x_n will have a $\mathfrak{p}$-component of large p-power order, and hence without loss of generality, we may assume that all the points x_n lie in $A[\mathfrak{p}^\infty]$ (the union of all the kernels of $\iota(\mathfrak{p}^\nu)$ for $\nu \to \infty$). In particular, the degrees $[k'(x_n): k']$ contain arbitrarily large powers of p, whence the fields $k'(x_n)$ contain arbitrarily large extensions $k'(\mu_{p^\nu})$. If $r \geqq 2$ and $\mathfrak{P}'$ is any prime ideal of k' dividing some prime $\mathfrak{p}_2, \ldots, \mathfrak{p}_r$, then $\mathfrak{P}'$ is unramified in $k'(x_n)$. But the ramification indices at all primes dividing p in k' tend to infinity as n tends to infinity. Hence again $r = 1$.

Now suppose that x_n has order $\mathfrak{p}^n$, meaning that $\mathfrak{p}^n$ is the kernel of the map

$$\alpha \mapsto \iota(\alpha)\, x_n .$$

We shall prove that $k'(x_n) = k'(A[\mathfrak{p}^n])$. We have an isomorphism

$$\mathfrak{o}/\mathfrak{p}^n \approx \iota(\mathfrak{o})\, x_n .$$

On the other hand, $A[\mathfrak{p}^n]$ is cyclic module over $\mathfrak{o}/\mathfrak{p}^n$, generated by an element z, so that $x_n = \iota(\alpha)\cdot z$ for some α. Then α must be a unit in the local ring of $\mathfrak{o}$ at $\mathfrak{p}$, whence in fact

$$\iota(\mathfrak{o})\, x = A[\mathfrak{p}^n] .$$

This proves that $k'(x_n) = k'(A[\mathfrak{p}^n])$.

Using arbitrarily large n, we conclude that $k'(A[\mathfrak{p}^\infty])$ is contained in $k'(\mu_{p^\infty})$. But according to Kubota [Ku], the Galois group Gal $(k'(A[\mathfrak{p}^\infty])/k')$ is a Lie group of dimension $\geqq 2$. Since the Galois group of the p-primary roots of unity is a Lie group of dimension 1, we have a contradiction, which concludes the proof.

III. A variant

Let us agree to call a scheme S *accessible* if there exists an absolutely finitely generated field K for which the set $S(K)$ of K-valued points of S is non-empty. Thus for example, if K is an absolutely finitely generated field, then for *any* subring $R \subset K$, Spec (R) is accessible (by the K-valued point $R \hookrightarrow K$); also any subring R' of the power-series ring $K[[X_1, \ldots, \ldots]]$ over K in any number of variables has Spec (R') accessible

$$(\text{by } R' \hookrightarrow K[[X_1, \ldots]] \xrightarrow{X \to 0} K).$$

On the other hand, the spectrum of a field F is accessible if and only if F is absolutely finitely generated.

Theorem 2. *Let S be a connected, locally noetherian scheme which is accessible. Let X/S be a proper and smooth S-scheme with geometrically connected fibres. Then the group* Ker (X/S) *is finite.*

Proof. We begin by reducing to the case when S is a finitely generated field. In view of the accessibility of S, this reduction results from the following simple lemma applied with $T = \text{Spec}(K)$.

Lemma 4. *Let X/S be proper and smooth with geometrically connected fibres over a connected locally noetherian scheme S. Given a connected locally noetherian S-scheme T, denote by X_T/T the inverse image of X/S on T, i.e. form the cartesian diagram*

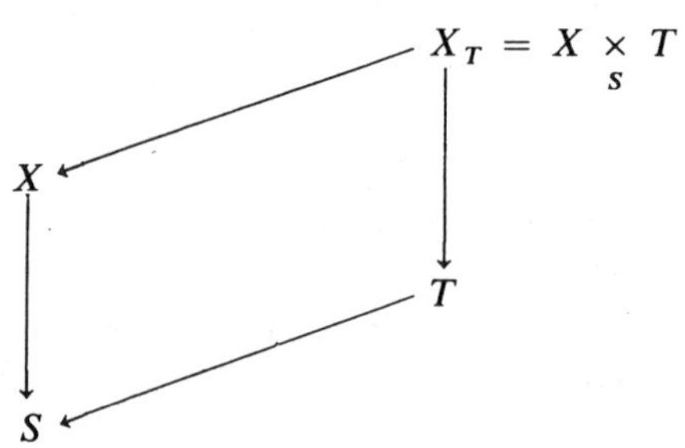

The natural map (cf. 1.5)

$$\mathrm{Ker}\,(X_T/T) \to \mathrm{Ker}\,(X/S)$$

is surjective.

Proof. Let t be a geometric point of T, s the image geometric point of S, and x a geometric point on the fibre X_s. The homotopy exact sequences (SGA I, Exp X, 1.4) for X/S and X_T/T sit in a commutative diagram

$$\begin{array}{ccccccc} \pi_1(X_s, x) & \longrightarrow & \pi_1(X_T, x) & \longrightarrow & \pi_1(T, t) & \longrightarrow & 0 \\ \| & & \downarrow & & \downarrow & & \\ \pi_1(X_s, x) & \longrightarrow & \pi_1(X, x) & \longrightarrow & \pi_1(S, s) & \longrightarrow & 0 \end{array}$$

Passing to the abelianizations yields the commutative diagram with exact rows

$$\begin{array}{ccccccc} \pi_1(X_s)^{ab} & \longrightarrow & \pi_1(X_T)^{ab} & \longrightarrow & \pi_1(T)^{ab} & \longrightarrow & 0 \\ \| & & \downarrow & & \downarrow & & \\ \pi_1(X_s)^{ab} & \longrightarrow & \pi_1(X)^{ab} & \longrightarrow & \pi_1(S)^{ab} & \longrightarrow & 0 \end{array}$$

whence we find

$$\pi_1(X_s)^{ab} \begin{array}{l} \twoheadrightarrow \mathrm{Ker}\,(X_T/T) = \text{image of } \pi_1(X_s)^{ab} \text{ in } \pi_1(X_T)^{ab}. \\ \qquad\qquad \downarrow \\ \twoheadrightarrow \mathrm{Ker}\,(X/S) \;\;= \text{image of } \pi_1(X_s)^{ab} \text{ in } \pi_1(X)^{ab}. \end{array}$$

QED

Thus we are reduced to proving the finiteness of Ker (X/K) when K is an absolutely finitely generated field, and X/K is proper, smooth, and geometrically connected. We have already proven this finiteness theorem when X/K is an abelian variety (cf. Remark (1) above). We will reduce to this case by making use of the theory of the Picard and Albanese varieties.

At the expense of replacing K by a finite extension, we may assume that X has a K-rational point x_0. The Picard scheme $Pic_{X/K}$ is then a commutative group-scheme locally of finite type over K, which represents the functor on {Schemes/K}

$$W \to \left\{ \begin{array}{l} \text{the group of } W\text{-isomorphism classes of pairs } (\mathscr{L}, \varepsilon) \text{ consisting} \\ \text{of an invertible sheaf } \mathscr{L} \text{ on } X \underset{K}{\times} W \text{ together with a} \\ \text{trivialization } \varepsilon \text{ of the restriction } \mathscr{L} \text{ to } \{x_0\} \underset{K}{\times} W \end{array} \right.$$

The subgroup-scheme $Pic^{\tau}_{X/K}$ of $Pic_{X/K}$ classifies those $(\mathscr{L}, \varepsilon)$ whose underlying $\mathscr{L}$ becomes τ-equivariant to zero when restricted to every geometric fibre of $X \times W/W$ (i.e. for each geometric point w of W, some multiple of $\mathscr{L} \mid X \times w$ is algebraically equivalent to zero). The identity component $Pic^{0}_{X/K}$ of $Pic_{X/K}$ classifies those $(\mathscr{L}, \varepsilon)$ whose $\mathscr{L}$ becomes algebraically equivalent to zero on each geometric fibre $X \times W/W$. The Picard *variety* $Pic^{0,\,\mathrm{red}}_{X/K}$ is an abelian variety over K, and it sits in an *f.p.p.f.* short exact sequence of commutative group schemes

$$0 \to Pic^{0,\,\mathrm{red}}_{X/K} \to Pic^{\tau}_{X/K} \to C \to 0 \tag{3.1}$$

in which the cokernel C is a finite flat group-scheme over K. This cokernel C should be thought of as the "scheme theoretic" torsion in the Neron-Severi group.

We denote by $\mathrm{Alb}_{X/K}$ the Albanese variety of X/K, *defined* to be the dual abelian variety to the Picard variety $Pic^{0,\,\mathrm{red}}_{X/K}$. We now recall the expression of $\pi_1(X \otimes \overline{K})^{ab}$ in terms of the Tate module of the Albanese, and a finite "error term" involving the Cartier dual $C^{\vee}$ of C.

LEMMA 5. *Let K be a field, and X/K a proper, smooth and geometrically connected K-scheme which admits a K-rational point. Then there is a canonical short exact sequence of* $\mathrm{Gal}(\overline{K}/K)$*-modules*

$$0 \to C^{\vee}(\overline{K}) \to \pi_1(X \otimes \overline{K})^{ab} \to T(\mathrm{Alb}_{X/K}(\overline{K})) \to 0\,. \tag{3.2}$$

Proof. By Kummer and Artin-Schreier theory, we have for each integer $N \geq 1$ a canonical isomorphism

$$\begin{aligned}&\mathrm{Hom}(\pi_1(X \otimes \overline{K})^{ab}, \mathbf{Z}/N\mathbf{Z})\\ &= H^1_{et}(X \otimes \overline{K}, \mathbf{Z}/N\mathbf{Z}) \xrightarrow{\sim} \mathrm{Hom}(\mu_N, (Pic^{\tau}_{X/K}) \otimes \overline{K})\,.\end{aligned}$$

in which the last Hom is in the sense of $\overline{K}$-group-schemes. Applying the functor $X \mapsto \mathrm{Hom}(\mu_N, X)$ to the short exact sequence

$$0 \to Pic^{0,\,\mathrm{red}} \to Pic^{\tau} \to C \to 0$$

gives a short exact sequence

$$\begin{aligned}0 &\to \mathrm{Hom}(\mu_N, (Pic^{0,\,\mathrm{red}}) \otimes \overline{K})\\ &\to \mathrm{Hom}(\mu_N, (Pic^{\tau}) \otimes \overline{K}) \to \mathrm{Hom}(\mu_N, C \otimes \overline{K}) \to 0\end{aligned} \tag{3.3}$$

(the final zero because over an algebraically closed field, the group $\mathrm{Ext}^1(\mu_N, A)$ vanishes for any abelian variety A, cf. the remark at the end of this section). We now "decode" its two end terms, using Cartier-Nishi duality for the first, and Cartier duality for the last.

The first is

$$\begin{aligned}
\mathrm{Hom}\,(\mu_N, (Pic^{0,\,\mathrm{red}}) \otimes \overline{K}) = {} & \mathrm{Hom}\,(\mu_N, (Pic^{0,\,\mathrm{red}})_N \otimes \overline{K}) \\
& \wr\downarrow \quad \text{Cartier-Nishi duality} \\
& \mathrm{Hom}\,(\mathrm{Alb}_{X/N})_N \otimes \overline{K}, \mathbf{Z}/N\mathbf{Z}) \\
& \wr\downarrow \quad \text{evaluation on } \overline{K}\text{-points} \\
& \mathrm{Hom}\,((\mathrm{Alb}_{X/K}(\overline{K}))_N, \mathbf{Z}/N\mathbf{Z}) \\
& \wr\downarrow \\
& \mathrm{Hom}\,(T\,(\mathrm{Alb}_{X/K}(\overline{K})), \mathbf{Z}/N\mathbf{Z})\,.
\end{aligned}$$

The last is

$$\begin{aligned}
\mathrm{Hom}\,(\mu_N, C \otimes \overline{K}) \xrightarrow[\text{Cartier duality}]{\sim} & \;\mathrm{Hom}\,(C^\vee \otimes \overline{K}, \mathbf{Z}/N\mathbf{Z}) \\
& \;\wr\downarrow \text{ evaluation} \\
& \;\mathrm{Hom}\,(C^\vee(\overline{K}), \mathbf{Z}/N\mathbf{Z})
\end{aligned}$$

"Substituting" into the exact sequence (3.2), we find a canonical short exact sequence

$$\begin{aligned}
\text{(3.4)} \qquad & 0 \to \mathrm{Hom}\,(T\,(\mathrm{Alb}_{X/K}(\overline{K})), \mathbf{Z}/N\mathbf{Z}) \\
& \to \mathrm{Hom}\,(\pi_1(X \otimes \overline{K})^{ab}, \mathbf{Z}/N\mathbf{Z}) \to \mathrm{Hom}\,(C^\vee(\overline{K}), \mathbf{Z}/N\mathbf{Z}) \to 0
\end{aligned}$$

Passing to the *direct* limit as N grows multiplicatively, we obtain a canonical short exact sequence

$$\begin{aligned}
\text{(3.5)} \qquad & 0 \to \mathrm{Hom}\,(T\,(\mathrm{Alb}_X(\overline{K}), \mathbf{Q}/\mathbf{Z}) \\
& \to \mathrm{Hom}\,(\pi_1(X \otimes \overline{K})^{ab}, \mathbf{Q}/\mathbf{Z}) \to \mathrm{Hom}\,(C^\vee(\overline{K}), \mathbf{Q}/\mathbf{Z}) \to 0\,.
\end{aligned}$$

Taking its Pontryagin dual, we find the required exact sequence (3.2). QED

To complete the reduction of Theorem 2 to the case of abelian varieties, we simply notice that the exact sequence of lemma 5 yields, upon passage to coinvariants, an exact sequence

$$(3.6) \qquad (C^\vee(\overline{K}))_{\mathrm{Gal}/\overline{K}/K)} \to \mathrm{Ker}\,(X/K) \to \mathrm{Ker}\,(\mathrm{Alb}_{X/K}/K) \to 0$$

whose first term, being a quotient of the finite group $C^\vee(\overline{K})$, is finite. QED

Remark. In the course of the proof of Lemma 5, we appealed to the "well-known" vanishing of $\mathrm{Ext}^1(\mu_N, A)$ over an algebraically closed field, for an abelian variety A and any integer $N > 1$. Here is a simple proof. It is enough to prove this vanishing when N is either prime to the characteristic p of K, or, in case $p > 0$, when $N = p$.

Suppose first N prime to p. Because the ground-field is algebraically closed, we have $\mu_N \simeq \mathbf{Z}/N\mathbf{Z}$, so it is equivalent to prove the vanishing of $\mathrm{Ext}^1(\mathbf{Z}/N\mathbf{Z}, A)$. We will prove that *this* group vanishes for every integer $N > 1$. Consider such an extension:

$$0 \to A \to E \to \mathbf{Z}/N\mathbf{Z} \to 0$$

Pass to $\overline{K}$-valued points

$$0 \to A(\overline{K}) \to E(\overline{K}) \to \mathbf{Z}/N\mathbf{Z} \to 0$$

and consider the endomorphism "multiplication by N". Because the group $A(\overline{K})$ is N-divisible, the snake lemma gives an exact sequence

$$0 \to A(\overline{K})_N \to E(\overline{K})_N \to \mathbf{Z}/N\mathbf{Z} \to 0$$

But a point in $E(\overline{K})_N$ which maps onto "1" $\in \mathbf{Z}/N\mathbf{Z}$ is precisely a splitting of our extension.

Next consider the case $N = p = \mathrm{char}\,(K)$. We give a proof due to Barry Mazur. Using the *f.p.p.f.* exact sequence

$$0 \to A_p \to A \to A \to 0.$$

to compute $\mathrm{Ext}\,(\mu_p, -)$, we obtain a short exact sequence

$$0 \to \mathrm{Hom}\,(\mu_p, A) \to \mathrm{Ext}^1(\mu_p, A_p) \to \mathrm{Ext}^1(\mu_p, A) \to 0$$

To prove that $\mathrm{Ext}^1(\mu_p, A) = 0$, we will show that the groups $\mathrm{Hom}\,(\mu_p, A)$ and $\mathrm{Ext}^1(\mu_p, A_p)$ are both finite, of the same order. Trivially, we have $\mathrm{Hom}\,(\mu_p, A) = \mathrm{Hom}\,(\mu_p, A_p)$. Because we are over an algebraically closed field, and A_p is killed by p, its toroidal biconnected-etale decomposition looks like

$$A_p \simeq (\mu_p)^a \times (\text{biconnected}) \times (\mathbf{Z}/p\mathbf{Z})^b; \qquad [\text{in fact } a = b].$$

Only the μ p's in A_p can "interact" with μ_p. Thus we are reduced to showing that $\mathrm{Hom}\,(\mu_p, (\mu_p)^a)$ and $\mathrm{Ext}^1(\mu_p, (\mu_p)^a)$ are both finite of the same cardinality p^a.

By Cartier duality, it is equivalent to show that both Hom $(\mathbf{Z}/p\mathbf{Z}, \mathbf{Z}/p\mathbf{Z})$ and $\mathrm{Ext}^1(\mathbf{Z}/p\mathbf{Z}, \mathbf{Z}/p\mathbf{Z})$ have order p, and this is obvious (resolve the "first" $\mathbf{Z}/p\mathbf{Z}$ by

$$0 \to \mathbf{Z} \xrightarrow{p} \mathbf{Z} \to \mathbf{Z}/p\mathbf{Z} \to 0).$$

For another proof in this case, cf. Oort, [10], 85.

IV. Absolute Finiteness theorems

THEOREM 3. *Let $\mathcal{O}$ be the ring of integers in a finite extension K of $\mathbf{Q}$. Let X be a smooth $\mathcal{O}$-scheme of finite type whose geometric generic fibre $X \otimes_{\mathcal{O}} \overline{K}$ is connected, and which maps surjectively to* Spec $(\mathcal{O})$ *(i.e. for every prime $\mathfrak{p}$ of $\mathcal{O}$, the fibre over $\mathfrak{p}$, $X \otimes_{\mathcal{O}} (\mathcal{O}/\mathfrak{p})$, is non empty). Then the group $\pi_1(X)^{ab}$ is finite.*

Proof. This follows immediately from Theorem 1 and global classfield theory, according to which $\pi_1(\mathrm{Spec}(\mathcal{O}))^{ab}$, the galois group of the maximal unramified abelian extension of K, is *finite*. QED

THEOREM 4. *Let $\mathcal{O}$ be the ring of integers in a finite extension K of $\mathbf{Q}$, $\mathfrak{p}_1, \ldots, \mathfrak{p}_n$ a finite set of primes of $\mathcal{O}$, $N = p_1 \ldots p_n$ the product of their residue characteristics, and $\mathcal{O}[1/\mathfrak{p}_1 \ldots \mathfrak{p}_n]$ the ring of "integers outside $\mathfrak{p}_1, \ldots, \mathfrak{p}_n$" in K. Let X be a smooth $\mathcal{O}[1/\mathfrak{p}_1 \ldots \mathfrak{p}_n]$-scheme of finite type, whose geometric generic fibre $X \otimes_{\mathcal{O}} \overline{K}$ is connected, and which maps surjectively to* Spec $(\mathcal{O}[1/\mathfrak{p}_1 \ldots \mathfrak{p}_n])$ *(i.e. for every prime $\mathfrak{p} \notin \{\mathfrak{p}_1, \ldots, \mathfrak{p}_n\}$, the fibre*

$$X \otimes (\mathcal{O}/\mathfrak{p})$$

is non-empty). Then the group $\pi_1(X)^{ab}$ is the product of a finite group and a pro-N group.

Proof. Again an immediate consequence of Theorem 1 and global classfield theory, according to which $\pi_1(\mathrm{Spec}(\mathcal{O}[1/\mathfrak{p}_1 \ldots \mathfrak{p}_n]))^{ab}$, the galois group of the maximal abelian, unramified outside $\{\mathfrak{p}_1, \ldots, \mathfrak{p}_n\}$-extension of K is finite times pro-N. QED

THEOREM 5. *Let* S *be a normal, connected noetherian scheme, whose function field* K *is absolutely finitely generated. Let* $f: X \to S$ *be a smooth surjective morphism of finite type whose geometric generic fibre is connected, and which admits a cross-section* $X \overset{\varepsilon}{\leftrightarrows} S$. *Then there are only finitely many connected finite etale* X*-schemes* Y/X *which are galois over* X *with abelian galois group of order prime to* char (K) *and which are completely decomposed over the marked section. If in addition we suppose* X/S *proper, we can drop the proviso "of order prime to* char (K)".

Proof. This is just the concatenation of Theorems 1 and 2 with the physical interpretation (1.3) of the group Ker (X/S) in the presence of a section. QED

V. APPLICATION TO l-ADIC REPRESENTATIONS

Let l be a prime number, $\overline{\mathbf{Q}}_l$ an algebraic closure of $\mathbf{Q}_l$. By an l-adic representation ρ of a topological group π, we mean a finite-dimensional continuous representation

$$\rho: \pi \to GL(n, \overline{\mathbf{Q}}_l)$$

whose image lies in $GL(n, E_\lambda)$ for some finite extension E_λ of $\mathbf{Q}_l$.

THEOREM 6. (cf. Grothendieck, *via* [2], 1.3). *Let* K *be an absolutely finitely generated field,* X/K *a smooth, geometrically connected* K*-scheme of finite type,* $\bar{x}$ *a geometric point of* $X \otimes \overline{K}$, x *the image geometric point of* $\bar{x}$ *in* X. *Let* l *be a prime number, and* ρ *an* l*-adic representation of* $\pi_1(X, x)$;

$$\rho: \pi_1(X, x) \to GL(n, \overline{\mathbf{Q}}_l).$$

Let G *be the Zariski closure of the image* $\rho(\pi_1(X \otimes \overline{K}, \bar{x}))$ *of the geometric fundamental group* $\pi_1(X \otimes \overline{K}, \bar{x})$ *in* $GL(n, \overline{\mathbf{Q}}_l)$ *and* G^0 *its identity component. Suppose that either* l *is different from the characteristic* p *of* K, *or that* X/K *is proper. Then:*

(1) *the radical of* G^0 *is unipotent, or equivalently:*

(2) *if the restriction of* ρ *to the geometric fundamental group* $\pi_1(X \otimes \overline{K}, \bar{x})$ *is completely reducible, then the algebraic group* G^0 *is semi-simple.*

Proof. By Theorem 1, for $l \neq p$, or by Theorem 2 if $l = p$ and X/K is *proper*, we know that the l-part of Ker (X/K) is finite i.e. (cf. Lemma 1) the image of $\pi_1 (X \otimes \overline{K}, \bar{x})$ in $\pi_1 (X)^{ab}$ is the product of a finite group and a group of order prime to l. Given this fact, the proof proceeds exactly as in (Deligne [2], 1.3).

QED

Remarks. (1) This theorem is the group-theoretic version of Grothendieck's local monodromy theorem (cf. Serre-Tate ([15], Appendix) for a precise statement, as well as the proof) with X/K "replacing" the spectrum of the fraction field E of a henselian discrete valuation ring with residue field K, and with $\pi_1 (X \otimes \overline{K})$ "replacing" the inertia subgroup I of Gal $(\overline{E}/E)$. The "extra" feature of the "local" case is that the quotient of I by a normal pro-p subgroup is abelian. Therefore any l-adic representation ρ of I, with $l \neq p$, becomes *abelian* when restricted to a suitable open subgroup of I, and hence the associated algebraic group G^0 is automatically abelian. In particular, the radical of G^0 is G^0 itself.

(2) If X/K is itself an abelian variety A/K, then $\pi_1 (A \otimes \overline{K}, \bar{x})$ is abelian. Therefore if l is any prime, and ρ any l-adic representation of $\pi_1 (A \otimes \overline{K}, \bar{x})$, the associated algebraic groups G and G^0 will be abelian; hence if ρ is the restriction to $\pi_1 (A \otimes \overline{K}, \bar{x})$ of an l-adic representation of $\pi_1 (A, x)$, then G^0 is unipotent, i.e. the restriction of ρ to an open subgroup of $\pi_1 (A \otimes \overline{K}, \bar{x})$ is *unipotent* (compare Oort [11], 2).

(3) Can one give an example of X/K proper smooth and geometrically connected over an absolutely finitely generated field K of characteristic $p > 0$ whose fundamental group $\pi_1 (X, x)$ admits an n-dimensional p-adic representation with $n \geqslant 2$ (resp. $n \geqslant 3$) for which the associated algebraic group G^0 is $SL(n)$ (resp. $SO(n)$)? Can we find an abelian scheme A over such an X, all of whose fibres have the same p-rank $n \geqslant 2$, for which the associated p-adic representation of $\pi_1 (X, x)$ has $G^0 = SL(n)$? (cf. Oort [11] for the case of p-rank zero).

REFERENCES

[0] ARTIN, M., A. GROTHENDIECK and J. L. VERDIER. *Théorie des Topos et Cohomologie Etale des Schémas* (SGA 4), Tome 3. Springer Lecture Notes 305, 1973.

[1] BLOCH, S. Algebraic K-theory and class-field theory for arithmetic surfaces. *To appear in Annals of Math.*

[2] DELIGNE, P. La conjecture de Weil II. *Pub. Math. IHES 52* (1980).

[3] GROTHENDIECK, A. *Revêtements Etales et Groupe Fondamental* (SGA I). Springer Lecture Notes, 224, 1971.

[4] LANG, S. *Diophantine Geometry*. Interscience Publishers, New York, 1962.

[5] —— On the Lefshetz principle. *Ann. of Math. 64* (1956), pp. 326-327.

[6] —— Unramified class field theory over function fields in several variables. *Ann. of Math. 64* (1956), pp. 286-325.

[7] —— Sur les séries L d'une variété algébrique. *Bull. Soc. Math. France 84* (1956), pp. 385-407.

[8] LANG, S. et J. P. SERRE. Sur les revêtements non ramifiés des variétés algébriques. *Amer. J. Math.* (1957), pp. 319-330.

[9] MUMFORD, D. *Abelian Varieties*. Oxford University Press, 1970.

[10] OORT, F. *Commutative group schemes*. Springer Lecture Notes 15, 1966.

[11] —— Subvarieties of Moduli Spaces. *Inv. Math. 24* (1974), pp. 95-119.

[12] PARSHIN, A. H. Abelian coverings of arithmetic schemes. *Doklady Akad. Nauk. Tome 243 No. 4* (1978), 855-858; English translation in *Soviet Mathematics, Doklady, Vol. 19* (1978), 1438-1442.

[13] SERRE, J.-P. Quelques propriétés des variétés abéliennes en car. *p. Amer. J. Math. vol. 80* (1958), pp. 715-739.

[14] —— *Groupes algébriques et corps de classes*. Hermann, Paris, 1959.

[15] SERRE, J.-P. and J. TATE. Good reduction of abelian varieties. *Annals Math. 88, No. 3* (1968), pp. 492-517.

[16] WEIL, A. *Courbes algébriques et variétés abéliennes*. Hermann, Paris, 1971.

[17] IMAI, H. A remark on the rational points of abelian varieties with values in cyclotomic $\mathbf{Z}_p$-extensions. *Proc. Japan Acad. 51* (1975), pp. 12-16.

[18] KUBOTA, T. On the field extension by complex multiplication. *Trans. AMS 118, No. 6* (1965), pp. 113-122.

[19] RIBET, K. Division fields of abelian varieties with complex multiplication. *Mémoire, Soc. Math. France, 2e Série, n° 2* (1980), pp. 75-94.

[20] MAZUR, B. Modular curves and the Eisenstein ideal. *Publ. IHES 47* (1977), 33-186.

(Reçu le 20 janvier 1981)

Nicholas M. Katz

Fine Hall
Department of Mathematics
Princeton University
Princeton, N.J. 08544, USA

Serge Lang

Mathematics Department
Box 2155 Yale Station
New Haven, Conn. 06520
USA

APPENDIX: TORSION POINTS OF ABELIAN VARIETIES IN CYCLOTOMIC EXTENSIONS

by Kenneth A. Ribet [1])

Let k be a number field, and let $\bar{k}$ be an algebraic closure for k. For each prime p, let K_p be the subfield of $\bar{k}$ obtained by adjoining to k all p-power roots of unity in $\bar{k}$. Let K be the compositum of all of the K_p, i.e., the field obtained by adjoining to k *all* roots of unity in $\bar{k}$.

Suppose that A is an abelian variety over k. Mazur has raised the question of whether the groups $A(K_p)$ are finitely generated [4]. In this connection, H. Imai [1] and J.-P. Serre [5] proved (independently) that the *torsion subgroup* of $A(K_p)$ is finite for each p. The aim of this appendix is to prove that more precisely one has the following theorem, cf. [3], §II, Remark 3.

THEOREM 1. *The torsion subgroup* $A(K)_{\text{tors}}$ *of* $A(K)$ *is finite.*

Let G be the Galois group Gal $(\bar{k}/k)$ and let H be its subgroup Gal $(\bar{k}/K)$. For each positive integer n, let $A[n]$ be the kernel of multiplication by n in $A(\bar{k})$. For each prime p, let V_p be the $\mathbf{Q}_p$-adic Tate module attached to A. If M is one of these modules, we denote by M^H the set of elements of M left fixed by H. Since H is normal in G, M^H is stable under the action of G on M.

Because of the structure of the torsion subgroup of $A(\bar{k})$, one sees easily that Theorem 1 is equivalent to the conjunction of the following two statements:

THEOREM 2. *For all but finitely many primes* p, *we have* $A[p]^H = 0$.

THEOREM 3. *For each prime* p, *we have* $V_p^H = 0$.

Indeed, Theorem 2 asserts the vanishing of the p-primary part of $A(K)_{\text{tors}}$, while Theorem 3 asserts the finiteness of this p-primary part.

[1]) Partially supported by National Science Foundation contract number MCS 80-02317.

In proving these statements, we visibly have the right to replace k by a finite extension of k. Therefore, using ([SGA 71], IX, 3.6) we can (and will) assume that A/k is semistable. Next, consider the largest subextension k' of K/k which is unramified at all finite places of k.

LEMMA. *For each prime p, let L_p be the largest extension of k in K which is unramified at all places of k except for primes dividing p and the infinite places of k. Then L_p is the compositum $k'K_p$.*

Proof. Let A be the Galois group Gal (K/k), viewed as a subgroup of $\hat{\mathbf{Z}}^*$. We consider $\hat{\mathbf{Z}}^*$ as the direct product of its two subgroups $\mathbf{Z}_p^*$ and $\prod_{l \neq p} \mathbf{Z}_l^*$. Let I (resp. J) be the subgroup of A generated by the inertia groups of A for primes of k which divide p (resp. which do not divide p). Then I is a subgroup of Z_p^*, while J is a subgroup of $\prod_{l \neq p} \mathbf{Z}_l^*$. The product $I \times J$ is the subgroup of A generated by all inertia groups of A. We have $J = \operatorname{Gal}(\bar{k}/L_p)$, $I \times J = \operatorname{Gal}(\bar{k}/k')$, and $\operatorname{Gal}(\bar{k}/K_p) = A \cap \left(\prod_{l \neq p} \mathbf{Z}_l^*\right)$. Now Gal $(\bar{k}/k'K_p)$ is the intersection of the two Galois groups Gal $(\bar{k}/k')$ and Gal $(\bar{k}/K_p)$. Putting these facts together, we prove the desired assertion.

We now replace k by its finite extension k'. With this replacement made, K_p becomes equal to L_p. Furthermore, for odd primes p, the largest extension of k in K which is unramified outside p and infinity and which has degree prime to p is the field obtained by adjoining to k the p-th roots of unity in $\bar{k}$.

Proof of Theorem 2. We shall consider only primes p which are odd, unramified in k, and such that A has good reduction at at least one prime of k dividing p. Let p be such a prime and v a prime of k over p at which A has good reduction. Suppose that the G-module $A[p]^H$ is non-zero, and let W be a simple G-submodule of this module. The algebra $\operatorname{End}_G W$ is a finite field $\mathbf{F}$, and the action of G on W is given by a character

$$\phi : G \to \mathbf{F}^*$$

since the action of G on $A[p]^H$ is abelian. (Here the point is simply that G/H is an abelian group.) In particular, the image of G in Aut $(A[p])$ has order prime to p. On the other hand, the character ϕ is unramified at primes of k not dividing p because A/k is semistable. By the discussion following the lemma, we know that ϕ factors through the quotient Gal $(k(\mu_p)/k)$ of G; here, μ_p denotes the group of p-th roots of unity. In particular, ϕ must have order dividing $p - 1$, so that its values lie in the prime field $\mathbf{F}_p$. Since W was chosen to be simple, its dimension over $\mathbf{F}_p$ must be 1; i.e., W is a group of order p.

Let $\chi : G \to \mathbf{F}_p^*$ be the mod p cyclotomic character, i.e., the character giving the action of G on μ_p. Since ϕ factors through Gal $(k(\mu_p)/k)$, we may write ϕ in the form χ^n, where n is an integer mod $(p-1)$. We claim that n can only be 0 or 1.

To verify this claim, it is enough to check that it is true after we replace G by an inertia group I in G for the prime v, since χ is totally ramified at v. We remark that W is the I-module associated to a finite flat commutative group scheme $\mathscr{W}$ over the ring of integers of the completion of k at v, since v is such that A has good reduction at v. Because $\mathscr{W}$ has order p, the classification of Tate-Oort ([8], especially pp. 15-16) applies to $\mathscr{W}$. Because v is absolutely unramified, the classification shows immediately that $\mathscr{W}$ is either étale or the dual of an étale group. In the former case, I acts trivially on W; in the latter case, I acts on W via χ. This completes the verification of the claim.

Thus, if Theorem 2 is false, there are infinitely many primes p for which $A[p]$ contains a G-submodule isomorphic to either $\mathbf{Z}/p\mathbf{Z}$ or to μ_p. Of course, the former case can occur only a finite number of times, since $A(k)$ is finite. One way to rule out the latter case is to argue that whenever μ_p is a submodule of $A[p]$, the group $\mathbf{Z}/p\mathbf{Z}$ is a quotient of the dual of $A[p]$, which is the kernel of multiplication by p on the abelian variety $A^\vee$ dual to A. In other words, if μ_p occurs as a submodule of $A[p]$, then there is an abelian variety isogenous to $A^\vee$ (and therefore in fact to A) which has a rational point of order p over k. Therefore p is a divisor of the order of a finite group that may be specified in advance, viz. the group of rational points of any reduction of A at a good unramified prime of k of residue characteristic different from 2. (See the appendix to Katz's recent paper [2] for a discussion of this point.)

Proof of Theorem 3. Suppose that p is a prime such that V_p^H is non-zero. We again choose W to be an irreducible G-submodule (i.e., $\mathbf{Q}_p[G]$-submodule) of V_p^H. Because the action of G on W is abelian, and because W is simple, each element of G acts semisimply on W. Since A/k is semistable, it follows that the homomorphism

$$\rho : G \to \mathrm{Aut}(W)$$

giving the action of G on W is unramified at all primes of k not dividing p. Therefore, ρ factors through Gal (K_p/k) in view of the lemma and the subsequent replacement $k \to k'$. In other words, starting from the hypothesis that the p-torsion subgroup of $A(K)$ is infinite, we have deduced that the p-torsion subgroup of $A(K_p)$ is infinite.

Of course, this situation is ruled out by the theorem of Imai and Serre mentioned above. Nevertheless, we will sketch for the reader's convenience an argument which leads to a contradiction. Let v be a place of k dividing p, and let $D \subset G$ be a decomposition group for v. By ([SGA 71], IX, Prop. 5.6), the D-module V_p is an extension of D-modules attached to p-divisible groups over the integer ring of the completion of k at v. Because of Tate's theory [7], the semisimplification V_p^{ss} of the D-module V_p has a Hodge-Tate decomposition. (Here we should remark that submodules and quotients of Hodge-Tate modules are again Hodge-Tate.) Since W is semisimple as a D-module (because semisimple and *abelian* as a G-module), W may be viewed as a submodule of V_p^{ss}. Therefore, W is a Hodge-Tate module.

By ([6], III, Appendix), we know that ρ is a locally algebraic abelian representation of G. Using this information, plus the fact that ρ factors through Gal (K_p/k), we find that there is an open subgroup G_0 of G with the following property: the restriction of ρ to G_0 is the direct sum of 1-dimensional representations, each described by an integral power χ_p^n of the standard cyclotomic character $\chi_p: G \to \mathbf{Z}_p^*$. After replacing k by a finite extension, we may assume that G_0 is G. Take a prime w of k which is prime to p and such that A has good reduction at w. Let $g \in G$ be a Frobenius element for w. The eigenvalues of $\rho(g)$ will be integral powers of $\chi_p(g)$, i.e., of the norm Nw of w. However, by a well known theorem of Weil, these eigenvalues all have archimedian absolute values equal to $(\mathrm{N}w)^{1/2}$. This contradiction completes the proof of Theorem 3.

REFERENCES

[1] IMAI, H. A remark on the rational points of abelian varieties with values in cyclotomic $\mathbf{Z}_p$ extensions. *Proc. Japan Acad. 51* (1975), 12-16.

[2] KATZ, N. Galois properties of torsion points on abelian varieties. *Invent. Math. 62* (1981), 481-502.

[3] KATZ, N. and S. LANG. Finiteness theorems in geometric classfield theory.

[4] MAZUR, B. Rational points of abelian varieties with values in towers of number fields. *Invent. Math. 18* (1972), 183-266.

[5] SERRE, J.-P. Letters to B. Mazur, January, 1974.

[6] —— *Abelian l-adic Representations and Elliptic Curves.* New York: Benjamin 1968.

[7] TATE, J. *p*-divisible groups. In: *Proceedings of a Conference on Local Fields.* Berlin-Heidelberg-New York: Springer-Verlag 1967.

[8] TATE, J. and F. OORT. Group schemes of prime order. *Ann. scient. Éc. Norm. Sup., 4e série 3* (1970), 1-21.

[SGA 71] *Groupes de Monodromie en Géométrie Algébrique* (séminaire dirigé par A. Grothedieck avec la collaboration de M. Raynaud et D. S. Rim). *Lecture Notes in Math. 288* (1972).

(Reçu le 20 janvier 1981)

Kenneth A. Ribet

U.C. Berkeley
Mathematics Department
Berkeley, Ca. 94720
USA

Sém. Delange-Pisot-Pitou, Paris 1981-1982
(Birkhäuser): 125-136

REPRÉSENTATIONS LOCALEMENT ALGÉBRIQUES DANS LES CORPS CYCLOTOMIQUES

S. LANG

Soit k un corps de nombres. On note k^a la clôture algébrique (même notation pour tout autre corps). Soit μ l'ensemble des racines de l'unité. Récemment, Ribet [Ri] a démontré que le groupe de torsion d'une variété abélienne dans le corps $k(\mu)$ est fini. Ceci généralise des résultats de Imai [Im] avec $k(\mu_{p^\infty})$ au lieu de $k(\mu)$ (et une condition de bonne réduction en plus) ; et de Katz-Lang [Ka-L] avec $k(\mu)$ mais supposant que la variété abélienne est de type CM . Nous allons axiomatiser ici ces résultats. Les démonstrations suivent Ribet [Ri].

Nous rappellerons en appendice ce qu'on veut dire par une représentation localement algébrique, au sens de Serre [Se]. Soit p un nombre premier, soit $G_k = \mathrm{Gal}(k^a/k)$, et soit

$$\rho : G_k \longrightarrow GL(V)$$

une représentation de G_k dans un espace vectoriel de dimension finie sur $\mathbf{Q}_p$. ("Représentation" signifie homomorphisme continu, pour la topologie de Krull sur G_k qui en fait un groupe compact, et la topologie p-adique sur $GL(V)$). On dira que ρ a <u>bonne semi-simplification</u> si les conditions suivantes sont satisfaites :

BSS 1. Si λ est une place finie de k ne divisant pas p, alors la semi-simplification $(V|D_\lambda)^{ss}$ par rapport au groupe de décomposition D_λ est non ramifiée en λ.

BSS 2. Si v est une place divisant p, alors $(V|D_v)^{ss}$ est localement algébrique (donc abélienne) sur le groupe d'inertie I_v.

Soit k' une extension finie de k. Par restriction, ρ donne une représentation de $G_{k'}$, et les deux conditions sont alors satisfaites relativement à k', si elles le sont relativement à k.

Nous aurons aussi besoin d'une forme faible de l'hypothèse de Weil-Riemann :

WRH. Il existe une place λ de k ne divisant pas p, telle que les valeurs propres de Frobenius $\rho(Fr_\lambda)$ ne sont pas des puissances entières de la norme absolue $\mathbf{N}\lambda$; et de même quand ρ est restreint à $G_{k'}$ pour toute extension finie k' de k.

THÉORÈME 1. <u>Soit</u> $H = G_{k(\boldsymbol{\mu})}$ <u>le groupe de Galois laissant</u> $k(\boldsymbol{\mu})$ <u>fixe</u>. <u>Supposons que</u> ρ <u>ait bonne semi-simplification et satisfasse à</u> WRH. <u>Soit</u> V^H <u>le sous-espace des éléments de</u> V <u>fixes par</u> H. <u>Alors</u> $V^H = 0$.

<u>Démonstration</u>. Supposons $V^H \neq 0$. Soit W un sous-espace G_k-simple de V^H. Alors ρ_W est une composante de la semi-simplification V^{ss}, et est abélien, se factorisant par $G/H = \mathrm{Gal}(k(\boldsymbol{\mu})/k)$. Sans perte de généralité, nous pouvons remplacer V par W, et ainsi supposer que $\rho = \rho_W$.

Si λ est une place de k ne divisant pas p, alors ρ étant abélienne on en déduit que ρ est semi-simple sur D_λ, donc non ramifiée en λ. D'autre part, pour v divisant p, puisque $(V|D_v)^{ss}$ est localement algébrique sur le groupe d'inertie I_v, il s'en suit que W est aussi localement algébrique. Soit

$$\varphi : U_v \longrightarrow GL(W)$$

l'homomorphisme sur les unités locales $U_v \subset k_v^*$ obtenu en composant l'application de réciprocité et ρ_W.

Soit $T_{k_v/\mathbb{Q}_p}$ le tore sur $\mathbb{Q}_p$ obtenu par restriction des scalaires, et soit

$$R : T_{k_v/\mathbb{Q}_p} \longrightarrow GL_W$$

l'homomorphisme algébrique tel que $\varphi = R$ sur un sous-groupe ouvert de U_v (voir l'appendice pour la notation $T_{k_v/\mathbb{Q}_p}$). Notons que φ se factorise par la norme locale

$$N_{k_v/\mathbb{Q}_p} : U_v \longrightarrow \mathbb{Z}_p^*$$

puisque ρ se factorise par le groupe de Galois cyclotomique localement et donc par le caractère cyclotomique. Il s'en suit que R se factorise par le quotient de $T_{k_v/\mathbb{Q}_p}$ par la clôture Zariskienne du groupe des éléments de norme 1, qui est connexe. Soit

$$N_{alg} : T_{k_v/\mathbb{Q}_p} \longrightarrow \mathbb{G}_m$$

l'homomorphisme algébrique déterminé par la norme locale. Alors il existe une représentation algébrique

$$h : \mathbb{G}_m \longrightarrow GL_W \quad \text{telle que} \quad R = h \circ N_{alg} .$$

Il en découle que

$$\rho = h \circ \chi_p \quad \text{sur un sous-groupe de} \quad I_v ,$$

où χ_p est le caractère cyclotomique. Mais d'autre part :

ρ et $h \circ \chi_p$ sont non-ramifiés en dehors de p ;

ρ et $h \circ \chi_p$ se factorisent par G/H .

Donc tous les deux se factorisent par $Gal(L/k)$ où L est la sous-extension maximale de $k(\mu)$ non ramifiée en dehors de p . Mais on vérifie facilement grâce à la structure des groupes d'inertie, presque partout égaux à $\mathbb{Z}_\ell^*$ que L est une extension finie de $k(\mu_{p^\infty})$. [Par exemple, si $k = \mathbb{Q}$, on sait par la théorie des nombres algébriques élémentaires que $L = \mathbb{Q}(\mu_{p^\infty})$. Une extension finie k n'apporte qu'une perturbation finie à la situation. Voir plus bas]. Donc l'image de I_v dans $\mathrm{Gal}(L/k)$ est ouverte, donc $h\circ\chi_p$ et ρ sont égaux sur un sous-groupe ouvert de G_k. En remplaçant k par une extension finie si nécessaire, on peut donc supposer sans perte de généralité que $h\circ\chi_p = \rho$. En particulier, pour $\sigma \in G_k$ les valeurs propres de $\rho(\sigma)$ sont des valeurs propres de $h(\chi_p(\sigma))$, et donc des puissances entières du nombre p-adique $\chi_p(\sigma)$. Soit λ une place de k ne divisant pas p et satisfaisant à la condition WRH. La conclusion qui précède s'applique à l'élément de Frobenius

$$\sigma = \mathrm{Fr}_\lambda ,$$

et montre que les valeurs propres de $\rho(\mathrm{Fr}_\lambda)$ sont des puissances intégrales de $\mathbf{N}\lambda$, contredisant WRH. Ceci termine la démonstration du théorème.

Pour la convenance du lecteur, je donne les détails de l'assertion employée plus haut en lemme comme dans [Ka-L].

LEMME. _Soit_ k _un corps de nombres_. _Il existe un entier positif_ m _tel que, si_ E _est une extension finie de_ k _ramifiée seulement en_ p, _et contenue dans_ $k(\mu)$, _alors_

$$E \subset k(\mu_{p^\infty}, \mu_m) .$$

Démonstration. Il existe un ensemble fini de nombres premiers S tel que

$$\mathrm{Gal}(k(\mu)/k) = G_S \times \prod_{\ell \notin S} G_\ell ,$$

ou $G_\ell \approx \mathbb{Z}_\ell^*$ et G_S contient un sous-groupe

$$H_S = \prod_{\ell \in S} H_\ell$$

avec H_ℓ ouvert dans $\mathbb{Z}_\ell^*$. Sans perte de généralité on peut supposer que S contient tous les nombres premiers qui se ramifient dans k. Si $\ell \notin S$ alors le groupe d'inertie en ℓ

contient G_ℓ (plongé comme composante dans le produit). Si $\ell \in S$ alors le groupe d'inertie en ℓ contient un sous-groupe H'_ℓ ouvert dans H_ℓ . Par conséquent le sous-groupe du groupe de Galois engendré par tous les groupes d'inertie aux nombres premiers $\ell \neq p$ contient

$$\prod_{\substack{\ell \in S \\ \ell \neq p}} H'_\ell \times \prod_{\ell \notin S} G_\ell .$$

Ceci démontre le lemme.

Pour la suite, soit $V[p]$ un espace vectoriel de dimension finie sur le corps premier $\mathbf{F}_p$, et soit

$$\rho_p : G_k \longrightarrow GL\ V[p]$$

une représentation. On dira que ρ_p a <u>bonne semi-simplification</u> si les conditions suivantes sont satisfaites :

BSS 1_p . Si λ est une place de k ne divisant pas p , alors $(V[p]|D_\lambda)^{ss}$ est non ramifié sur le groupe de décomposition D_λ .

BSS 2_p . Si v est une place de k divisant p , alors chaque composante simple de $(V[p]|D_v)^{ss}$ s'étend à un schéma en groupe fini et plat sur l'anneau des entiers de k_v .

Nous supposerons donnée une famille $\{\rho_p\}$ pour un ensemble infini de nombres premiers p impairs, telle que chaque ρ_p ait bonne semi-simplification. On dira que cette famille <u>satisfait à une faible hypothèse de Weil-Riemann</u> si la condition suivante est satisfaite :

WRH. Il existe une place λ de k et un polynôme $P_\lambda(T) \in \mathbf{Z}[T]$ n'ayant pas $\mathbf{N}\lambda$ pour racine, tels que pour n'importe quel premier p avec $\lambda \nmid p$, les valeurs propres de $\rho_p(Fr_\lambda)$ sont des racines de $P_\lambda(T)$ mod p , et ces racines ne sont pas égales à $\mathbf{N}\lambda$. De même quand k est remplacé par une extension finie k', et ρ_p est remplacé par sa restriction à $G_{k'}$.

THÉORÈME 2. <u>Supposons satisfaites les conditions ci-dessus. De plus, supposons que</u> $V[p]^{G_k}=0$ <u>pour presque tout</u> p, <u>et de même quand</u> k <u>est remplacé par une extension finie</u>. <u>Soit</u> $H = G_{k(\mu)}$. <u>Alors</u> $V[p]^H = 0$ <u>pour presque tout</u> p.

<u>Démonstration</u>. Soit k' la plus grande sous-extension non-ramifiée de k dans $k(\mu)$. Alors pour chaque premier p, $k'(\mu_{p^\infty})$ est la plus grande sous-extension de $k'(\mu)$ non ramifiée en dehors de p. On le vérifie facilement en regardant le corps fixé par le groupe d'inertie, qui est un sous-groupe de Z_p^*. Donc on peut remplacer k par k' sans perte de généralité. Il s'en suit que $k(\mu_p)$ est la plus grande sous-extension de $k(\mu)$ non-ramifiée en dehors de p et de degré premier à p.

Soit maintenant p tel que $V[p]$ ait bonne semi-simplification, dans la famille donnée. Supposons que $V[p]^H \neq 0$ et soit W un sous-espace simple pour G_k. Alors W est un constituant de $V[p]^{ss}$ et satisfait donc aux conditions BSS. On voit comme avant que W est non ramifié en dehors de p. Soit m la dimension de W sur F_p. Démontrons que $m=1$. L'espace $F_p^a \otimes W$ se décompose en une somme directe d'espaces de dimension 1 sur F_p^a, et sur chaque tel sous-espace, G_k opère via un caractère $G_k \longrightarrow (F_p^a)^*$. Puisque W est G_k-simple, ces caractères sont permutés transitivement par $\mathrm{Gal}(F_p^a/F_p)$. Puisque chaque tel caractère est non-ramifié en dehors de p et d'ordre premier à p, chaque caractère est une puissance du caractère cyclotomique

$$\chi : G_k \longrightarrow F_p^*$$

provenant de l'action de G_k sur μ_p. Puisqu'un tel caractère prend ses valeurs dans F_p^*, il est fixe par $\mathrm{Gal}(F_p^a/F_p)$. Ceci démontre que $m=1$.

Il en résulte que W est de dimension 1 sur F_p et que G_k opère sur W via χ_p^n pour un exposant entier n convenable, déterminé mod $p-1$. Soit v comme dans BSS 2_p.

Alors W s'étend à un schéma en groupe fini et plat W' sur l'anneau des entiers de k_v . Omettons les p tels que p est ramifié en k_v . La classification de Oort-Tate [O-T] montre que W' est étale ou bien que son dual de Cartier est étale. Si W' est étale, alors le groupe d'inertie I_v opère trivialement sur W , donc $n = 0$ et W est G_k-isomorphe à $\mathbb{Z}/p\mathbb{Z}$. Si le dual de Cartier est étale, alors le même argument montre que G_k opère trivialement sur le groupe dual de W , et donc que G_k opère via χ_p sur W , donc $n = 1$. Dans ce cas, W est G_k-isomorphe à μ_p .

Donc si $V[p]^H \neq 0$ il existe une infinité de premiers p tels que $V[p]$ a un G_k-sous-espace isomorphe à $\mathbb{Z}/p\mathbb{Z}$ ou à μ_p . Le cas $\mathbb{Z}/p\mathbb{Z}$ ne peut se produire qu'un nombre fini de fois par l'une des hypothèses du théorème, donc on a μ_p pour presque tout p .

Soit finalement λ comme dans la condition WRH, et considérons une infinité de p comme dans cette condition. Alors Fr_λ opère en élevant les éléments de μ_p à la puissance $N\lambda$, donc les valeurs propres de $\rho_p(Fr_\lambda)$ sont précisément égales à $N\lambda$ mod p . Donc pour une infinité de p on a

$$P(N\lambda) \equiv 0 \bmod p ,$$

et par conséquent $P(N\lambda) = 0$. Ceci contredit WRH, et termine la démonstration du Théorème 2.

Application aux variétés abéliennes

Le théorème 1 appliqué aux variétés abéliennes dit que pour chaque p il n'existe qu'un nombre fini de points d'ordre une puissance de p rationnels dans le corps cyclotomique. Pour cela, on prend la représentation de G_k sur le module p-adique de Tate (Deuring-Weil). Le théorème 2 appliqué dans le même cas dit que pour presque tout p , il n'y a pas de point d'ordre p rationnel dans le corps cyclotomique. La conjonction des deux théorèmes implique que le nombre de points de torsion rationnels dans $k(\mu)$ est fini.

Le fait que les représentations provenant des points de torsion d'une variété abélienne satisfont aux conditions de bonne semi-simplification énoncées plus haut est, comme l'a remarqué Raynaud [SGA 7I], après une extension finie du corps de base, une conséquence du théorème de réduction semi-stable de Grothendieck-Mumford, et de la théorie de Tate sur les modules de Hodge-Tate [Ta].

On peut s'attendre à ce que les représentations p-adiques provenant de la cohomologie d'une variété algébrique satisfassent aux axiomes BSS 1 et BSS 2 . Par contre, BSS 2_p a été formulé ad hoc et on ne s'attend pas à ce qu'il s'applique à la cohomologie en dimensions > 1 . Néanmoins, Ribet attire mon attention sur une question de Serre [Se 2], p. 278, suggérant que les exposants des caractères de l'inertie opérant sur une composante simple d'une représentation cohomologique sont bornés. Etant donné cette possibilité, en tenant compte de la première partie de la démonstration du Théorème 2, on peut remplacer l'axiome BSS 2_p par la condition que si G_k opère sur W via χ_p^n , alors les exposants n sont bornés ; et l'on remplace la condition WRH en spécifiant que le polynôme caractéristique P_λ n'a aucune des valeurs $\mathbb{N}\lambda^n$ pour racine, avec ces valeurs bornées de n . La même démonstration s'applique alors sans changement. En tous cas, à ce que je sache, pour le moment on ne connait pas d'exemples auxquels les axiomes s'appliquent autre que les variétés abéliennes.

Le théorème peut être appliqué aussi de la façon suivante, qui en est même à sa source, voir [Ka-L]. Soit A une variété abélienne sur un corps de nombre k . On veut démontrer que le revêtement maximal défini sur k , et tel que le groupe d'automorphismes soit abélien sur k , est fini. Soit $\pi : B \longrightarrow A$ un revêtement abélien sur k . Son groupe de Galois peut s'identifier à un sous-groupe des points rationnels de B dans k . Le dual de ce revêtement est un revêtement

$$ {}^t\pi : {}^tA \longrightarrow {}^tB $$

et le noyau de ${}^t\pi$ est dual du noyau de π , donc est un μ-sous-groupe de ${}^tA(k^a)$. (Un μ-sous-groupe est un sous-groupe dont chaque élément est un vecteur propre pour l'action du groupe de Galois correspondant au caractère cyclotomique ; autrement dit, il est engendré par des μ-points, un point d'ordre n étant un μ-point si $\sigma x = \chi(\sigma)x$, où χ est le caractère cyclotomique

$$ \chi : \mathrm{Gal}(k^a/k) \longrightarrow (\mathbb{Z}/n\mathbb{Z})^* $$

tel que $\sigma\zeta = \zeta^{\chi(\sigma)}$ si ζ est une racine n-ième de l'unité).

Les μ-points forment un sous-groupe de $A(k(\mu))_{tor}$ et sont donc en nombre fini d'après les théorèmes précédents. Ceci démontre qu'il n'y a qu'un nombre fini de revêtements comme ci-dessus.

Dans [Ka-L] on procède en sens inverse, en donnant une démonstration directe de ce résultat réduisant modulo p , et réduisant le théorème au cas du corps de classes sur les corps finis. On en déduit que le groupe des μ-points est fini, ceci ayant donné lieu à la conjecture que $A(k(\mu))_{tor}$ est fini, maintenant démontrée en général.

Appendice

Nous rappelons ici brièvement la notion de représentation localement algébrique. Soit A une algèbre commutative sur un corps k de caractéristique 0 pour simplifier. On suppose que A possède un élément unité 1 , et une base finie $\{w_1,\ldots,w_d\}$ sur k . Si $x = (x_1,\ldots,x_d)$ sont des variables, et $y = (y_1,\ldots,y_d)$ aussi, on peut définir une multiplication dans l'espace affine de dimension d par

$$ xy = (f_1(x,y),\ldots,f_d(x,y)) , $$

où les $f_i(x,y)$ sont les polynômes dans $k[x,y]$ tels que

$$(x_1w_1+\ldots+x_dw_d)(y_1w_1+\ldots+y_dw_d) = \sum_{i=1}^{d} f_i(x,y)w_i \ .$$

Ecrivons

$$1 = \sum_{\nu=1}^{d} e_\nu w_\nu \ .$$

Alors $(e_1,\ldots,e_d)$ est un élément unité pour la multiplication. Il existe un polynôme $D(x)$ tel que pour toute valeur spéciale de x dans une k-algèbre, le système d'équations $f_\nu(x,y) = e_\nu$ $(\nu = 1,\ldots,d)$ est résoluble si et seulement si $D(x) \neq 0$; donc que l'élément correspondant

$$x_1 \otimes w_1 + \ldots + x_d \otimes w_d$$

est inversible si et seulement si $D(x) \neq 0$. On note $T = T_A$ le groupe algébrique défini sur k , comme l'ouvert de Zariski formé des éléments x tels que $D(x) \neq 0$, avec la multiplication définie comme ci-dessus. Pour toute k-algèbre B , l'ensemble des points $T(B)$ est alors un groupe.

On notera que $T(k) = A^*$ est le groupe des unités dans A . Plus généralement, $T(B) = (B \otimes_k A)^*$ est le groupe des unités dans l'algèbre obtenue en étendant les scalaires de k à B .

En particulier, soit K une extension finie de k . Alors K est une algèbre comme ci-dessus, et on note $T = T_{K/k}$ le groupe algébrique associé comme ci-dessus. On dit aussi que T est le tore associé à l'extension K/k par restriction des scalaires. On identifie K^* au groupe $T_{K/k}(k) = T_K$ (si k est fixe).

Soit U un sous-groupe de K^* . Soit

$$R : U \longrightarrow GL(V)$$

une représentation de U dans un espace vectoriel V de dimension finie sur k . On dira que R est <u>algébrique</u> s'il existe un homomorphisme algébrique, défini sur k ,

$$f : T_K \longrightarrow GL_V$$

tel que $R = f$ sur U, c'est-à-dire que $R(x) = f(x)$ pour tout x dans U. Dans les applications, U est Zariski dense dans T, par conséquent f est uniquement déterminé par sa restriction à U.

Si K est une extension finie de E, alors la norme ordinaire

$$N_{K/E} : K^* \longrightarrow E^*$$

donne lieu à un homomorphisme algébrique

$$N_{K/E,\mathrm{alg}} : T_K \longrightarrow T_E$$

défini en fonction des coordonnées x par le produit

$$N_{\mathrm{alg}}(x) = \prod_{\sigma} (x_1 w_1^{\sigma} + \ldots + x_n w_n^{\sigma})$$

pris sur tous les plongements de K sur E. On exprime le membre de droite comme combinaison linéaire d'une base de E sur k, et les coefficients donnent les coordonnées de la norme algébrique de x.

Enfin, soient $k = \mathbb{Q}_p$ et K une extension finie de $\mathbb{Q}_p$. Soit U_K le groupe des unités dans K. D'après la théorie du corps de classes, il existe un isomorphisme canonique

$$\varphi_K : U_K \longrightarrow I_K^{ab}$$

des unités avec le groupe d'inertie abélien. (Différentes conventions prendront peut être son inverse, ça n'a aucune importance ici). Soit

$$\rho : G_K \longrightarrow GL(V)$$

une représentation abélienne de G_K dans un espace vectoriel de dimension finie V sur $\mathbb{Q}_p$. Le composé $\rho \circ \varphi_K$ donne une représentation de U_K dans $GL(V)$. Si cette représentation est algébrique sur un sous-groupe ouvert de U_K, on dit que ρ est <u>localement algébrique</u>. On notera que si U' est un sous-groupe ouvert de U_K alors U' est Zariski dense dans T_K. La représentation algébrique ci-dessus est donc uniquement déterminée par ρ.

Bibliographie

[Im] H. IMAI.- A remark on the rational points of abelian varieties with values in cyclotomic Z_p extensions. Proc. Japan Acad. 51 (1975) pp. 12-16.

[Ka-L] N. KATZ and S. LANG.- Finiteness theorems in geometric class field theory. L'Enseignement Mathématique (1981) pp. 285-314.

[O-T] F. OORT and J. TATE.- Group schemes of prime order. Ann. Scient. Ecole Normale Sup. 4e série t. 3 fasc. 1 (1970) pp. 1-21.

[Ri] K. RIBET.- Torsion points of abelian varieties in cyclotomic extensions. Appendix to [Ka-L] loc. cit. pp. 315-319.

[Se] J.-P. SERRE.- Abelian ℓ-adic representations and elliptic curves. W.A. Benjamin, 1968.

[Se 2] J.-P. SERRE.- Propriétés galoisiennes des points d'ordre fini des courbes elliptiques. Invent. Math. 15 (1972) pp. 259-231.

[Ta] J. TATE.- p-divisible groups, in Proc. Conference on Local Fields. Springer Verlag 1967 (Driebergen Conference).

BULLETIN (New Series) OF THE
AMERICAN MATHEMATICAL SOCIETY
Volume 6, Number 3, May 1982

UNITS AND CLASS GROUPS IN NUMBER THEORY AND ALGEBRAIC GEOMETRY

BY SERGE LANG[1]

CONTENTS

§1. Introductory remarks

Determining unramified coverings over various base spaces is a classical activity, which can take place in many contexts: topological, complex analytic, algebraic, and arithmetic. The abelian coverings are simpler to handle than the non-abelian ones, and in these lectures, we shall concentrate on abelian cases. Furthermore, the base space will have mostly dimension 1.

It turns out that the study of the arithmetic case is inextricably intertwined with that of the other cases, in many ways. Thus even though I (for example) start being motivated by the arithmetic case, I come eventually to a consideration of the other cases.

Presented as the Colloquium Lectures at the eighty-fifth summer meeting of the American Mathematical Society, in Pittsburgh, August 17–20, 1981; received by the editors July 8, 1981.

1980 *Mathematics Subject Classification*. Primary 12A35, 12A90; Secondary 14G25, 10D12.

[1]Supported by NSF Grant.

"Motivation", by the way, is a very relative term. What is motivation for one is an irrelevancy for another. I hope that somewhere in the combination of different problems which I shall list, most people will find some motivation for themselves.

One source of motivation for many is that the problems which are considered have their roots in 19th century mathematics. I personally don't make a fetish of this particular item, but it is indeed the case for the arithmetic problems which we shall encounter here. The solutions (as far as they have gone today) however, lie in contemporary mathematics, including the vast panoply of algebraic geometry-commutative algebra developed by Grothendieck and his school, ranging over several thousand pages.

For the topologist, unramified coverings of a space are classified by the fundamental group, and the abelian ones are classified by the first homology group. Although I restrict these lectures to number theory and algebraic geometry, I cannot refrain from mentioning an application of cyclotomic theory to free actions of finite groups on spheres in §2.

Before passing to the arithmetic, which by definition will be our prime source of interest, I would like to mention briefly the general setting for the considerations of algebraic geometry and complex analytic geometry of curves which will play an important role. This may provide motivation for some analysts or geometers.

Let X be a (compact, oriented) Riemann surface, and let $D = D(X)$ be the free abelian group generated by the points on X. This is called the group of **divisors**. A divisor

$$\mathfrak{a} = \sum_P n_P(P)$$

is called **linearly equivalent** to 0 if it is the divisor of a (meromorphic) function on X. This means that there exists a function f on X such that $n_P = \operatorname{ord}_P f$ is the order of the zero (pole) of f at P. Since the number of zeros of a function is equal to the number of poles, the divisor of a function is contained in the group of divisors of **degree** 0, that is divisors such that $\sum n_P = 0$. The factor group

$$D_0/D_l = \operatorname{Pic}(X)$$

of divisors of degree 0 by divisors linearly equivalent to 0 is called the **Picard group** of X, or also the group of **divisor classes** of degree 0. If f is a function, we let (f) denote its associated divisor.

Over the complex numbers, it is known (theorem of Abel-Jacobi) that $\operatorname{Pic}(X)$ is isomorphic to a complex torus of dimension g, where g is the genus of X, that is

$$\operatorname{Pic}(X) \approx \mathbf{C}^g/\Lambda,$$

where Λ is a lattice in $\mathbf{C}^g$. In particular, the structure of its torsion subgroup is clear. For each positive integer N, let $\operatorname{Pic}_N(X)$ denote the subgroup of elements of order N. Then

$$\operatorname{Pic}_N(X) \approx (\mathbf{Z}/N\mathbf{Z})^{2g}.$$

This group is directly related to unramified coverings as follows. Let $\mathfrak{a}$ be a representative divisor for an element of Pic_N, so there exists a function f such that

$$N\mathfrak{a} = (f).$$

Let $K = \mathbf{C}(X)$ be the function field of X, i.e. the field of all meromorphic functions on X. Then

$$K(f^{1/N})$$

is the function field of an unramified covering. Taking roots in this manner for all elements of Pic_N yields a maximal unramified abelian covering of exponent N (meaning that if T is an element of the group of covering transformations, then $T^N = \mathrm{id}$). Furthermore, if Y/X is this maximal covering, and G is its Galois group (group of covering transformations), then it can easily be shown that as finite abelian groups, G and Pic_N are canonically dual to each other (Kummer theory), and in particular are isomorphic (non-canonically), so have the same order. Thus the abelian coverings of X are determined in terms of divisor classes, and conversely, divisor classes are interpreted in terms of abelian coverings. The description of the Kummer theory is so simple that I give it.

Let $c \in \mathrm{Pic}_N$ and let T be a covering transformation. Let $\mathfrak{a}$ be a divisor representing c with $N\mathfrak{a} = (f)$ as above. Let φ be a meromorphic function on the covering such that $\varphi^N = f$ (so loosely speaking, $\varphi = f^{1/N}$, but two N-th roots of f differ by an N-th root of unity). Let $\boldsymbol{\mu}_N$ denote the group of all N-th roots of unity. Then the association

$$(T, c) \mapsto T\varphi/\varphi \in \boldsymbol{\mu}_N$$

defines a pairing between the group G and Pic_N, which is easily shown to give the above mentioned duality.

The Riemann surface X may be associated with an algebraic curve defined over the rational numbers. In that case, we would reserve the letter X to denote the curve, and the complex analytic manifold of its complex points would then be denoted by $X_{\mathbf{C}}$ or $X(\mathbf{C})$. The curve itself may be defined in projective space, or an affine open set may be defined by a single polynomial equation

$$\Phi(x, y) = 0,$$

where Φ is a polynomial in the two affine coordinates x, y. If the coefficients of Φ lie in a field k, we say that the curve is **defined** over k. Important curves for us later will be defined over $\mathbf{Q}$ or over other interesting algebraic number fields, namely finite extensions of the rational numbers. Among these curves will be the so-called modular curves, discussed at greater length in §3. Certain subgroups of the divisor class group for these curves at first present a striking analogy with ideal class groups in algebraic number theory, and recent discoveries have actually established precise connections which will provide much of the substance of these lectures.

Let X be a curve defined over a number field K. Associated with this curve is what is known as its Jacobian variety $J = J(X)$, which gives an algebraic representation of the Picard group Pic(X). Indeed, the analytic manifold $\mathrm{Pic}(X)_{\mathbf{C}} \approx \mathbf{C}^g/\Lambda$ (complex torus of dimension g) admits a projective embedding so that the group law on the torus corresponds to an algebraic group law. The image of $\mathbf{C}^g/\Lambda$ in projective space is called the **Jacobian variety**. Of course, there exist many such projective embeddings, and all such are algebraically isomorphic. Let us suppose that X has a rational point O in K (that is, X is coordinatized, defined over K, and there is a point all of whose coordinates lie in K). Then there is a projective model for J which is defined over K, and an embedding

$$X \to J$$

such that O (in X) goes to the origin in J. The embedding of X in J is in fact given by the map

$$P \mapsto \text{class of } (P) - (O) \text{ in } \mathrm{Pic}(X) \approx J;$$

furthermore X, identified with its image in J, generates J.

Having coordinatized J in that fashion, we may then speak of **rational points of J in some extension L of K**. They are the points whose affine coordinates lie in L. Such points form a group, denoted by J_L. The group of complex points $J_{\mathbf{C}}$ is complex analytically isomorphic to $\mathbf{C}^g/\Lambda$. If L is a number field, the theorem of Mordell-Weil asserts that J_L is finitely generated. We are interested here in the subgroup of torsion points J_{tor}.

For a given positive integer N, the group of points of order N in J will be denoted by J_N or $J[N]$. This is a finite subgroup, consisting of points which are algebraic over K (all their coordinates are algebraic over K). Let us denote by

$$K(J_N)$$

the field generated by the coordinates of all points in J_N over K. Then $K(J_N)$ is a Galois extension of K, and the effect of any automorphism $\sigma \in \mathrm{Gal}(K(J_N)/K)$ is determined by its effect on the points in J_N. Since we have an isomorphism $J_N \approx (\mathbf{Z}/N\mathbf{Z})^{2g}$, where g is the genus of X, we obtain a representation of this Galois group in $GL_{2g}(\mathbf{Z}/N\mathbf{Z})$. To determine the image of this representation in general is a fundamental problem relating number theory and algebraic geometry. When $g = 1$, so $X = J$, fundamental results have been obtained by Serre [**Se**], but here we want to concentrate on other cases which affect the theory of cyclotomic fields, and give rise to abelian unramified extensions.

In fact, it is necessary to consider certain special subgroups $\mathfrak{g}$ of J. We denote by $K(\mathfrak{g})$ the extension obtained by adjoining to K all coordinates of all points of $\mathfrak{g}$. For special choice of curve X and group $\mathfrak{g}$, we get interesting extensions. This is part of the general framework of giving explicit irrationalities via algebraic geometric objects for the generators of extensions predicted from purely internal structures of algebraic number theory, like ideal class groups. In other words, we want to construct (parametrize) explicitly the algebraic extensions of a given number field. For our purposes, we limit ourselves to abelian extensions, and even to special number fields like cyclotomic fields, for instance fields generated by roots of unity over the rational numbers, since the general problems are already very substantial in those cases.

In §2, §3, §4, §5, and §7 you will see examples of a situation with a group $\mathfrak{g}$ of order p, in a Jacobian variety or in the multiplicative group, admitting a group of automorphisms of order $p-1$. This situation is represented in different (but related) contexts, with a Galois group having a representation by 2×2 matrices

$$\begin{pmatrix} a & b \\ 0 & d \end{pmatrix}, \quad \text{with } a,\ d \in (\mathbf{Z}/p\mathbf{Z})^*,\ b \in \mathbf{Z}/p\mathbf{Z}.$$

Thus the study of abelian coverings leads to non-abelian coverings.

One particular way of obtaining interesting subgroups $\mathfrak{g}$ as mentioned above is the following, which ties up divisor class groups with units in the context of Riemann surfaces.

Let X be a Riemann surface again. Let S be a finite non-empty set of points. Let R be the ring of functions on X which have no poles outside S, that is functions whose poles lie in S. Then the points of S may be viewed as points at infinity, and we may say that the elements of R are the functions with poles only at infinity. Let R^* denote the group of units in R (invertible elements). Then R^* is the group of functions whose zeros and poles are at infinity. Consider the map

$$f \mapsto (f) = \sum_{P \in S} (\operatorname{ord}_P f)(P).$$

Let $s = |S|$ be the cardinality of S. The above map is a homomorphism of R^* into a free group of rank s generated by the points in S. Its kernel is precisely the group of constant functions (a function without zeros and poles on a compact Riemann surface is constant); and its image is contained in the group of divisors of degree 0, so its image has rank at most $s-1$. "Usually" this image will have much smaller rank. We shall meet later a special situation when for suitable choice of X and S, this rank is precisely $s-1$. Let $D^S = D^S(X)$ be the group of divisors with support in S. Then the group

$$D_0^S / D_l^S = \operatorname{Pic}^S(X) \subset \operatorname{Pic}(X)$$

is the subgroup of the divisor classes with support in S, or at infinity. When the above rank is $s-1$, then this subgroup is finite, and provides a very interesting object. For the modular curves considered in §2, they provide geometric counterparts for the ideal class groups in algebraic number theory. The interplay of the algebraic number theory and the theory of divisor classes for certain special number fields and special curves is precisely the topic of these lectures. These modular curves will be obtained as a quotient of the upper half plane by a subgroup of $SL_2(\mathbf{Z})$, of finite index. Such a quotient turns out to be the "affine part" of the complex points on a curve, and the points at infinity, classically called the cusps, furnish the set S. In this way, both the classical and contemporary theory of modular functions enter the picture.

Remark. The word "usually" used above can be made more precise as follows. The **Manin-Mumford conjecture** asserts that fixing a point O on X, there is only a finite number of points P such that the divisor $(P) - (O)$ is of finite order in the Picard group $\text{Pic}(X)$. Essentially this gives a bound on the size of the set S for which the S-units have maximal rank. The conjecture was reduced to a Galois theoretic statement in [**L 3**], and Shimura observed that this statement can be proved in a case known as the complex multiplication case, so the conjecture is true in that case. For a partial result in the general direction, cf. Bogomolov [**Bo**]. [*Raynaud recently proved the conjecture.*]

The above concepts are partly topological and partly geometric, over the complex numbers. However, they can be viewed as being merely the "local" concepts associated with more global concepts of geometric objects over rings of finite type over the integers. In that case, reduction mod primes p, or the associated objects over p-adic fields give other local objects, and gives the possibility for p-adic local results. Putting all these results together for various p and also the archimedean places gives rise to global results. So far, a few results and conjectures are known in various directions. Starting from a very classical situation, I shall expand the range of considerations throughout these lectures to arrive at a more encompassing outlook, pointing to broader directions in which each one of these considerations merely represents one facet of what eventually will become one huge theory.

I shall start with what I hope is a self contained and reasonably accessible level of exposition for a broad audience. As things move forward, the level will rise unavoidably. The actual lectures covered only the first part of the material. There was no reason why the written exposition should abide by the same limitations as the oral exposition.

§2. Cyclotomic Fields

(a) **The basic objects.** As good a point as any to start is the Fermat problem about the solutions to the equation

$$x^N - y^N = 1.$$

The left hand side factorizes as

$$\prod_{\varsigma} (x - \varsigma y)$$

where the product is taken over all N-th roots of unity. This immediately leads into the study of **cyclotomic fields**, namely the field $\mathbf{Q}(\boldsymbol{\mu}_N)$, where $\boldsymbol{\mu}_N$ denotes the group of N-th roots of unity. By a number field we shall mean a finite extension of the rational numbers. If K is a finite extension of $\mathbf{Q}$, we denote by $\mathfrak{o}_K$ (or $\mathfrak{o}$ if the reference to K is clear) the subring of algebraic integers. When $K = \mathbf{Q}(\boldsymbol{\mu}_N)$, then

$$\mathfrak{o}_K = \mathbf{Z}[\boldsymbol{\mu}_N]$$

is the ring generated over the ordinary integers **Z** by the roots of unity. If R is any ring, we let R^* be the group of units (invertible elements) in that ring. In the early study of the Fermat curve, many mistakes were made because people thought that ring had unique factorization into irreducible (prime) elements, which turned out to be false. For instance, a recent issue of the *Mathematical Intelligencer* (1979) reproduces a page of the *Compte Rendu des Séances de l'Académie des Sciences* of 1 March 1847 where Lamé announces a proof of the Fermat problem. (Cf. the Lenstra article, p. 6.) This announcement is followed by critical remarks of Liouville, pointing out that Lamé's paper is deficient in not taking into account the lack of unique factorization.

The obstruction to unique factorization of course lies in the ideal class group, which is defined as follows. Let $\mathfrak{a}, \mathfrak{b}$ be two ideals ($\neq 0$) of $\mathfrak{o}$. We say that $\mathfrak{a}$ and $\mathfrak{b}$ are **linearly equivalent**, or lie in the same **ideal class**, if there exists an element α of K such that $\mathfrak{a} = \alpha\mathfrak{b}$. Under this equivalence relation, the ideals form a group called the **ideal class group** $\mathrm{Cl}(K)$ of K, and this group is finite. Its order is denoted by h_K and is called the **class number**.

Kummer proved that if the class number of $\mathbf{Q}(\boldsymbol{\mu}_p)$ is not divisible by p, then the Fermat problem is solved affirmatively for p. We shall describe below other deeper considerations of Kummer concerning this class number and the Fermat problem.

A fundamental problem about number fields is the determination of the class number and of the structure of the ideal class group. For instance: how large is h_K? By what primes is it divisible? How does it behave asymptotically with N when $K = \mathbf{Q}(\boldsymbol{\mu}_N)$? Or $K = \mathbf{Q}(\boldsymbol{\mu}_{p^n})$ with a fixed prime p and n variable?

The ideal class group is directly related to unramified coverings as follows. For any extension L of K, the primes $\mathfrak{p}$ of K may decompose in this extension:

$$\mathfrak{p}\mathfrak{o}_L = \mathfrak{P}_1^{e_1} \cdots \mathfrak{P}_r^{e_r},$$

where the right hand side is the unique factorization of $\mathfrak{p}\mathfrak{o}_L$ into power products of prime *ideals*. If the exponents $e_1, \ldots, e_r$ are all equal to 1, we say that $\mathfrak{p}$ is **unramified**. If every prime $\mathfrak{p}$ of K is unramified, and if in addition every embedding of K into the real numbers extends only to real embeddings of L, then we say that L is **unramified** over K. A fundamental fact of **class field theory** asserts:

Let L be the maximal abelian unramified extension of K. Then there is a canonical isomorphism

$$C_K \approx \mathrm{Gal}(L/K)$$

of the ideal class group of K with the Galois group of L over K.

Another difficulty encountered in analyzing the Fermat curve, besides the ideal class group, was due to the units of $\mathfrak{o}_K$. Kummer had understood the deeper significance of the units, and of a distinguished subgroup of the units which he could write down immediately. Indeed, let ζ be a primitive N-th root of unity, and let i, j be positive integers prime to N. Then

$$\frac{1-\varsigma^i}{1-\varsigma^j}$$

is a unit. This is easily seen. First we note that

$$\frac{1-\varsigma^i}{1-\varsigma} = 1 + \varsigma + \cdots + \varsigma^{i-1},$$

so $1 - \varsigma$ divides $1 - \varsigma^i$. But we can find a positive integer a such that $aj \equiv 1 \bmod N$ since j is assumed relatively prime to N, so we get also that $1 - \varsigma^j$ divides $1 - \varsigma$, as asserted.

Let $K = \mathbf{Q}(\boldsymbol{\mu}_N)$. We let $E_{\mathrm{cyc}}(K)$ be the group generated by all roots of unity in K and all the units of the form

$$\prod_{1 \le a < N} (1 - \varsigma^a)^{n(a)}.$$

We call $E_{\mathrm{cyc}}(K)$ the group of **cyclotomic units**. It is a subgroup of the units $E(K)$. A classical theorem of Dirichlet asserts that $E(K)$ (so $E_{\mathrm{cyc}}(K)$) are finitely generated, of the same rank. In fact, we have

$$\operatorname{rank} E(K) = \frac{[K : \mathbf{Q}]}{2} - 1,$$

where $[K : \mathbf{Q}]$ is the degree of K over $\mathbf{Q}$, which is known to be equal to $\phi(N)$ (ϕ is the Euler function). Although one knows the group $E_{\mathrm{cyc}}(K)$, the structure of $E(K)$, and especially the factor group E/E_{cyc} remains a mystery.

It is clear that the essential aspects of the units have to do with the real subfield K^+ of K. Indeed, let E^+ denote the group of units in K^+. With Hasse, define the index

$$Q = (E : \boldsymbol{\mu}_K E^+).$$

Then it is easy to show that $Q = 1$ if N is a prime power, and $Q = 2$ if N is composite. In addition, concerning the cyclotomic units, suppose for simplicity that N is odd. If ς is an N-th root of unity, we can write $\varsigma = \lambda^2$ for some N-th root of unity λ, and then

$$1 - \lambda^2 = \lambda(\lambda^{-1} - \lambda).$$

This immediately shows that up to roots of unity, the cyclotomic units are real.

In the formula for the rank given above, the factor of $1/2$ is due to the fact that up to roots of unity, the units come from the real subfield; and we subtract 1 because there is a relation analogous to the property in function fields that a function has the same number of zeros and poles, counting multiplicities. In the present instance, the places at infinity are simply the embeddings of a number field into the complex numbers. If v denotes such an embedding, and $|\ |_v$ denotes the absolute value induced by the embedding, then one has the relation for any unit ϵ:

$$\sum_v \log |\epsilon|_v = 0.$$

If r_1 denotes the number of real embeddings and $2r_2$ the number of complex embeddings, then it is clear that the rank of the units is at most $r_1 + r_2 - 1$, and Dirichlet's theorem implies that the rank is precisely equal to this number. In the cyclotomic case $K = \mathbf{Q}(\mu_N)$ with $N \geq 3$ there is of course no real embedding of K.

We shall now see more precisely what is known and what is conjectured about the ideal class group and unit group in the cyclotomic fields.

We must view the various groups we have introduced as representation spaces for the Galois group

$$G = \mathrm{Gal}(\mathbf{Q}(\mu_N)/\mathbf{Q}).$$

This group G is isomorphic to $(\mathbf{Z}/N\mathbf{Z})^*$, under the association

$$a \mapsto \sigma_a.$$

If a is an integer prime to N, then σ_a is the automorphism of $\mathbf{Q}(\mu_N)$ such that

$$\sigma_a \zeta = \zeta^a.$$

The group of principal ideals is stable under G, so G acts on C_K with $K = \mathbf{Q}(\mu_N)$. It is also clear that the group of cyclotomic units is stable under G, so G acts on E/E_{cyc}. What is the structure of these groups as G-modules?

In any representation theory, one tries to decompose representation spaces into eigenspaces for the characters of the group. The first immediate such decomposition that we deal with is that arising from complex conjugation. Note the element σ_{-1} in G such that

$$\sigma_{-1}\zeta = \zeta^{-1}.$$

If A is any G-module we let A^+ be the $(+1)$-eigenspace for σ_{-1}, that is the subgroup of elements fixed by σ_{-1}. We let A^- be the (-1)-eigenspace, that is the subgroup of elements x such that $\sigma_{-1}x = -x$ for $x \in A$ (writing the operation of A additively). If multiplication by 2 is invertible on A, then we have a direct sum decomposition

$$A = A^+ \oplus A^-.$$

Consider the group of ideal classes $C = C_K$. Let K^+ denote the real subfield of K. An ideal $\mathfrak{a}$ in K^+ lifts to an ideal of K, namely

$$\mathfrak{a} \mapsto \mathfrak{a}\mathfrak{o}_K.$$

This induces a homomorphism on the ideal class groups

$$C_{K^+} \to C_K$$

which can be proved to be injective. Likewise, we have a norm map

$$N_{K/K^+} : C_K \to C_{K^+}$$

given by $c \mapsto cc^\rho$, where $\rho = \sigma_{-1}$ is complex conjugation. It can be shown that this norm map is surjective. By definition, the kernel of the norm map consists of those ideal classes c such that $c^{1+\rho} = 1$, or in other words $c^\rho = c^{-1}$. This kernel is precisely the (-1)-eigenspace for complex conjugation. Thus we have an exact sequence

$$1 \to C_K^- \to C_K \to C_{K^+} \to 1.$$

(For proofs, cf. [**L 1**], Chapter 3, §4.) We let:

$$h^+ = \text{order of } C^+ \quad \text{and} \quad h^- = \text{order of } C^-.$$

Then h^+ is the class number of the real subfield, and we have

$$h = h^- h^+.$$

We are thus led to study h^+ and h^- separately.

(b) **Plus eigenspaces.** Let us begin with h^+. First there is a relation of this number with units. We assume for simplicity that $N = p$ is a prime ≥ 3 and $K = \mathbf{Q}(\boldsymbol{\mu}_p)$. We then have a coincidence of orders:

Theorem 2.1. $h^+ = (E : E_{\text{cyc}})$.

This theorem is due to Kummer. Its proof today is viewed as elementary, and results from the factorization of the zeta function into Dirichlet L-series. There is no need to go into the proof here. Kummer made the first basic discoveries concerning these class numbers, and made the first basic conjectures. Especially:

Conjecture 1. *For* $\mathbf{Q}(\boldsymbol{\mu}_p)$, h^+ *is prime to* p.

The history of that conjecture is interesting. Kummer made it in no uncertain terms in a letter to Kronecker dated 28 December 1849. Kummer first tells Kronecker off for not understanding properly what he had previously written about cyclotomic fields and Fermat's equation, by stating "so liegt hierin ein grosser Irrthum deinerseits..."; and then he goes on (Collected Works, Vol. 1, p. 84):

> Deine auf dieser falschen Ansicht berühenden Folgerungen fallen somit von selbst weg. Ich gedenke vielmehr den Beweis des Fermatschen Satzes auf folgendes zu grunden:
>
> 1. Auf den noch zu beweisenden Satz, dass es für die Ausnahmszahlen λ stets Einheiten giebt, welche ganzen Zahlen congruent sind für den Modul λ, ohne darum λte Potenzen anderer Einheiten zu sein, oder was dasselbe ist, dass hier niemals D/Δ durch λ theilbar wird.

In our notation: $\lambda = p$ and $D/\Delta = h^+$. Writing h^+ in this form is explainable in terms of the expression in Theorem 2.1. The quotient D/Δ represents the order of the factor group E/E_{cyc}. Thus Kummer rather expected to prove the conjecture. According to Barry Mazur, who reviewed Kummer's complete works when they were published recently by Springer-Verlag, Kummer never mentioned the conjecture in a published paper, but he mentioned it once more in another letter to Kronecker on 24 April 1853 (loc cit p. 93):

> Hierein hängt auch zusammen, dass eines meiner Haupresultate auf welches ich seit einem Vierteljahre gebaut hatte, dass der zweite Faktor der Klassenzahl D/Δ niemals durch λ theilbar ist, falsch ist oder wenigstens unbewiesen...Ich werde also vorlaufig hauptsachlich meinen Fleiss nur auf die Weiterführung der Theorie der complexen Zahlen wenden, und dann sehen ob etwas daraus entsteht, was auch uber jene Aufgabe Licht verbreitet.

So the situation was less clear than Kummer thought at first. Much later, Vandiver made the same conjecture, and wrote [**Va 1**]:

> ...However, about twenty-five years ago I conjectured that this number was never divisible by p [referring to h^+]. Later on, when I discovered how closely the question was related to Fermat's Last Theorem, I began to have my doubts, recalling how often conjectures concerning the theorem turned out to be incorrect. When I visited Furtwängler in Vienna in 1928, he mentioned that he had conjectured the same thing before I had brought up any such topic with him. As he had probably more experience with algebraic numbers than any mathematician of his generation, I felt a little more confident...

Vandiver, like others before him, wanted to have such a result for the application to what is called the *first case* of Fermat's theorem: there are no solutions other than the trivial ones to the equation

$$x^p + y^p = z^p$$

in relatively prime integers x, y, z which are also prime to p. Many years ago, Feit was unable to understand a step in Vandiver's "proof" that $p \nmid h^+$ implies the first case of Fermat's Last Theorem [**Va 2**], and stimulated by this, Iwasawa found a precise gap which is such that there is no proof.

In number theory, or elsewhere, when two groups arising naturally in the course of a theory have the same order, one immediately asks whether this coincidence is not due to the fact that the groups are isomorphic. Iwasawa has an example showing that C^+ and E/E_{cyc} are not isomorphic as G-modules. It is generally believed that the two groups are not always isomorphic, even as abelian groups. However, in this direction, one has the next best thing.

Theorem 2.2. *For any prime l not dividing the degree $[K : \mathbf{Q}]$ with $K = \mathbf{Q}(\boldsymbol{\mu}_p)$ the l-primary parts of C^+ and E/E_{cyc} have isomorphic semisimplications.*

Recall that the l-primary part $A^{(l)}$ of a torsion abelian group A is the subgroup generated by all elements whose order is a power of l.

We also recall briefly the definition of semisimplification. Let A be a representation module for G. We say that A is **simple** (or gives a simple, or irreducible representation of G) if $A \neq \{0\}$, and if A has no G-submodules other than 0 or itself. The module A may of course have a coefficient ring or field acting on it, commuting with the action of G. If A is a finite group, we may take this ring to be $\mathbf{Z}$. Suppose that A is finite, or finite dimensional over a field. Then A has a chain of submodules

$$A = A_0 \supset A_1 \supset \cdots \supset A_r$$

such that every A_i/A_{i+1} is simple. The direct sum

$$\bigoplus_{i=0}^{r-1} A_i/A_{i+1}$$

is uniquely determined as a G-module, up to isomorphism, and is called the **semisimplification**. In earlier terminology, the simple components A_i/A_{i+1} are called the **Jordan-Hölder** factors.

Theorem 2.2 was conjectured for several years, stemming from the work of Leopoldt [**Le 2**]. The conjecture is made explicit in Gras [**Gra**]. A related conjecture appears as a "Remark" in Coates-Lichtenbaum [**Co-L**], p. 520. Greenberg [**Gr 2**] showed that the statement of Theorem 2.2 was a consequence of a standard conjecture concerning certain infinite extensions, now proved by Mazur-Wiles, see below Theorem 2.10.

By going up the tower of p^n-th roots of unity, one can make a conjecture concerning the limiting behavior of units and ideal classes as modules over the infinite Galois group. This is more elaborate to state, and we shall touch on that aspect of the question at the end of this section.

The group theoretic situation in the case when N is composite was unclear for a long time. Sinnott [**Si 1**], [**Si 2**] made a breakthrough when he discovered the appropriate group-ring formulation needed to handle the analogue of Theorem 2.1 in the composite case. In this case, the reader should be warned of at least one important new phenomenon: the presence of a power of 2 in the relation of Theorem 1.1, between h^+ and the index $(E : E_{\text{cyc}})$, namely

$$(E : E_{\text{cyc}}) = (E^+ : E^+_{\text{cyc}}) = 2^\nu h^+,$$

where $\nu = 2^{t-2}+1-t$, and t is the number of prime factors of N, whenever $t \geq 2$. Thus when $t = 1$ or 2, $2^\nu = 1$, but $\nu > 0$ otherwise.

(c) **A topological interlude.** Let G be a finite group operating freely on the n-sphere S^n with $n \geq 3$, and let S^n/G be the quotient space. Then $G = \pi_1(S^n/G)$ is the fundamental group. Such operations therefore give rise to maximal unramified coverings, which are even finite. One wishes to classify such actions up to various equivalence relations. Two representations R_1, R_2 of a group G in the group of topological automorphisms of a space X are called topologically conjugate if there exists a topological automorphism T of X such that $TR_1(g)T^{-1} = R_2(g)$ for all $g \in G$. A folklore conjecture asserts that:

> a topologically free action of a finite group on the 3-sphere S^3
> is topologically conjugate to a free linear action.

Seifert must have known this possibility. Milnor drew attention to the problem in the late fifties, and solved a special case. The problem is stated in Thurston's Lecture Notes from Princeton (in circulation). For background material, see J. Hempel's book on 3-Manifolds (Ann. of Math. Studies 96, 1976).

In higher dimensions, there is of course always the linear action of finite subgroups of $O(n)$ on S^n. For such linear theory, cf. for instance J. Wolf, *Spaces of Constant Curvature*, 2nd Edition, Publish or Perish. I am indebted to C. B. Thomas for drawing my attention to the above literature, and to forthcoming papers of his, concerning the classification of free actions of finite groups on S^n which C. T. C. Wall has already shown to depend in part on the 2-primary component of the ideal class group in real cyclotomic fields $\mathbf{Q}(\mu_N)^+$ for suitable N.

In particular, Thomas tells me that given any free action of a finite group G on S^n with $n \geq 5$, there exist infinitely many distinct topological conjugacy classes of actions of G on S^n, and there are only finitely many topological conjugacy classes of linear actions. Using the algebraic background of a paper of Wall [**Wa**], applied to the surgery exact sequence, Thomas gives examples for the binary dihedral group D_{4p} of order $4p$ operating freely on S^{4k-1}

with $k \geq 2$, when p is a prime such that h_p^+ is odd, e.g. $p < 163$. These operations are even homotopically distinct from the classical linear actions. [For the convenience of the reader, I reproduce a definition of D_{4p}. It can be represented as the group generated by the matrices

$$\begin{pmatrix} 0 & -1 \\ 1 & 0 \end{pmatrix} \quad \text{and} \quad \begin{pmatrix} \varsigma & 0 \\ 0 & \varsigma^{-1} \end{pmatrix}$$

where ς is a primitive p-th root of unity. It is one of the extensions of the cyclic group of order p by the cyclic group of order 4. In the above representation, it acts linearly on $\mathbf{C}^2$, whence on S^3 which is naturally contained in $\mathbf{C}^2$.]

Furthermore, according to Thomas, there exist free actions by D_{4p} which can be topologically distinguished only by an invariant in the 2-primary part of the ideal class group of $\mathbf{Q}(\boldsymbol{\mu}_p)^+$. A paper of Kubert [**Ku 1**] shows the existence of a large 2-primary component in $\mathbf{Q}(\boldsymbol{\mu}_N)$ when N is divisible by many primes, but says nothing about the 2-primary part for $\mathbf{Q}(\boldsymbol{\mu}_p)$.

In the above context, the ideal class group (2-primary part) intervenes. I am indebted to J. Milgram for pointing out to me the existence of a substantial literature which relates the existence of free action of finite groups on spheres to units in cyclotomic fields, and especially cyclotomic units. For example in [**Mi 1**], [**Mi 2**] Milgram reduces questions of which groups act on S^n $(n > 3)$ to explicit questions involving the structure of such units. Since my main interest here was only to point out briefly a connection of units and ideal class groups with topology, and constitute an aside for the main topics of these lectures, I refer interested readers to the discussions and bibliographies at the end of Milgram's papers for further information.

(d) **The minus eigenspaces.** So much for the plus part of the class number and the units at this time. Let us look at the minus part. The situation is now radically different. We are seeking all relations for the ideal classes C^- in the group ring $R = \mathbf{Z}[G]$. Classical relations are provided by a construction of Kummer and Stickelberger, as follows.

Given a real number $\bmod \mathbf{Z}$, say $x \in \mathbf{R}/\mathbf{Z}$, we let $\langle x \rangle$ denote its unique representative in $\mathbf{R}$ such that

$$0 \leq \langle x \rangle < 1.$$

We define the first **Bernoulli polynomial**

$$\mathbf{B}_1(X) = X - \frac{1}{2}.$$

(There will be a second Bernoulli polynomial later.) We then define the **Stickelberger element** in $\mathbf{Q}[G]$:

$$\theta = \sum_{a=1}^{p-1} \mathbf{B}_1\left(\frac{a}{p}\right)\sigma_a^{-1} = \sum_{a \in \mathbf{Z}(p)^*} \mathbf{B}_1\left(\left\langle \frac{a}{p} \right\rangle\right)\sigma_a^{-1}.$$

This element θ has rational coefficients in the group algebra. How to integralize it? Note that for any positive integer c odd and prime to p, the rational numbers

$$\left\langle \frac{ca}{p} \right\rangle - c\left\langle \frac{a}{p} \right\rangle \quad \text{and} \quad \frac{c-1}{2}$$

are integers. From this it follows at once that $(\sigma_c - c)\theta \in \mathbf{Z}[G]$. In other words, let I be the ideal in the group ring $\mathbf{Z}[G]$ generated by all elements $\sigma_c - c$ with c prime to $2p$. Then

$$I\theta \subset \mathbf{Z}[G],$$

and in fact, if we let $R = \mathbf{Z}[G]$, then

$$R\theta \cap R = I\theta.$$

Furthermore, $\theta = \theta^-$. We call $I\theta$ the **Stickelberger ideal** S.

Theorem 2.3. *The elements of the Stickelberger ideal annihilate C^-. (Actually, they also annihilate C.)*

This theorem of Kummer and Stickelberger is proved by showing that for any ideal $\mathfrak{a}$, the ideal $\mathfrak{a}^\alpha$ is principal when $\alpha = (\sigma_c - c)\theta$, and by exhibiting explicitly the algebraic number generating this ideal, which is a quotient of Gauss sums. We do not go into this explicit determination here since we want to emphasize other aspects of the theory. For a general context, see Conjecture 8.3. Kummer had already proved the theorem by getting the relation for prime ideals of degree 1 over $\mathbf{Q}$, see [**Kum 2**], p. 628, and by using the "Stickelberger element" in special cases.

Iwasawa [**Iw 3**] proved:

Theorem 2.4. $(R^- : S) = h^-$.

As for the cyclotomic units, this relation holds without extra factor when dealing with N equal to a prime power, and with an extraneous power of 2, as determined by Sinnott, when N is composite. We stick to the prime case for simplicity.

Iwasawa and Leopoldt emphasized repeatedly the problem of determining the relation between the factor module R^-/S and C^-.

Theorem 2.5. *Let $K = \mathbf{Q}(\mu_p)$. For any prime l not dividing the degree $[K : \mathbf{Q}]$, the l-primary parts of C^- and R^-/S have isomorphic semi-simplifications. (Here we may have $l = p$.)*

Theorem 2.5 was conjectured by Leopoldt [**Le 2**], [**Le 3**], and like its plus counterpart, follows from the Mazur-Wiles general theorems. For the p-primary part, one has in addition:

Conjecture 2. *The p-primary part of C^- is cyclic over the group ring, namely it is generated by one element, and consequently there is an isomorphism*

$$(R^-/S)^{(p)} \approx C^{-(p)},$$

where the superscript (p) indicates p-primary part.

In this manner, the study of the minus p-primary part of the class group would be reduced to the study of the Stickelberger ideal.

Note. Conjecture 2 stems from the work of Iwasawa and Leopoldt, but neither have explicitly stated it as a conjecture. They certainly drew attention to its possibility, and for convenience, it may be useful to refer to it as the Iwasawa-Leopoldt conjecture. If it is true, they should get the credit, and if it is false, I should get the blame.

When studying the eigenspaces for the characters of G (after a suitable extension of scalars), one encounters the character values

$$\chi(\theta) = \sum_{a=1}^{p-1} \mathbf{B}_1\left(\frac{a}{p}\right)\bar{\chi}(a) = B_{1,\bar{\chi}} \quad \text{by definition,}$$

identifying σ_a with a in $(\mathbf{Z}/p\mathbf{Z})^*$. Following Iwasawa, Leopoldt, and Mazur, this sum can be written as an integral

$$B_{1,\bar{\chi}} = \int_{\mathbf{Z}/p\mathbf{Z}} \bar{\chi}\, d\mu_{\mathbf{B}_1},$$

but we do not discuss this aspect of the question. However, we note that the divisibility properties of h^- depend on the divisibility properties of these sums $\chi(\theta)$, which are still very difficult to determine. Indeed, it is shown classically also by elementary L-series considerations that one has the explicit formula:

Theorem 2.6. $h^- = w \prod_{\chi\ odd} -\frac{1}{2}B_{1,\chi}.$

Here w is the number of roots of unity in $\mathbf{Q}(\boldsymbol{\mu}_p)$, namely $2p$. The product is taken over odd characters χ, meaning characters such that $\chi(-1) = -1$. Thus the divisibility properties of h^- are determined by the divisibility properties of the numbers $B_{1,\chi}$, the "generalized Bernoulli numbers" of Leopoldt [**Le 1**] who investigated their congruence properties, picking things up where Kummer left them a century before.

So we look at characters χ of $(\mathbf{Z}/p\mathbf{Z})^*$, even or odd since we are going to deal with the cyclotomic fields and the modular curves. The group $(\mathbf{Z}/p\mathbf{Z})^*$ is merely the multiplicative group of the prime field $\mathbf{Z}/p\mathbf{Z}$, and consists of $(p-1)$th roots of unity. On the other hand, the p-adic integers $\mathbf{Z}_p$ contain the $(p-1)$th roots of unity as a subgroup of $\mathbf{Z}_p^*$, and reduction mod p gives an isomorphism

$$\boldsymbol{\mu}_{p-1} \to (\mathbf{Z}/p\mathbf{Z})^*$$

of the group of $(p-1)$th roots of unity in characteristic 0 with the $(p-1)$th roots of unity in characteristic p. The inverse of this isomorphism,

$$\omega : (\mathbf{Z}/p\mathbf{Z})^* \to \boldsymbol{\mu}_{p-1}$$

is called the **Teichmuller character**. So $\omega(a)$ is the unique root of unity congruent to $a \bmod p$.

Instead of the group ring $\mathbf{Z}[G]$, let us look at the group ring $\mathbf{Z}_p[G]$ over the p-adic integers. We lose nothing in so widening the coefficients when we look at representations of G in finite abelian groups whose order is a power of p; and we gain a lot, because the usual idempotent projecting on the χ-eigencomponents can now be written with coefficients in $\mathbf{Z}_p$, namely

$$e_\chi = \frac{1}{|G|} \sum_{\sigma \in G} \bar{\chi}(\sigma)\sigma.$$

As usual, $|G|$ denotes the order of G, namely $p-1$ in the present case. Note that $p-1$ is invertible in $\mathbf{Z}_p^*$, so $e_\chi \in \mathbf{Z}_p[G]$. Let M be a G-module on which multiplication by $p-1$ is invertible. Then $e_\chi M$ is the χ-eigenspace, namely the subgroup of all elements x such that

$$\sigma x = \chi(\sigma)x.$$

This eigenspace will be denoted by $M(\chi)$.

Here we let M be the co-Stickelberger module,

$$M = \mathbf{Z}_p[G]^- / \mathbf{Z}_p S.$$

According to Conjecture 2, we have $M \approx C^{-(p)}$. In any case, one can ask two questions: what is the structure of M, and how closely does it approximate $C^{-(p)}$.

Theorem 2.7. (i) *If* $\chi = \omega$, *then* $M(\chi) = 0$.
(ii) χ *is an odd character,* $\chi \neq \omega$ *then* $M(\chi)$ *is cyclic,*

$$M(\chi) \approx \mathbf{Z}_p / B_{1,\bar{\chi}} \mathbf{Z}_p.$$

In particular, $\operatorname{ord}_p M(\chi) = \operatorname{ord}_p B_{1,\bar{\chi}}$.

This theorem is an immediate consequence of the definitions. Theorem 2.5 then implies that

$$\operatorname{ord}_p C^{-(p)}(\chi) = \operatorname{ord}_p B_{1,\bar{\chi}},$$

and Conjecture 2 predicts that $C^{-(p)}(\chi)$ is cyclic, for an odd character $\chi \neq \omega$, of order $p^{m(\chi)}$, where $m(\chi) = \operatorname{ord}_p B_{1,\bar{\chi}}$. Thus the study of M is reduced to the study of p-divisibility of the Bernoulli numbers.

The above conjectures and theorems constitute an essential part of the present vision of what the structure of units and class numbers should be like. It can be shown that the Kummer-Vandiver conjecture h^+ prime to p is true implies all the theorems and conjectures of this section. Thus the conjectures are related, and in fact are related via Kummer theory and class field theory as follows.

Consider the maximal abelian unramified extension of $K = \mathbf{Q}(\boldsymbol{\mu}_p)$ which is of exponent p, meaning that if $\mathcal{G}$ is its Galois group, then $\sigma^p = 1$ for all $\sigma \in \mathcal{G}$. By Kummer theory, if L is a cyclic subextension of degree p, then

$$L = K(\epsilon^{1/p})$$

for some element ϵ. The Kummer-Vandiver conjecture implies that ϵ can be taken to be a cyclotomic unit. But then the representation of $\mathrm{Gal}(K/\mathbf{Q})$ on the cyclotomic units is sufficiently explicit so that by mixing Kummer theory and class field theory one obtains the cyclicity of C^- over the group ring. Actually, the situation is more involved, because as described above, we are only dealing with the subgroup of elements of order p, and to get the full structure, one has to consider higher cyclotomic fields of p^n-th roots of unity, for n tending to infinity. In other words, one has to go up the cyclotomic tower. This requires a more elaborate foundation and the introduction of new concepts which are more technical. In any case, one is led to consider double decked extensions

$$\begin{array}{c} K(\epsilon^{1/p}) \\ | \\ K = \mathbf{Q}(\mu_p) \\ | \\ \mathbf{Q} \end{array}$$

such that each layer is abelian, but the composite layer is not; and similarly when p is replaced by p^n for arbitrarily large n. Attempts to recognize directly which extensions $K(\epsilon^{1/p})$ are unramified over $\mathbf{Q}(\mu_p)$ have failed for a century. The newly developed methods via algebraic geometry replace the ordinary p-th root of a unit by another p-th root, namely the root taken relative to the group law on Jacobian varieties, where the additional structure is more complicated, richer and leads to results which so far were unobtainable otherwise.

In the tower, we should note a theorem of Washington [**Wash**].

Theorem 2.8. *Let C_n^- be the minus part of the ideal class group in $\mathbf{Q}(\mu_{p^{n+1}})$. Let l be a prime $\neq p$. Then the powers of l dividing the orders of C_n^- are bounded.*

One asks not only for the behavior of class numbers "vertically", that is in the tower of fields $\mathbf{Q}(\mu_{p^n})$ for n tending to infinity, but one also asks for the "horizontal" behavior of these class numbers. This amounts to the horizontal behavior of $B_{1,\chi}$ from the point of view of divisibility, when χ is a character of $(\mathbf{Z}/p\mathbf{Z})^*$, and p is viewed as variable. As far as I know, there are no results in this direction. Computations indicate a perturbation of random behavior. For instance, Trotter has done some computations for the class number

$$h = -B_{1,\chi}$$

of an imaginary quadratic field $F = \mathbf{Q}(\sqrt{-p})$, where χ is the associated quadratic character. Consider the primes $\equiv 3 \bmod 4$, with $7 \leq p < 48{,}611$. There are 2,512 such primes. We have mod 3:

$h \equiv 0$ for 964 (out of 2,512) or 38.4 o/o,

$h \equiv 1$ for 761 (out of 2,512) or 30.3 o/o,

$h \equiv 2$ for 787 (out of 2,512) or 31.3 o/o.

As Trotter remarks, the evidence is strong that $h \equiv 0 \bmod 3$ occurs more often than a third of the time, quite strong that $h \equiv 0 \bmod 5$ occurs more than one fifth of the time, and definitely suggests that $h \equiv 0 \bmod 7$ occurs more than one seventh of the time. One awaits more precise conjectures and proofs.

However, the distribution over all quadratic fields $\mathbf{Q}(\sqrt{-D})$ appears to be random, cf. Davenport and Heilbronn [**Da-H**], cf. see also Kuroda-Leopoldt in [**Zi**], p. 42.

When the quadratic subfield F is contained in the cyclotomic field $\mathbf{Q}(\boldsymbol{\mu}_p)$, then the ideal class group of F occurs as a quotient group of the ideal class group of $\mathbf{Q}(\boldsymbol{\mu}_p)$, by considerations of class field theory, so the class numbers of subfields of $\mathbf{Q}(\boldsymbol{\mu}_p)$ are important components of the class numbers of the full cyclotomic field itself.

(e) **The cyclotomic p-tower.** In this last part I shall summarize briefly the way one formulates results in the infinite tower of p-extensions, first investigated by Iwasawa. The algebra becomes a little heavier. The reader might omit this part at first, in order to minimize the obstacles preceding the discussion of geometric connections of number theory and groups of finite order on Jacobian varieties, given in §3 and §4 for $\mathbf{Q}(\boldsymbol{\mu}_p)$.

I limit the discussion to the standard cyclotomic case, so let p be an odd prime, and let

$$K_n = \mathbf{Q}(\boldsymbol{\mu}_{p^{n+1}}), \qquad K_\infty = \bigcup_{n=1}^{\infty} K_n.$$

Let:

$\mathcal{G}_n = \mathrm{Gal}(K_n/K_0)$,

$\mathcal{G}_\infty = \mathrm{Gal}(K_\infty/K_0) =$ projective limit of the groups $\mathcal{G}_n$. Then

$$\mathcal{G}_\infty \approx \mathbf{Z}_p.$$

$\gamma =$ topological generator of $\mathcal{G}_\infty$, for instance $\gamma = \sigma_{1+p}$.

$\Lambda =$ projective limit of $\mathbf{Z}_p[\mathcal{G}_n]$. Then there is a unique isomorphism

$$\Lambda \approx \mathbf{Z}_p[[X]]$$

with the power series in one variable over $\mathbf{Z}_p$, such that $1+X$ corresponds to the chosen generator γ. This means that the image of $1+X$ in $\mathbf{Z}_p[\mathcal{G}_n]$ is equal to the image of γ in $\mathcal{G}_n$ for all n. We call Λ the **Iwasawa algebra**. Cf. [**Iw 1**] and [**Se 2**]. Let:

$\mathcal{A}_n = p$-primary part of $\mathrm{Cl}(K_n)^-$ and

$\mathcal{A} = \varprojlim \mathcal{A}_n$.

Then $\mathcal{A}$ is a Λ-module, which can be proved to be torsion and finitely generated, and $\mathcal{A}$ is also a topological γ-module. Cf. Iwasawa [**Iw 1**] and Serre [**Se 2**]. Let further:

$G_n = \mathrm{Gal}(K_n/\mathbf{Q})$.

Then $\mathcal{A}$ is a G_0-module. Since $\mu_{p-1} \subset \mathbf{Z}_p$, if χ denotes odd characters of G_0, we may form the decomposition

$$\mathcal{A} = \bigoplus_{\chi} \mathcal{A}(\chi).$$

For each n there is a Stickelberger ideal S_n, and the **co-Stickelberger module at level** n:

$$M_n = \mathbf{Z}_p[G_n]^- / \mathbf{Z}_p S_n.$$

We let the **co-Stickelberger module** be the projective limit

$$M = \varprojlim M_n.$$

If the Iwasawa-Leopoldt conjecture is true, then we have an isomorphism for each odd $\chi \neq \omega$:

$$\mathcal{A}(\chi) \approx M(\chi).$$

We wish to describe the Mazur-Wiles theorem concerning the structure of $\mathcal{A}$ as Λ-module.

Two modules M_1 and M_2 are called **quasi-isomorphic** if there is a homomorphism $h : M_1 \to M_2$ with finite kernel and finite cokernel. It is a theorem of Serre [**Se 2**] that any finitely generated torsion module over Λ is quasi-isomorphic to a direct sum

$$\bigoplus_i \Lambda/p^{r_i}\Lambda \oplus \bigoplus_j \Lambda/f_j\Lambda$$

where each f_j is a **Weierstrass polynomial**, namely a polynomial in X with leading coefficient 1, and all other coefficients congruent to $0 \bmod p$. Such a module will be said to be of **Jacobian type** if there are no factors of type $\Lambda/p^r\Lambda$.

Theorem 2.9. (**Ferrero-Washington**) *The module $\mathcal{A}$ and the co-Stickelberger module are of Jacobian type.*

The proof in [**Fe-W**] relies on p-adic measure theoretic considerations, and we do not go into it here. If f is a Weierstrass polynomial, then $\Lambda/f\Lambda$ is free of dimension $\deg f$ over $\mathbf{Z}_p$. For a module of Jacobian type, we define the **characteristic polynomial** to be

$$\prod f_j.$$

It is the characteristic polynomial of $\gamma - 1$ acting on the $\mathbf{Q}_p$-vector space obtained by tensoring with $\mathbf{Q}_p$.

Theorem 2.10. (**Mazur-Wiles**) *For any odd character $\chi \neq \omega$, $\chi \neq \bar{\omega}$, the modules $\mathcal{A}(\chi)$ and $M(\chi)$ have the same characteristic polynomials. Equivalently, the $\mathbf{Q}_p$-vector spaces $\mathbf{Q}_p\mathcal{A}(\chi)$ and $\mathbf{Q}_p M(\chi)$ have isomorphic semi-simplifications as Λ-modules.*

A generator of the ideal formed with the characteristic polynomial is determined only up to a unit in Λ. Another generator of independent interest arises in the theory and illuminates Theorem 2.10, namely a power series g_χ having the following property. Let ψ be a character of $1+p\mathbf{Z}_p$, and let n be the smallest positive integer such that $1 + p^n\mathbf{Z}_p$ is contained in the kernel of ψ. Such p^n is called the conductor of ψ. One can define generalized Bernoulli numbers $B_{1,\chi\psi}$. Identifying $\mathcal{G}_\infty$ with $1 + p\mathbf{Z}_p$ by using the topological generator γ, we may view ψ as a character on $\mathcal{G}_\infty$. Then $\psi(\gamma)$ is a primitive p^{n-1}-th root of unity. There exists a unique power series $g_\chi = g_{\chi,\gamma}$ (of Kubota-Leopoldt) such that for all ψ, we have

$$g_\chi(\psi(\gamma) - 1) = -B_{1,\bar{\chi}\bar{\psi}}.$$

The Ferrero-Washington theorem is equivalent with the property that the coefficients of g_χ are not all divisible by p, and hence that

$$g_\chi(X) = c_0 + c_1X + \cdots + c_\lambda X^\lambda + \text{higher terms},$$

where $c_\nu \equiv 0 \bmod p$ for $\nu < \lambda$, and c_λ is a p-adic unit. The Weierstrass preparation theorem states that g_χ differs from a Weierstrass polynomial by a unit in $\mathbf{Z}_p[[X]]$. The following theorem is a direct consequence of the definitions and p-adic interpolation, belonging to the basic theory of p-adic L-functions due to Iwasawa.

Theorem 2.11. *The characteristic polynomial of $\gamma - 1$ on the co-Stickelberger module $M(\chi)$ is the Weierstrass polynomial of the Kubota-Leopoldt power series $g_{\chi,\gamma}$.*

Although the Mazur-Wiles theorem does not completely elucidate the module structure of $\mathcal{A}$, or at the first level of $\mathrm{Cl}^-(K_0)^{(p)}$, it is sufficient to imply consequences for the orders of these groups. We shall state such a consequence in a later section in connection with the algebraic-geometric considerations entering in its proof.

The Iwasawa-Leopoldt conjecture would be more precise than the Mazur-Wiles theorem for the classical cyclotomic tower that we have considered. It is related to the simplicity of the roots of the characteristic polynomials involved.

On the other hand, Mazur-Wiles treat more general ground fields than the rationals, namely any abelian field (a subfield of a cyclotomic field); and thereby they deal with a more general character decomposition than that of the group $\mathrm{Gal}(K_0/\mathbf{Q})$. For these more general ground fields, the analogue of the Iwasawa-Leopoldt conjecture is definitely not always true. It is still a problem, even in the most classical tower over $\mathbf{Q}$, to determine the extent to which the modules $M(\chi)$ and $\mathcal{A}(\chi)$ differ, for instance: is $\mathcal{A}(\chi)$ quasi-cyclic, or equivalently are $M(\chi)$ and $\mathcal{A}(\chi)$ quasi-isomorphic? To what extent are the characteristic roots simple?

One should not miss the importance of having these more general ground fields and characters, and I want to add a few words about that. Let F be an abelian extension of the rationals, contained in some cyclotomic field. For each prime number p, the Galois group $\mathrm{Gal}(\mathbf{Q}(\boldsymbol{\mu}_{p^\infty})/\mathbf{Q})$ has a finite torsion subgroup, whose fixed field Z_p is a $\mathbf{Z}_p$-extension of $\mathbf{Q}$, that is

$$\mathrm{Gal}(Z_p/\mathbf{Q}) \approx \mathbf{Z}_p.$$

Let FZ_p be the composite. Then FZ_p is called the **cyclotomic $\mathbf{Z}_p$-extension** of F. Let γ_p be a topological generator of $\mathrm{Gal}(FZ_p/F)$. For odd character χ with conductor divisible at most by the first power of p, one may form the projective limits $\mathcal{A}_{F,p}(\chi)$ and $M_{F,p}(\chi)$ as before using the appropriate generalized Bernoulli numbers. The Mazur-Wiles theorem asserts that these two Λ_p-modules have the same characteristic polynomial $f_{p,\chi}$ for $\gamma_p - 1$.

For almost all p we can make a canonical choice of γ_p. Indeed, if p is odd, then $\mathrm{Gal}(Z_p/\mathbf{Q})$ is generated by σ_{1+p}, and if $p = 2$, then by σ_{1+4}. For almost all p, F is disjoint from Z_p, so σ_{1+p} may be viewed as a generator of $\mathrm{Gal}(FZ_p/F)$. For this choice of generator one may then ask more refined questions concerning the coefficients and the roots of the characteristic polynomial, for a given character χ, or after taking a product over χ for the characteristic polynomial f_p of $\sigma_{1+p} - 1$ on $\mathbf{Q}_p M_{F,p}$. One may also ask questions concerning the behavior of f_p for varying p, for instance: are the degrees bounded as a function of p; how do they vary with p; what is the nature of the roots; what is the distribution of p for which the roots are simple; etc. Some of these questions are now being thought about by those active in the field, but nothing is known at present. This leads into the consideration of extensions of type

$$FZ_{p_1} \cdots Z_{p_t}$$

for a finite set of primes $p_1, \ldots, p_t$, and in general passing to the limit. Such extensions have recently been considered by Greenberg and Friedman following Iwasawa.

Finally, I should also emphasize that one can take much more general ground fields than abelian fields, and one can define a Stickelberger element and ideal via the zeta function instead of doing it ad hoc as we did here with Bernoulli polynomials. Cf. §8, where this will be done in a different context. Similar questions then arise for arbitrary $\mathbf{Z}_p$-extensions: we are faced with (at least) two modules, the projective limit of the ideal class group (p-primary part) and the co-Stickelberger module, so that the investigation of their relation can be posed as a problem in this generality. Cf. [**Co 1**].

§3. Modular curves

Let $\mathfrak{H}$ be the upper half plane, that is the set of complex numbers $\tau = x + iy$ with $y > 0$. Let $\Gamma(1) = SL_2(\mathbf{Z})$ be the **modular group**, that is the group of matrices

$$\gamma = \begin{pmatrix} a & b \\ c & d \end{pmatrix}$$

with integer coefficients, determinant 1. Then $\Gamma(1)$ operates on $\mathfrak{H}$ by

$$\tau \mapsto \frac{a\tau + b}{c\tau + d} = \gamma(\tau),$$

and ± 1 operates trivially, so we get a faithful representation of $\Gamma(1)/\pm 1$. The coset space $\Gamma(1)\setminus\mathfrak{H}$ has a representative fundamental domain which has the well-known shape pictured below.

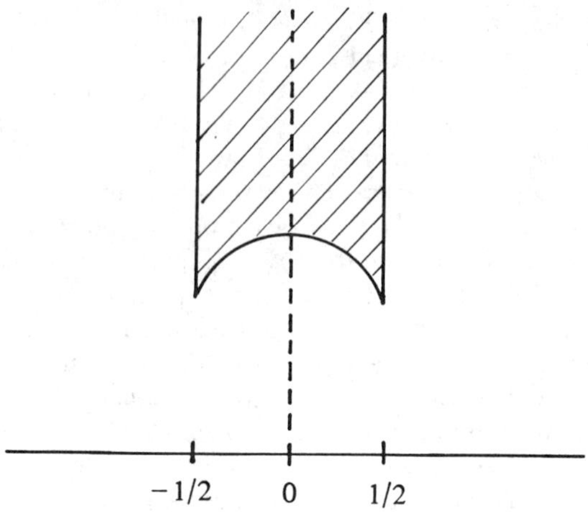

There is a classical function, holomorphic on $\mathfrak{H}$ and invariant under $\Gamma(1)$, called the j-function, which gives a complex analytic isomorphism

$$j : \Gamma(1)\setminus\mathfrak{H} \to \mathbf{P}^1_{\mathbf{C}} - \{\infty\}$$

with the affine line (projective one-dimensional space from which infinity is deleted). If one takes $q = e^{2\pi i\tau}$ as a local uniformizing parameter at infinity, then one can compactify $\Gamma(1)\setminus\mathfrak{H}$ by adjoining one point at infinity, thus obtaining a compact Riemann surface namely $\mathbf{P}^1_{\mathbf{C}}$. In terms of q, the function j has a Laurent expansion

$$j = \frac{1}{q} + 744 + 196884q + \text{higher terms.}$$

One can characterize j analytically by stating that $j(i) = 1728$ and $j(e^{2\pi i/3}) = 0$, while $j(\infty) = \infty$. This is rather ad hoc. A better way to conceive of j is in terms of isomorphism classes of complex toruses as follows.

Let $\Lambda = [\omega_1, \omega_2]$ be a lattice in $\mathbf{C}$, with basis ω_1, ω_2 over the integers. This means that Λ is the abelian group generated by ω_1, ω_2, and that these two elements are linearly independent over the real numbers. In addition, we shall always suppose that ω_1/ω_2 lies in the upper half plane, so we put $\omega_1/\omega_2 = \tau$. Then the invariance of j under $SL_2(\mathbf{Z})$ shows that the value $j(\tau)$ is independent of the choice of basis as above, and in addition, is the same if we replace $[\omega_1, \omega_2]$ by $[c\omega_1, c\omega_2]$ for any complex number $c \neq 0$. Thus we may define

$$j(\Lambda) = j(\tau),$$

and we have $j(c\Lambda) = j(\Lambda)$. But $\mathbf{C}/\Lambda$ is a complex torus of dimension 1, and the above arguments show that j is the single invariant for isomorphism classes of such toruses. The value 1728 is selected for usefulness in arithmetic applications. One can give many analytic expressions for j, arising from the theory of elliptic functions. For instance, associated with the lattice are the two invariants

$$g_2(\Lambda) = 60 \sum \omega^{-4} \quad \text{and} \quad g_3(\Lambda) = 140 \sum \omega^{-6},$$

where the sums are taken for $\omega \in \Lambda$, $\omega \neq 0$. Then

$$j = 1728 g_2^3/(g_2^3 - 27g_3^2).$$

Now let Γ be a subgroup of $\Gamma(1)$, of finite index. Then $\Gamma \setminus \mathfrak{H}$ is a finite (possibly ramified) covering of $\Gamma(1) \setminus \mathfrak{H}$. We shall be specifically interested in some very special subgroups Γ, which we now describe. Let N be a positive integer. We define:

$$\Gamma(N) = \text{subgroup of elements } \gamma \equiv \begin{pmatrix} 1 & 0 \\ 0 & 1 \end{pmatrix} \bmod N;$$

$$\Gamma_1(N) = \text{subgroup of elements } \gamma \equiv \begin{pmatrix} 1 & b \\ 0 & 1 \end{pmatrix} \bmod N \text{ with arbitrary } b;$$

$$\Gamma_0(N) = \text{subgroup of elements } \gamma \equiv \begin{pmatrix} a & b \\ 0 & d \end{pmatrix} \bmod N \text{ with arbitrary } a, b.$$

Since $\det \gamma = 1$ and a, b, c, d are integers, we must have

$$d \equiv a^{-1} \bmod N,$$

and a, d are prime to N. It is then easily seen that

$$\Gamma_0(N)/\Gamma_1(N) \approx (\mathbf{Z}/N\mathbf{Z})^*.$$

In fact we get two exact sequences:

$$1 \to \Gamma(N) \to \Gamma(1) \to SL_2(\mathbf{Z}/N\mathbf{Z}) \to 1$$

where the right hand map is reduction mod N; and

$$1 \to \Gamma_1(N) \to \Gamma_0(N) \to (\mathbf{Z}/N\mathbf{Z})^* \to 1$$

where the right hand map is the projection on $a \bmod N$.

If $\Gamma_1 \subset \Gamma_2$ then we have a covering (possibly ramified):

$$\Gamma_1 \setminus \mathfrak{H} \to \Gamma_2 \setminus \mathfrak{H}.$$

For any Γ, the projective embedding j of $\Gamma(1) \setminus \mathfrak{H}$ can be lifted to a projective embedding

$$\Gamma \setminus \mathfrak{H} \xrightarrow{f_\Gamma} \text{some projective space}$$

such that the image of f_Γ is an affine curve, denoted by $Y(\Gamma)_{\mathbf{C}}$. The coordinate functions of f_Γ can be chosen in many ways, suited for different applications, and exhibiting different properties of this affine curve, which is called a modular curve. The discussion of such functions becomes technical, and will be omitted, except that in §6 we discuss one possible generator for projective coordinates. Cf. Klein [**Kl**] and [**Ku-L 1**].

When Γ is one of the three groups defined above, then the corresponding affine curve is denoted by

$$Y(N), \quad Y_1(N), \quad Y_0(N) \text{ respectively,}$$

with the subscript $\mathbf{C}$ when we refer to the Riemann surface of its complex points. Thus we have a commutative diagram of coverings:

$$\begin{array}{ccc} \Gamma_1(N)\backslash\mathfrak{H} & \to & Y_1(N)_{\mathbf{C}} \\ \downarrow & & \downarrow \\ \Gamma_0(N)\backslash\mathfrak{H} & \to & Y_0(N)_{\mathbf{C}} \end{array}$$

To slide into the algebraic terminology, we shall speak more systematically of elliptic curves rather than complex toruses $\mathbf{C}/\Lambda$. The curves $Y_1(N)$ and $Y_0(N)$ have an interpretation as parametrizing certain isomorphism classes of objects, similar to $Y(1)$ (that is $\Gamma(1)\backslash\mathfrak{H}$) parametrizing isomorphism classes of elliptic curves, as follows:

$Y_0(N)$ parametrizes isomorphism classes of pairs (A, Z), where A is an elliptic curve (complex torus) and Z is a cyclic subgroup of order N.

$Y_1(N)$ parametrizes isomorphism classes of pairs (A, P), where A is an elliptic curve and P is a point of order exactly N.

In terms of the analytic objects, if $A = \mathbf{C}/[\tau, 1]$, we take P to be the point represented by $1/N$, and we take Z to be the subgroup generated by P. The invariance under the groups $\Gamma_0(N)$ and $\Gamma_1(N)$ gives the bijection between $\Gamma_0(N)\backslash\mathfrak{H}$ resp. $\Gamma_1(N)\backslash\mathfrak{H}$ and isomorphism classes of pairs as stated above.

However, using the algebraic language and formulation has one advantage: it can be used in an arithmetic context, because relative to a suitable choice of coordinatization, the curves $Y_1(N)$ and $Y_0(N)$ are defined over the rational numbers.

In terms of the above representation, the covering map $Y_1(N) \to Y_0(N)$ associates to each pair (A, P) the pair (A, Z) where Z is the cyclic group generated by P.

Now let us look at the points at infinity.

The affine curve $Y(\Gamma)$ can be compactified, and the points in the inverse image of ∞ on the j-line are called the **points at infinity**, or the **cusps**. The projective curve consisting of $Y(\Gamma)$ and the points at infinity is denoted by $X(\Gamma)$. Thus we have

$$X(\Gamma) = Y(\Gamma) \cup X^\infty(\Gamma)$$

where $X^\infty(\Gamma)$ is the set of cusps.

The cusps have a simple model as follows. Let

$$\mathfrak{H}^* = \mathfrak{H} \cup \mathbf{Q} \cup \{\infty\}.$$

Then $SL_2(\mathbf{Z})$ operates on $\mathbf{Q} \cup \{\infty\}$. One can give a topology and complex analytic structure to $\mathfrak{H}^*$ such that $\Gamma \backslash \mathfrak{H}^*$ is a compact Riemann surface. A typical neighborhood of a rational number r is a disc in the upper half plane tangent to r; and a typical neighborhood of ∞ is the part of the upper half plane lying above a horizontal line, as shown in the figure.

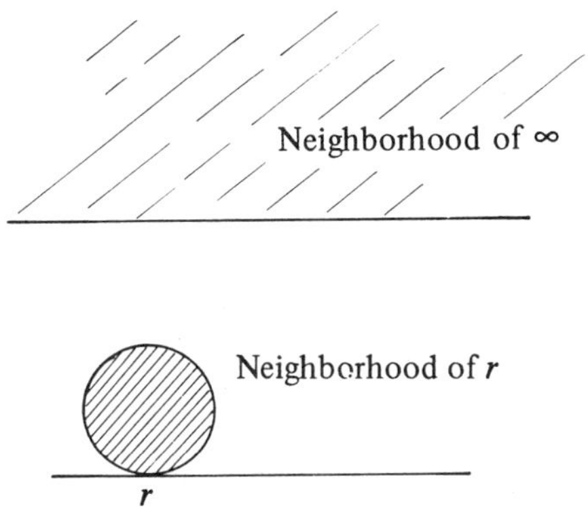

Then $X^\infty(\Gamma)$ is the set of equivalence classes of $\mathbf{Q} \cup \{\infty\}$ with respect to the action of Γ.

Again, when Γ is one of the three special subgroups defined above, the corresponding projective curve is denoted by

$$X(N), \quad X_1(N), \quad X_0(N) \text{ respectively.}$$

The (ramified) covering $Y_1(N) \to Y_0(N)$ extends to a (ramified) covering

$$X_1(N) \to X_0(N).$$

If, as before, we let $G \approx (\mathbf{Z}/N\mathbf{Z})^*$, then this covering has a group of covering transformations $G/\pm 1$, under the association

$$a \mapsto \gamma_a$$

where γ_a is any element of $\Gamma_0(N)$ satisfying

$$\gamma_a \equiv \begin{pmatrix} a & 0 \\ 0 & a^{-1} \end{pmatrix} \bmod N.$$

Now take $N = p$ *prime* ≥ 3. It turns out first that there are precisely two cusps on $X_0(N)$ lying above $j = \infty$. The degree of the covering is given by

$$[X_1(N) : X_0(N)] = \frac{N-1}{2},$$

and it is easily seen that the covering is unramified over the cusps. Therefore, if we denote by Q_0 and Q_1 the two cusps of $X_0(N)$, then there are precisely $(N-1)/2$ distinct points on $X_1(N)$ lying above each of these two points. We now let:

$$S = \text{set of } (N-1)/2 \text{ points on } X_1(N) \text{ lying above } Q_0.$$

We take this as a particularly interesting set of points at infinity, according to the general situations mentioned in the introductory remarks. More precisely, Q_0 and the set of points in S are chosen to be the rational cusps (rational points in $\mathbf{Q}$).

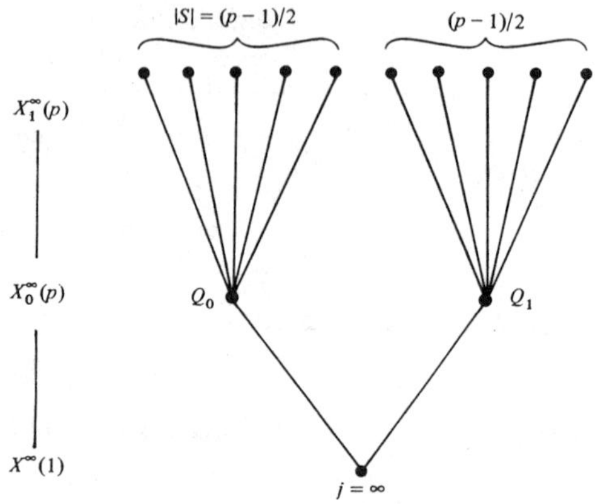

Just as in the case of ideal classes, we seek all relations in the cuspidal divisor class groups $\mathcal{C}^\infty(\Gamma) = \text{Pic}^\infty(X(\Gamma))$ for the Γ introduced above. The finiteness of this group was originally proved by Manin-Drinfeld [**Dr**], but the investigation of its structure in more explicit form was begun and carried out in the Kubert-Lang series of papers. Cf. also [**Ku-L 1**]. Here we shall limit ourselves to describing some results from that series which give a striking analogy with the cyclotomic case, and lead into the deeper connections established subsequently by Wiles and Mazur-Wiles. These results pertain to the group $\mathcal{C}_1^S(N)$ for N prime, where $\mathcal{C}_1^S(N) = \text{Pic}^S(X_1(N))$.

We are dealing again with the group ring $\mathbf{Z}[G]$, or rather $\mathbf{Z}[G/\pm 1]$. Let

$$\mathbf{B}_2(x) = x^2 - x + \frac{1}{6}$$

be the **second Bernoulli polynomial**. Note that $\mathbf{B}_2(1-x) = \mathbf{B}_2(x)$ for $0 \leq x \leq 1$. We let

$$\theta^{(2)} = N \sum_{a \in \mathbf{Z}(N)^*/\pm 1} \frac{1}{2} \mathbf{B}_2\left(\left\langle \frac{a}{N} \right\rangle\right) \sigma_a^{-1}$$

be the **Stickelberger element of order** 2. Let:

$I^{(2)}$ = ideal of $\mathbf{Z}[G/\pm 1]$ generated by the elements $\sigma_c - c^2$ with c prime to $6N$.

$R = \mathbf{Z}[G/\pm 1]$ = the group ring.

R_0 = ideal of elements in R of degree 0, and $I_0 = I \cap R_0$.

We let the **Stickelberger ideal** $S^{(2)}$ be the ideal

$$S^{(2)} = R\theta^{(2)} \cap R_0 = I_0^{(2)}\theta^{(2)},$$

and we then get the **co-Stickelberger module** $R_0/S^{(2)}$. For the rest of this section we omit the superscripts (2) since we deal only with them. Otherwise, we have to have a notation to distinguish the present situation from that dealing with the first Bernoulli polynomials.

Theorem 3.1. (i) *There is a natural isomorphism*

$$R_0/S \approx C_1^S(N).$$

(ii) *We have the class number formula*

$$(R_0 : S) = N \prod_{\chi \neq 1} \pm \frac{1}{2} B_{2,\chi}$$

where

$$B_{2,\chi} = N \sum_a \mathbf{B}_2\left(\left\langle \frac{a}{N} \right\rangle\right)\chi(a).$$

Here χ ranges over the characters of $(\mathbf{Z}/N\mathbf{Z})^*/\pm 1$, or what is the same, the even characters of $(\mathbf{Z}/N\mathbf{Z})^*$.

When N is a prime power, the situation is somewhat more complicated, since the Stickelberger ideal needs to be refined, cf. the [**Ku-L**] series; and when N is composite, Jing Yu [**Yu**] has applied the Sinnott methods to get corresponding structure theorems for $C_1^S(N)$ and $C^\infty(N)$.

It is at the moment a problem to give similar geometric interpretations for the co-Stickelberger module formed with the higher degree Bernoulli polynomials, defined by the generating series

$$\frac{te^{tX}}{e^t - 1} = \sum \mathbf{B}_k(X)\frac{t^k}{k!}.$$

Cf. Kubert-Lang, as in [**L 1**], Chapter 2 and Bergelson [**Be**] for purely algebraic index computations of Stickelberger ideals involving such polynomials.

The proof of Theorem 3.1 is done by a complete characterization of the group of functions (modulo constants) which have zeros and poles only at the cusps, or in the set S as stated above. Such functions can be constructed explicitly in terms of "modular forms", but such a discussion gets more technical and we wish to proceed with our general survey rather than go into these more elaborate constructions.

In any case, what was Conjecture 2 in the cyclotomic theory is a theorem in the context of the cuspidal divisor class group on the modular curve. The analogue of the theorem on eigenspces, namely the analogue of Theorem 2.7, is then true not only for the group ring modulo the Stickelberger ideal, but for the cuspidal group itself. We suppose that $p \geq 5$.

Theorem 3.2. *Let χ be an even character of G and let*

$$M = \mathbf{Z}_p[G]_0/S_p^{(2)} \quad \textit{where } S_p^{(2)} = \mathbf{Z}_p S^{(2)}.$$

Then

$$C_1^S(p)^{(p)} \approx M,$$

and we have the following eigenspace descriptions.

(i) *If χ is trivial, or $\chi = \omega^2$ then $M(\chi) = 0$.*
(ii) *If $\chi \neq 1$ and $\chi \neq \omega^2$ then*

$$M(\chi) \approx \mathbf{Z}_p/B_{2,\bar{\chi}}\mathbf{Z}_p,$$

so this group is cyclic of order $p^{n(\chi)}$, where $n(\chi) = \operatorname{ord}_p B_{2,\bar{\chi}}$.

§4. The Wiles and Mazur-Wiles connection

So far, we have described two analogous theories. The possibility that the geometric theory would affect the cyclotomic theory was immediately apparent (cf. [**L 1**], p. 53), but it was Wiles who first showed precisely how this connection would come about, following work of Ribet. The situation is now going to get more complicated, and we have to expand very considerably the range of notions which intervene.

We are concerned with the p-primary component of the group of ideal classes in $\mathbf{Q}(\boldsymbol{\mu}_p)$, say; by class field theory this corresponds to an abelian unramified extension. Can one obtain this extension "geometrically", by means of certain finite groups on appropriately selected Jacobians of curves as in the introduction? Indeed, let us go back to Theorem 2.10, which implies a conjectural existence of a certain abelian unramified extension of $K = \mathbf{Q}(\boldsymbol{\mu}_p)$, of order p^m where

$$m = m(\chi) = \operatorname{ord}_p B_{1,\bar{\chi}},$$

for $\chi \neq \omega$. Ribet [**Ri**] proved that if p divides $B_{1,\bar{\chi}}$, then there exists a cyclic unramified abelian extension of degree p, of the appropriate character, so that p divides $|C^{-(p)}(\chi)|$.

He introduced the theory of modular functions, and especially the curve $X_1(p)$, via a fundamental theorem of Shimura [**Sh 1**], Theorem 7.14. Ribet showed that by selecting an appropriate finite subgroup of torsion points in a quotient of the Jacobian $J_1(p)$ of $X_1(p)$, he could generate a cyclic extension of degree p,

$$\mathbf{Q}(\boldsymbol{\mu}_p)(\mathfrak{g}),$$

by adjoining the coordinates of the points in this group. However, more has now been proved:

Theorem 4.1. *Suppose that χ is an odd character, and $\chi \neq \omega, \chi \neq \bar{\omega}$. Then the order of $C^{-(p)}(\chi)$ is exactly $p^{m(\chi)}$.*

With the additional hypothesis that $C^{-(p)}(\chi)$ is cyclic, this theorem was proved by Wiles [**Wi**]. By an extension of the methods, it was proved as stated by Mazur-Wiles [**Ma-W**]. These methods link the algebraic geometry of the modular curves in §2 with the arithmetic of the cyclotomic fields, and are based extensively on deep, original and far reaching results of Mazur concerning the arithmetic properties of these modular curves and their Jacobians, via his theory of the "Eisenstein ideal" in [**Ma 1**]. I am going to try to explain how this comes about, following the introduction and first section of Wiles' paper [**Wi**].

Note that if χ is an odd character $\neq \omega$, $\neq \bar{\omega}$, then $\chi\omega$ is an even character $\neq 1$, ω^2. Let Z be the cyclic cuspidal group

$$Z = \mathcal{C}_1^S(p)(\chi\omega)^{(p)} \approx \mathbf{Z}_p/B_{2,\overline{\chi\omega}}\mathbf{Z}_p.$$

In other words, Z is the p-primary part of the cuspidal group on $X_1(p)$, with support in the set of cusps S described in §2, and forming the $\chi\omega$-eigenspace. The algebra of endomorphisms of $J_1(p)$ contains a subalgebra $\mathbf{T}$, called the **Hecke algebra**, which we shall describe very briefly below. For purposes of this section, the Hecke algebra $\mathbf{T}$ is assumed to have coefficients in $\mathbf{Z}_p$.

Theorem 4.2. *Let χ be an odd character $\neq \omega$, $\neq \bar{\omega}$. Let $\mathbf{I}_{\chi\omega} = \mathbf{I}$ be the ideal in $\mathbf{T}$ annihilating the cyclic group Z above, and let $\mathfrak{g}$ be the finite group of zeros of $\mathbf{I}$ in $J_1(p)$, namely the group of points $x \in J_1(p)$ such that $\mathbf{I}x = 0$. Let $p^n = |Z|$ be the order of Z. Then*

$$\mathbf{Q}(\boldsymbol{\mu}_{p^n})(\mathfrak{g})$$

is an unramified abelian extension of $\mathbf{Q}(\boldsymbol{\mu}_{p^n})$, of degree p^n.

This theorem describes how to construct certain abelian unramified extensions by means of torsion points on the Jacobian of the modular curve. Even though Z is a cuspidal group, $\mathfrak{g}$ is not. Note that Theorem 4.2 constructs an abelian extension of $\mathbf{Q}(\boldsymbol{\mu}_{p^n})$ but not of $\mathbf{Q}(\boldsymbol{\mu}_p)$, which is what we wanted in the first place. Thus it is still a complicated matter to "descend" the above construction back to $\mathbf{Q}(\boldsymbol{\mu}_p)$. The matter is sufficiently complicated that I shall limit myself to describe in general terms what is done in the papers of Wiles and Mazur. This involves the following steps.

(i) For arbitrary "levels", that is for arbitrary curves $X_1(p^\nu)$ and appropriate cuspidal divisor class groups, construct abelian unramified extensions of cyclotomic fields $\mathbf{Q}(\boldsymbol{\mu}_{p^n})$ with suitable, and arbitrarily large prime powers p^n, in a way similar to the construction of Theorem 4.2.

(ii) Express the limiting result (injective limit or projective limit) in the category of continuous modules over the Galois group $\mathrm{Gal}(\mathbf{Q}(\boldsymbol{\mu}_{p^\infty})/\mathbf{Q}(\boldsymbol{\mu}_p))$; or alternatively of modules over the "Iwasawa algebra", equal to the projective limit of the group rings formed at finite levels, namely

$$\Lambda = \lim \mathbf{Z}_p[\mathcal{G}_n]$$

where $\mathcal{G}_n = \mathrm{Gal}(\mathbf{Q}(\boldsymbol{\mu}_{p^{n+1}})/\mathbf{Q}(\boldsymbol{\mu}_p))$; and do this for the eigenspaces relative to the characters of $\mathrm{Gal}(\mathbf{Q}(\boldsymbol{\mu}_p)/\mathbf{Q})$, as in §2.

(iii) Elaborate the algebraic theory of the "twist" which allows one to shift from eigencomponents of even characters, arising from the algebraic geometry, to the odd characters arising from the ideal-class-group theory; and follow this twist by taking Galois invariants to recover the structure (as far as possible) of the ideal class group of $\mathbf{Q}(\boldsymbol{\mu}_p)$. See [**L 1**], Chapter 2, §7; and [**Ma-W**], as well as the end of [**Wi**], p. 33 where Wiles goes from "p^n" to "p^m".

For these lectures, I preferred to emphasize the connection between the algebraic number theory and the algebraic geometry; and to give the flavor of the geometric construction used to obtain unramified abelian extensions of cyclotomic fields, rather than to enter into the relatively heavy algebra needed to describe more precisely the above three steps, except for the last part of §2.

We finish this section with a few words concerning the Hecke algebra. It is generated by endomorphisms denoted T_l (for $l \neq p$), $U_p, \langle a \rangle$ (for a prime to p), and w_ς. Their descriptions can be given in various contexts, especially as correspondences on the modular curve $X_1(p)$, whose points are viewed as pairs (E, P) where E is an elliptic curve, and P is a point of order p; or in analytic terms, as function of the variable τ in the upper half plane. To keep this discussion brief, we use the first context.

T_l is the correspondence

$$(E, P) \mapsto \sum_B (E/B, (P+B)/B)$$

which to each pair (E, P) associates the formal sum shown above, taken over all subgroups B of order l in E.

U_p is the correspondence

$$(E, P) \mapsto \sum_B (E/B, (P+B)/B)$$

where the sum is now taken over the subgroups of order p but unequal to the subgroup generated by P itself.

$\langle a \rangle$, for an integer in $(\mathbf{Z}/p\mathbf{Z})^*$, is the correspondence

$$(E, P) \mapsto (E, aP).$$

Finally, to describe w_ς, one needs to know that there is a canonical pairing

$$e_p : E_p \times E_p \to \boldsymbol{\mu}_p,$$

which makes E_p self dual, and is alternating. Fix a primitive p-th root of unity ς. To each point P of order p, we let P' be a point (well determined in $E/(P)$ by the alternating property) such that

$$e_p(P, P') = \varsigma.$$

Then w_ς is the correspondence such that

$$(E, P) \mapsto (E/(P), P').$$

Mazur [**Ma 1**] had studied extensively this algebra in relation to the cuspidal points on $X_0(p)$. By extending this study to $X_1(p)$, Wiles was led to his theorems, subsequently completed and extended by Mazur-Wiles.

For the convenience of the reader, I also give explicitly generators for the ideal $\mathbf{I}_{\chi\omega}$ (the Eisenstein ideal, better called the $(S, \chi\omega)$-**cuspidal ideal.**)

Theorem 4.3. *Let χ be an odd character $\neq \omega$, $\neq \bar{\omega}$. Then the cuspidal ideal $\mathbf{I}_{\chi\omega}$ of Theorem 4.2 is generated by the elements*

$$T_l - (1 + l\langle l \rangle), \quad U_p - 1, \quad \langle a \rangle - \chi\omega(a), \quad p^n$$

where p^n is the exact power of p dividing $B_{2,\overline{\chi\omega}}$.

Remark. The above generators for the cuspidal ideal correspond to the choice of rational cusps which we have made, and is therefore a slight variation of the choice made in [**Wi**] relative to the other set of cusps lying above Q_1.

The situation is obviously deep and complicated. I want to emphasize that the techniques and ideas used to solve certain concrete classical problems bring out the full panoply of commutative algebra-algebraic geometry over rings of Grothendieck and his school, put to work in concrete contexts (especially of modular curves). In particular, to show that extensions obtained by adjoining coordinates of certain torsion points are unramified, one needs the general theory of commutative group schemes of Oort-Tate and Raynaud [**O-T**], [**Ray**]; the general theory of modular schemes over $\mathbf{Z}$, as in Deligne-Rapoport [**De-Rap**]; all based extensively on assorted EGA and SGA to the tune of several thousand pages, cf. for instance the bibliography at the end of Wiles' paper listing four of these volumes. Not only that, but Mazur-Wiles also use the Langlands Antwerp paper [**Lgds**] based on representation theory-trace formula techniques to be able to handle the ramification for $J_1(p^\nu)$ for arbitrarily large powers p^ν. However, the applicability of these last techniques is itself based on a good understanding of the algebraic geometry of the various moduli schemes involved. The appeal to representation theory can be replaced by a direct appeal to the work of Katz-Mazur [**Ka-M**] on these moduli schemes.

My listing of the above items is not meant to discourage anyone from reading the Mazur-Wiles papers. I merely did not want to hide what was involved. However, it should also be understood that few people have a complete grasp of all the elements put together by Mazur-Wiles. Some people may have the ability to take much for granted; to make just the right selection of items culled from more extensive works, and be comfortable with this selection; and to develop their intuition from carefully worked out special cases, for instance $X_0(37)$ and $X_1(37)$. To each his own.

§5. Geometric class field theory

It is possible to give a geometric context for some problems of class field theory. I shall select here a special case which can be easily formulated, and I follow a joint paper with Katz [**Ka-L**]. (For varieties over finite fields, cf. [**L 4**] and [**L 5**].) Although [**Ka-L**] deals with varieties in general, I limit myself to curves here for simplicity of language. (In higher dimensions, the Jacobian must be replaced by the Albanese variety or its dual.)

Let X be a projective non-singular curve defined over a number field k. Unramified coverings of X are usually defined over an algebraic extension of k. One may ask for those which are defined over k, and which, furthermore, are abelian over k: the elements of the group of covering transformations are defined over k also. The following theorem is an analogue of the finiteness of the class number.

Theorem 5.1. *Let x_0 be a rational point of X in k. There exists a maximal abelian unramified covering*

$$\pi : X' \to X,$$

also defined over k, such that x_0 splits completely in the covering, that is $\pi^{-1}(x_0)$ consists of d distinct k-rational points, where $d = [X' : X]$ is the degree of the covering.

Note that implicit in the statement of Theorem 5.1 is the fact that the maximal abelian unramified covering in which x_0 splits completely is actually finite. The proof is done by reducing mod good primes, to the case of curves over finite fields, where the situation is known by more classical class field theory.

We can generalize the notion of covering to include constant field extensions of k, which are then regarded as "unramified" over k. Then from **[Ka-L]** we have:

Theorem 5.2. *The maximal abelian unramified covering of $k(X)$ is the composite of the maximal abelian extension k^{ab} of k and the geometric covering whose existence is asserted in Theorem* 5.1.

This geometric covering is in fact obtained by pull-back from the Jacobian: there exists a corresponding covering $J' \to J$ of the Jacobian of X, defined over k, such that X' is the pull-back of J':

$$\begin{array}{ccc} X' & \to & J' \\ \downarrow & & \downarrow \\ X & \to & J. \end{array}$$

In the canonical map of X in J, the point x_0 is assumed to map on the origin of J which splits completely in J'. Thus the problem of determining unramified abelian extension of X is reduced to the same problem over the Jacobian. The group of covering transformations $\mathrm{Aut}(J'/J)$ can be identified with a group of translations by rational points of J' in k; that is an element $T \in \mathrm{Aut}(J'/J)$ acts as T_a for some $a \in J'_{\mathrm{tor}}(k)$, where

$$T_a(y) = y + a.$$

Then J can be viewed as the quotient of J' by a finite subgroup $\mathfrak{g} \subset J'(k)$ of k-rational points of J'.

By an elementary duality, the group of rational points on a covering J' of J correspond to a certain group on J itself as follows. Let $x \in J_N$ be a point of order N on J, rational over some finite extension of k. Let k^{a} denote the algebraic closure of k (the field of all algebraic numbers), and $\mathrm{Gal}(k^{\mathrm{a}}/k) = G_k$ the Galois group of k^{a} over k. We shall say that x is a **$\boldsymbol{\mu}$-point** if the cyclic group generated by x is G_k-isomorphic to the group $\boldsymbol{\mu}_N$ of all N-th roots of unity. In particular, x is rational over $k(\boldsymbol{\mu}_N)$. A finite subgroup of J is called a **$\boldsymbol{\mu}$-group** if all its points are $\boldsymbol{\mu}$-points. Then it can be shown by duality that finite k-rational subgroups of points on a covering J' of J as above are in bijection with $\boldsymbol{\mu}$-groups on J.

Theorem 5.3. *The maximal $\boldsymbol{\mu}$-subgroup is finite.*

This statement is equivalent with the finiteness of the maximal geometric unramified abelian covering of J (or X) over k, and follows from Theorem 5.1. This led to the conjecture that not only is the maximal $\boldsymbol{\mu}$-subgroup finite, but so is the group of torsion points of J in the maximal cyclotomic extension of k. This conjecture was proved in [**Ka-L**] in the case of complex multiplication, and was extended to the general case by Ribet (see the appendix to [**Ka-L**]). Since Theorem 5.3 can thus be proved ab ovo, it provides an alternative proof for the finiteness of the maximal covering in Theorem 5.1.

For each curve X it is then interesting to determine precisely the nature of this maximal $\boldsymbol{\mu}$-type group, especially for the modular curves which have proved so important in other contexts. Let us return to the ramified covering

$$X_1(p) \to X_0(p)$$

discussed in §3. It is cyclic of degree $(p-1)/2$. Let

$$n = \text{numerator of } \frac{p-1}{12}.$$

Let G be the group of covering transformations. Then G has a unique factor group of order n, which corresponds to an intermediate covering denoted by $X_2(p)$, so $X_1(p) \to X_2(p)$ is cyclic of degree $(p-1)/2n$, and the covering

$$\pi : X_2(p) \to X_0(p),$$

which is cyclic of degree n, is called the **Shimura covering**. Mazur had determined the maximal $\boldsymbol{\mu}$-subgroup of $J_0(p)$ in [**Ma 1**]. In the general framework of [**Ka-L**], this was interpreted as an explicit determination for $X_0(p)$ of the notion arising in Theorem 5.1:

Theorem 5.4. *The Shimura covering is the maximal abelian unramified covering of $X_0(p)$, defined over $\mathbf{Q}$, in which the rational cusp at infinity splits completely.*

On the other hand, one may consider models for $X_0(p)$ (or an arbitrary curve X) over the ring of integers of k. For the modular scheme as above, it is remarked in [**Ka-L**] that there is total ramification over one of the components at p, and therefore that scheme-theoretically, there is no unramified covering as in Theorem 5.4. One may also ask for the decomposition laws of "primes"– in this case, maximal ideals locally, in rings of dimension 2–in such coverings. Such decomposition laws would amount to higher dimensional "reciprocity laws", cf. [**L 4**] and [**L 5**], where the problem is raised in a general context of schemes over the integers (not yet called by that name). For instance in [**L 5**] I pointed out the surjectivity of the reciprocity law mapping from 0-cycles of the base space into the Galois group. Recently, Spencer Bloch [**B**] has made a great advance in this direction, by discovering how to formulate such laws for curves having everywhere non-degenerate reduction, following work of Kato [**Kato**] and Parshin [**Pa**]. Bloch formulates the decomposition laws in terms of the K-group of the base space. It remains to extend his results to more general cases (the modular curves have degenerate reduction at p, for instance), and to make his results explicit in the case of the modular curves. In any case, the maximal μ-subgroup can then be interpreted as a "class number", the number of classes being those in a suitable group of K-theory, in the unramified case of Bloch, and ultimately in the ramified case with conductor.

§6. Modular units

In this section, we describe briefly how one constructs the units in the modular function field, i.e. the meromorphic functions on $X(N)$ which have zeros and poles only at the points at infinity. We then show how one obtains units in the rings of integers in abelian extensions of imaginary quadratic fields.

(a) The function field

First consider an arbitrary lattice Λ in the complex plane $\mathbf{C}$. Then $\mathbf{C}/\Lambda$ is a complex torus, of complex dimension 1, real dimension 2. It admits a projective embedding in the projective plane, by means of the classical Weierstrass $\wp$-function and its derivative, which provide affine coordinates:

$$z \mapsto (\wp(z), \wp'(z)),$$

where

$$\wp(z) = \frac{1}{z^2} + \sum_{\omega \neq 0} \left[\frac{1}{(z-\omega)^2} - \frac{1}{\omega^2}\right],$$

$$\wp'(z) = \sum_{\omega} \frac{-2}{(z-\omega)^3}$$

and the sums are taken for $\omega \in \Lambda$, or $\omega \neq 0$ as indicated. The series for the $\wp$-function is concocted so as to make $\wp$ periodic with respect to Λ, and to insure convergence. Lattice points go to the point at infinity. If we put $x = \wp(z)$ and $y = \wp'(z)$, then

$$y^2 = 4x^3 - g_2 x - g_3,$$

where g_2, g_3 were already mentioned in §3.

We shall describe another type of projective coordinate.

First, we want an entire function which has a zero of order 1 at each lattice point and no other zero. The simplest normalization is that of the **sigma function** of Weierstrass,

$$\sigma(z,\Lambda) = z \prod_{\omega \neq 0} \left(1 - \frac{z}{\omega}\right) \exp\left(\frac{z}{\omega} + \frac{1}{2}\left(\frac{z}{\omega}\right)^2\right).$$

We then define the **Weierstrass zeta function** to be the logarithmic derivative

$$\zeta(z,\Lambda) = \sigma'/\sigma(z,\Lambda).$$

This function has the property that for any period $\omega \in \Lambda$ we have

$$\zeta(z+\omega,\Lambda) - \zeta(z,\Lambda) = \eta(\omega,\Lambda),$$

where $\eta(z,\Lambda)$ is a function which is **R**-linear in the variable z, and homogeneous of degree —1 in the pair (z,Λ), that is

$$\eta(cz, c\Lambda) = c^{-1}\eta(z,\Lambda) \quad \text{for } c \in \mathbf{C},\ c \neq 0.$$

We then define the **Klein form**

$$\mathfrak{k}(z,\Lambda) = e^{-\eta(z,\Lambda)z/2}\sigma(z,\Lambda),$$

which is homogeneous of degree 1 in (z,Λ). This function will be used to parametrize algebraic numbers analogous to the cyclotomic numbers $e^{2\pi i z} - 1$ when z ranges over division values of the lattice: rational multiples of $2\pi i$ in the cyclotomic case; rational multiples of elements in Λ in the elliptic case. Note that $\mathfrak{k}(z,\Lambda)$ is not holomorphic in z.

We recall that

$$\Delta = g_2^3 - 27g_3^2.$$

There is a natural 12th root, which is also denoted by $\eta(\tau)$, as a function of a variable in the upper half plane, and is called the **Dedekind** eta function (not to be confused with the Weierstrass eta function introduced above). The Dedekind eta function has a q-expansion given by the product

$$\eta(\tau) = q^{1/24} \prod_{n=1}^{\infty} (1 - q^n)$$

where as before, $q = e^{2\pi i \tau}$. Let ω_1, ω_2 be a basis for the lattice Λ over the integers **Z**. Then we can write any complex number z as a linear combination

$$z = a_1\omega_1 + a_2\omega_2,$$

with real numbers a_1, a_2. We let $a = (a_1, a_2)$ be a pair in $\mathbf{Q}^2$, and we suppose that $a \notin \mathbf{Z}^2$. We define the function

$$g_a(\tau) = \eta(\tau)^2 \mathfrak{k}_a(\tau),$$

where

$$\mathfrak{k}_a(\tau) = \mathfrak{k}(a_1\tau + a_2, [\tau, 1])$$

and $[\tau, 1]$ is the lattice generated by τ and 1 over the integers. Then g_a is a meromorphic function on $X(2N^2)$ if

$$a \in \frac{1}{N}\mathbf{Z}^2 \quad \text{but } a \notin \mathbf{Z}^2.$$

In [**Ku-L 1**] all the modular units are shown to be suitable power products of these functions g_a, which have zeros and poles only at the points at infinity, and one can describe precisely which power products have exact level N, that is, are invariant under the group of automorphisms $\Gamma(N)$ (or $\Gamma_1(N)$ for that matter). This precise determination leads to the structure of the cuspidal divisor class group $C_1^S(N)$ mentioned in §3.

(b) Fields of complex multiplication

Ever since the last century, it has been realized that abelian extensions of an imaginary quadratic field behave in a manner analogous to that of cyclotomic fields over the rationals, and basic theorems of class field theory were known in this case before they were known over arbitrary number fields. The reason for this was that one has explicit algebraic and analytic parametrizations for such abelian extensions. I shall summarize some aspects and concentrate on a selection of problems fitting into the general pattern considered in these lectures.

Suppose that K is an imaginary quadratic field, say $K = \mathbf{Q}(\sqrt{-D})$ where D is a positive integer, assumed square free, and let $\mathfrak{o} = \mathfrak{o}_K$ be the ring of algebraic integers in K. For example, if $K = \mathbf{Q}(\sqrt{-1})$, then $\mathfrak{o} = \mathbf{Z}[i]$. If $K = \mathbf{Q}(\sqrt{-3})$, then $\mathfrak{o} = \mathbf{Z}[\mu_3]$. Take $\Lambda = \mathfrak{o}$. Then $\mathbf{C}/\mathfrak{o}$ admits as endomorphisms multiplication by elements of $\mathfrak{o}$. Indeed, if $\alpha \in \mathfrak{o}$, then $\alpha\mathfrak{o} \subset \mathfrak{o}$ so we get an induced map

$$\alpha : \mathbf{C}/\mathfrak{o} \rightarrow \mathbf{C}/\mathfrak{o}$$

sending $z \mapsto \alpha z (\text{mod } \mathfrak{o})$. More generally, if $\mathfrak{a}$ is any (non zero) ideal of $\mathfrak{o}$, $\mathbf{C}/\mathfrak{a}$ again admits endomorphisms as above. We shall refer to the case when Λ is equal to such an ideal as the **complex multiplication** case.

In the two special cases of $\mathbf{Q}(i)$ and $\mathbf{Q}(\sqrt{-3})$ mentioned above, the ring of endomorphisms is generated by automorphisms of the curve. For instance,

$$y^2 = x^3 + ax$$

admits as automorphism $(x, y) \mapsto (-x, iy)$; while the curve

$$y^2 = x^3 + b$$

admits as automorphism $(x, y) \mapsto (\varsigma x, y)$ where $\varsigma^3 = 1$.

The values $\wp(z)$ and $\wp'(z)$ generate an abelian extension of $K(g_2, g_3)$ when z is a point of finite period with respect to $\mathfrak{a}$, or equivalently stated, when z is a torsion point in $\mathbf{C}/\mathfrak{a}$. However, the coordinates given by the Weierstrass functions are defined by means of "additive" expressions (just look at the series) and are inappropriate for the construction of units in abelian extensions of K. For that purpose, one has to use the Klein forms.

We must summarize some facts about ramified abelian extensions of K. Let $\mathfrak{f}$ be an ideal $\neq \mathfrak{o}$. There is a unique abelian extension $K(\mathfrak{f})$ of K, characterized by the following property. A prime ideal $\mathfrak{p}$ of K, $\mathfrak{p} \nmid \mathfrak{f}$, splits completely in $K(\mathfrak{f})$ if and only if $\mathfrak{p}$ is principal, $\mathfrak{p} = (\alpha)$, and a generator α (a priori determined only up to a unit of $\mathfrak{o}_K$) exists which satisfies $\alpha \equiv 1 \bmod \mathfrak{f}$. This extension $K(\mathfrak{f})$ is called the **ray class field**, of **conductor** $\mathfrak{f}$. "Split completely" in the present case means that if $\mathfrak{O}$ is the ring of algebraic integers in $K(\mathfrak{f})$, then $\mathfrak{p}\mathfrak{O}$ decomposes into a product of $[K(\mathfrak{f}) : K]$ distinct prime ideals:

$$\mathfrak{p}\mathfrak{O} = \mathfrak{P}_1 \cdots \mathfrak{P}_r \quad \text{where } r = [K(\mathfrak{f}) : K].$$

We are going to define a group of units in $K(\mathfrak{f})$ analogous to the cyclotomic units in $\mathbf{Q}(\boldsymbol{\mu}_N)$.

Although their origin lies in Kronecker's limit formula, the construction of such units was revived and extended in Siegel's Tata Institute Notes [**Sie**], and in the subsequent paper of Ramachandra [**Ra**]. A key step forward was made by Robert [**Ro 1**], who saw how to enlarge the known group by taking roots, so that the index of this unit group in the group of all units in the prime power case became essentially equal to the class number of $K(\mathfrak{f})$, (up to relatively large powers of 2 and 3) in analogy to the formula relating h^+ and E/E_{cyc} in the cyclotomic case. Here we shall follow the procedure developed by Kersey, Kubert and myself. See [**Ku-L**], Chapters 12 and 13, as well as [**Ke 2**].

Let $\mathbf{I}$ be the free abelian group on the ideals, so an element $\mathbf{a}$ of $\mathbf{I}$ can be written as a formal linear combination

$$\mathbf{a} = \sum n(\mathfrak{a})\mathfrak{a}$$

with integer coefficients $n(\mathfrak{a})$, almost all of which are equal to 0. Given $\mathfrak{f}$, and such $\mathbf{a}$ so that if $n(\mathfrak{a}) \neq 0$ then $\mathfrak{f}$ does not divide $\mathfrak{a}$, we define

$$\mathfrak{k}_{\mathfrak{f}}(\mathbf{a}) = \prod_{\mathfrak{a}} \mathfrak{k}(1, \mathfrak{f}\mathfrak{a}^{-1})^{n(\mathfrak{a})}.$$

This is an analogue to the cyclotomic numbers $\prod(\mathbf{e}(a/f) - 1)^{n(a)}$, but in order to obtain the "right" algebraic numbers, we have to impose some conditions. Let the degree of $\mathbf{a}$ be

$$\deg \mathbf{a} = \sum n(\mathfrak{a}).$$

We shall always require that $\deg \mathbf{a} \equiv 0 \bmod w$, where w is the number of roots of unity in K. Let $N(\mathfrak{f})$ be the smallest positive integer lying in $\mathfrak{f}$. The most important condition that we require is that

$$\sum n(\mathfrak{a})\mathbf{N}\mathfrak{a} \equiv 0 \bmod N(\mathfrak{f}).$$

As usual in number theory, $\mathbf{N}\mathfrak{a}$ denotes the absolute norm of $\mathfrak{a}$, namely the index $(\mathfrak{o} : \mathfrak{a})$. In addition to that, we require additional technical conditions to take care of problems with the prime 2 which we do not make explicit here. The abelian group generated by elements $\mathbf{a}$ satisfying these conditions is denoted by $\mathbf{I}_w(\mathfrak{f})$.

In addition to a formal linear combination $\mathbf{a}$ as above, we require other similar elements as follows. Let:

$S =$ a finite set of primes $\mathfrak{p}$ relatively prime to 6,

$R(\mathfrak{p}) =$ a set of representatives for $(\mathfrak{o}/\mathfrak{p})^*/\mathfrak{o}^*$;

for each $\mathfrak{p} \in S$ let $\mathbf{a}_\mathfrak{p} \in \mathbf{I}_w(\mathfrak{p})$.

If $\alpha \in \mathfrak{o}$ and $\mathbf{a}_\mathfrak{p} = \sum n_\mathfrak{p}(\mathfrak{a})\mathfrak{a}$, we denote $\alpha\mathbf{a}_\mathfrak{p} = \sum n_\mathfrak{p}(\mathfrak{a})(\alpha\mathfrak{a})$. Let

$$\beta = \mathfrak{k}_\mathfrak{f}(\mathbf{a}) \prod_{\mathfrak{p}\in S} \prod_{\alpha \in R(\mathfrak{p})} \mathfrak{k}_\mathfrak{p}(\alpha\mathbf{a}_\mathfrak{p}).$$

If we assume that the total degree in this product is 0, that is

$$\deg \mathbf{a} + \sum_{\mathfrak{p}\in S} \frac{\mathbf{N}\mathfrak{p} - 1}{w} \deg \mathbf{a}_\mathfrak{p} = 0,$$

then the element β above is an algebraic number, which can be shown to lie in $K(\mathfrak{f})$. The group generated by all such numbers is called the group of **modular numbers** in $K(\mathfrak{f})$, and denoted by $\mathfrak{K}(\mathfrak{f})$. The group generated by the roots of unity $\boldsymbol{\mu}_{K(\mathfrak{f})}$ and by the subgroup of units in $\mathfrak{K}(\mathfrak{f})$ is called the group of **modular units** in $\mathfrak{K}(\mathfrak{f})$. Thus

$$E_{\text{mod}}(K(\mathfrak{f})) = \boldsymbol{\mu}_{K(\mathfrak{f})}(\mathfrak{K}(\mathfrak{f}) \cap E),$$

where E denotes the group of all units.

If H denotes a subfield of $K(\mathfrak{f})$ containing K, then essentially one defines the group of modular units $E_{\text{mod}}(H)$ in H to be the group generated by $\boldsymbol{\mu}_H$ and by the norms down to $H \cap K(\mathfrak{g})$ of the modular units $E_{\text{mod}}(K(\mathfrak{g}))$, for all $\mathfrak{g} \mid \mathfrak{f}$.

The definition is somewhat elaborate and some aspects are fairly technical. The main point is that we define a rather large group of units in a canonical manner, expressed in terms of values of a single function $\mathfrak{k}(z, \Lambda)$, and containing all groups defined by other authors as mentioned above in a similar context. This latter inclusion property may be viewed as one possible reason for calling the group of modular units "large". Another is given more precisely by Kersey's index computation [**Ke 2**], the most precise result known today.

Theorem 6.1. $(E(K(\mathfrak{f})) : E_{\text{mod}}(K(\mathfrak{f}))) = \lambda h_{K(\mathfrak{f})}$ *where* λ *is a power of* 2 *depending on the number of prime factors of* $\mathfrak{f}$*;* $\lambda = 1$ *if* $\mathfrak{f}$ *is a prime power, relatively prime to* 6*, in particular if* $\mathfrak{f} = (1)$.

Kersey's proof is based in part on the same group theoretical considerations as Sinnott [Si] in the cyclotomic case, but the situation he faces is much more complicated for a variety of reasons: more complicated analytic functions to express units; more complicated structure of the Galois group; etc.

§7. The Coates-Wiles connection

Instead of starting from the diophantine problem posed by the higher degree Fermat equations, let us start over again now with the diophantine problem arising from cubic equations, say

$$y^2 = x^3 + ax + b,$$

where a, b are integers. We assume that the discriminant of the right hand side is not equal to 0. Then the set of complex points of this equation, together with one point at infinity, form a Riemann surface of genus 1, isomorphic to a complex torus. The curve is equal to its own Jacobian, taking as origin the point at infinity. Let A denote the curve, and $A(\mathbf{Q})$ its group of rational points. Classical questions concerning $A(\mathbf{Q})$ ask:

What is the structure of the group of torsion points $A(\mathbf{Q})_{\text{tor}}$?

Is there a rational point of infinite order? If so, what is the rank over $\mathbf{Z}$ of $A(\mathbf{Q})$ (which is finitely generated according to a celebrated theorem of Mordell)?

Again, I don't want to go into extensive terminology, and among an open ended choice of topics, I shall select one which establishes a relation between these diophantine questions and units in suitable number fields.

I shall also limit myself to even more special curves of type

$$y^2 = x^3 + b,$$

where b is an integer. We have seen in the last section that these curves admit μ_3 as a group of automorphisms, and belong to the "complex multiplication" category. Over the complex numbers, any two such curves become isomorphic, but over the rational numbers, they may exhibit completely different behavior. For instance, one curve may have rational points other than ∞, while another may not. We are interested to give a criterion when there is a rational point of infinite order, and we shall relate this question to the existence of certain (ramified) extensions of number fields.

Let $K = \mathbf{Q}(\sqrt{-3}) = \mathbf{Q}(\mu_3)$. Let p be a prime number relatively prime to $6b$. We shall suppose in the sequel that p splits completely in K, that is

$$p\mathfrak{o}_K = \mathfrak{p}\bar{\mathfrak{p}}, \qquad \mathfrak{p} \neq \bar{\mathfrak{p}}.$$

Since $\mathfrak{o}_K$ has unique factorization, there is a generator π for $\mathfrak{p}$, that is $\mathfrak{p} = \pi\mathfrak{o}$, and π is well-defined up to a root of unity. We have $p = \pi\bar{\pi}$. Furthermore, $K_{\mathfrak{p}} = \mathbf{Q}_p$, where $K_{\mathfrak{p}}$ is the $\mathfrak{p}$-adic completion of K.

For an element $\alpha \in \mathfrak{o}$ we let $[\alpha] = [\alpha]_A$ be the endomorphism of A induced by α. Let $\mathbf{Q}^{\mathrm{a}}$ denote the algebraic closure of $\mathbf{Q}$. For any field F, A_F or $A(F)$ denotes the group of points on A rational over F.

We assume that A has good reduction at $\mathfrak{p}$, and that the generator π is selected so that the reduction mod $\mathfrak{p}$ of $[\pi]_A$ is the Frobenius endomorphism $\mathrm{Fr}_{\mathfrak{p}}$. This means that for any point $(x, y) \in A(\mathbf{Q}^{\mathrm{a}})$ and any prime $\mathfrak{P}$ of $\mathbf{Q}^{\mathrm{a}}$ over $\mathfrak{p}$, we have

$$[\pi](x, y) \equiv (x^p, y^p) \bmod \mathfrak{P}.$$

Let A_π be the group of points t in $A(\mathbf{Q}^{\mathrm{a}})$ such that

$$[\pi]t = 0.$$

We could also write $A_\pi = A[\pi]$ or $A[\mathfrak{p}]$, to be the set of points t such that $[\alpha]t = 0$ for all $\alpha \in \mathfrak{p}$. Then A_π is a cyclic group of order p, and in fact, $K(A_\pi)$ is a cyclic extension of K, of degree $p - 1$ with Galois group $G \approx (\mathbf{Z}/p\mathbf{Z})^*$ in analogy with the cyclotomic theory. The action of G is determined by the representation on A_π, namely for $a \in \mathfrak{o}$,

$$\sigma_a t = [a]t,$$

analogous to the formula $\sigma_a \varsigma = \varsigma^a$ if $\varsigma \in \boldsymbol{\mu}_p$.

The theory of the extension $K(A_\pi)$ over K can be carried out in analogy with the p-adic theory of Kummer for $\mathbf{Q}(\boldsymbol{\mu}_p)$ over $\mathbf{Q}$. As usual, we have to look at eigenspaces. We let χ be the character such that for $\sigma \in G$ we have

$$\sigma t = \chi(\sigma)t, \quad \text{with } \chi(\sigma) \in (\mathbf{Z}/p\mathbf{Z})^*.$$

One may view χ as a character $\chi : \mathrm{Gal}(K^{\mathrm{a}}/K) \to (\mathbf{Z}/p\mathbf{Z})^*$, where K^{a} is the algebraic closure of K, and χ factors through $G = \mathrm{Gal}(K(A_\pi)/K)$.

Suppose, as in Coates-Wiles, that A has a rational point P of infinite order, and take P not in $[\pi]A(K)$. Then one may extract a π-th root of P, namely a point

$$Q \in \pi^{-1}(P)$$

such that $[\pi]Q = P$. This is analogous to extracting a p-th root in the multiplicative group when dealing with the cyclotomic theory. Then we have another example of double-decked extensions:

$$\begin{array}{c} K(A_\pi, \pi^{-1}(P)) \\ | \\ K(A_\pi) \\ | \\ K \end{array}$$

Let t be a generator for A_π over $\mathbf{Z}/p\mathbf{Z}$. Then we obtain a 2-dimensional representation of $\mathrm{Gal}(K^{\mathrm{a}}/K)$ arising from the action:

$$\sigma t = \chi(\sigma)t$$
$$\sigma Q = b(\sigma)t + Q$$

with $b(\sigma) \in \mathbf{Z}/p\mathbf{Z}$, thus giving the matrix representation

$$\sigma \mapsto \begin{pmatrix} \chi(\sigma) & b(\sigma) \\ 0 & 1 \end{pmatrix}.$$

Conversely, one may consider Galois extensions of K cyclic of degree p over $K(A_\pi)$ whose Galois group admits such a matrix representation with character χ, and ask if the existence of such extensions can arise only through the above construction with a point of infinite order. We shall say that p is **special** if $\pi + \bar{\pi} = 1$. Otherwise, we call p **non-special**.

Conjecture 7.1. *Assume that for infinitely many non-special primes p, there exists a Galois extension of K over $K(A_\pi)$ whose Galois group admits a representation in $GL_2(\mathbf{Z}/p\mathbf{Z})$, with matrices as above, and character χ. Then there exists a point P of infinite order in A_K, and all but a finite number of these extensions and representations are obtained as described above, by extracting a π-th root of some such P.*

The conjecture can be made somewhat more quantitative. For instance, extracting π-th roots of $A_K/[\pi]A_K$ gives rise to r independent p-extensions of $K(A_\pi)$, where r is the rank of A_K (for all but a finite number of primes p). We leave this aside, but we note that the conjecture essentially asserts that the existence of such extensions can only be explained as arising from a point of infinite order as described in the Coates-Wiles way, except in a finite number of cases. These "exceptional" cases are related to other more subtle invariants of the elliptic curve (something called the Tate-Shafarevitch group, which is beyond the level at which I wish to keep this exposition).

We turn to the local considerations of Coates-Wiles. Let:

$K_0 = K(A_\pi)$;

$F_0 =$ completion of K_0 at some prime ideal $\mathfrak{P}$ lying above $\mathfrak{p}$.

Then $F_0 = K_\mathfrak{p}(A_\pi) = \mathbf{Q}_p(A_\pi)$.

$U_0 =$ group of $\mathfrak{p}$-adic units in F_0.

$\mathcal{E}_0 =$ group of modular units in K_0.

$\overline{\mathcal{E}}_0 =$ closure of $\mathcal{E}_0$ in U_0.

Since the $(p-1)$th roots of unity are contained in $\mathbf{Z}_p$, and a fortiori in $\mathbf{Q}_p$, it follows that the cyclic extension F_0 of $\mathbf{Q}_p$, which is of degree $p-1$, is a Kummer extension, that is

$$F_0 = \mathbf{Q}_p(w_0) \quad \text{where } w_0^{p-1} \in \mathbf{Q}_p.$$

The lattice of fields is shown on the figure.

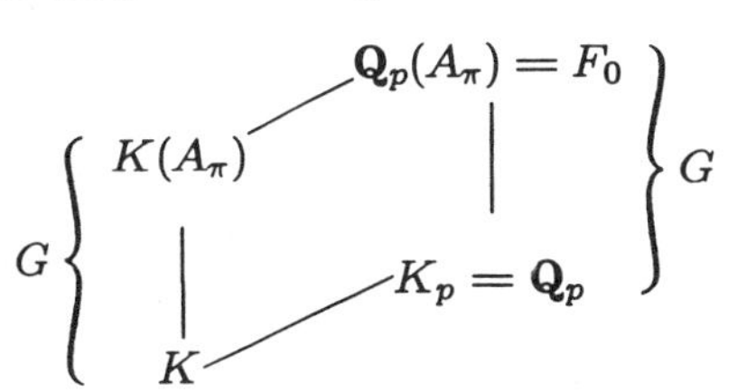

It might happen that $\mathbf{Q}_p(A_\pi) = \mathbf{Q}_p(\boldsymbol{\mu}_p)$. It is an easy technical matter to show that this happens if and only if $\pi + \bar{\pi} = 1$, in other words, π is special (provided $p > 5$). It can be shown that the set of special primes has density zero, so there are plenty of non-special primes. We assume p is non-special.

In dealing with the local extension over $\mathbf{Q}_p$, we can now view the character χ as the Kummer character of G, namely the character such that

$$\sigma w_0 = \chi(\sigma) w_0.$$

The double-decked extension can be viewed locally by extending the base field to $K_\mathfrak{p}$:

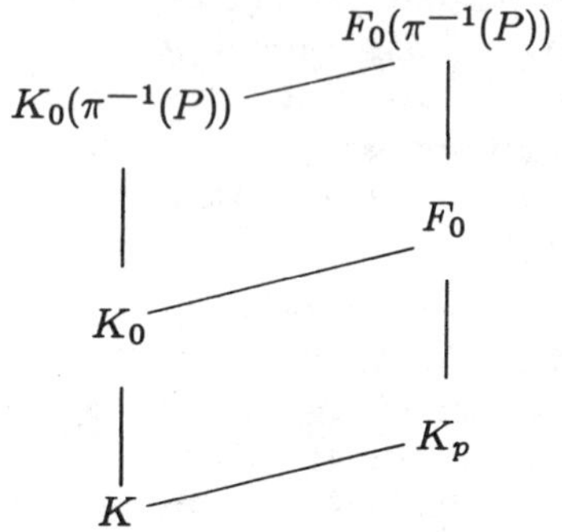

The extension $K_0(\pi^{-1}(P))$ of K_0 is a cyclic extension of degree p, closely related by class field theory to the factor group of local units modulo the closure of the global units.

If C is an abelian group on which G operates, we denote:

$C(p) = C/C^p$ (writing C multiplicatively);

$C(p, \chi) = (C/C^p)(\chi) = \chi$-eigenspace of $C(p)$.

Kummer had already studied the p-adic properties of the cyclotomic units, which in modern language amounts to the factor group of local units in $\mathbf{Q}(\boldsymbol{\mu}_p)$ by the closure of the cyclotomic units. A similar study can be made in the present context, but I shall limit myself to only one aspect. Motivated by the "Birch-Swinnerton-Dyer Conjecture", Coates-Wiles arrive at the following result.

Theorem 7.2. *Suppose that p is non-special and that there exists a rational point P of infinite order in $A(K)$. Then*

$$(U_0/\bar{\mathcal{E}}_0)(p, \chi) \neq 0.$$

The proof of that theorem is carried out by methods of class field theory, related to a p-adic "Kummer criterion", see [**Co-W**], [**Co 2**]. Robert [**Ro 2**] had also considered the analogue of Kummer's criterion in the context of elliptic curves.

Conjecture 7.3. *Conversely, assume that $(U_0/\bar{\mathcal{E}}_0)(p, \chi) \neq 0$ for infinitely many non-special primes. Then there exists a rational point on A of infinite order.*

This conjecture is a transformation of a special case of the Birch-Swinnerton Dyer conjecture. By class field theory, the hypothesis guarantees the existence of certain abelian extensions (subject to a certain eigenspace condition) of $K(A_\pi)$.

In any case, we see here once more how (possibly conjecturally) abelian extensions of number fields can be parametrized by extracting "roots", except that the roots occur on the group of an elliptic curve as well as the multiplicative group.

Of course, everything we have stated for K_0 can be extended by taking powers of π, and looking at the extensions

$$K_n = K(A[\pi^{n+1}]) \quad \text{and} \quad \Omega_n = K_n(\pi^{-(n+1)}P)$$

if P is a point of infinite order. Even to consider K_0 and Ω_0 as we have done above, it is necessary to pass to the limit in this tower of extensions to apply Iwasawa theory of $\mathbf{Z}_p$-extensions, which leads to the "right" statements concerning the quantitative description of the extensions one obtains by dividing not only torsion points but points of infinite order. This is a massive undertaking which goes beyond the scope of our discussion, but I want to emphasize that the formalism of the characteristic polynomial as in the appendix of §2, and the questions related to it, can be formulated in the present context also. For example, let C_n be the p-primary part of the ideal class group in K_n. We can form the same type of projective limit as in the cyclotomic $\mathbf{Z}_p$-extension. Is the projective limit of Jacobian type (analogue of the Ferrero-Washington theorem)? The answer is not known today.

§8. Stark units

Let us go back to cyclotomic fields, and to the cyclotomic units. Consider the real subfield $K = \mathbf{Q}(\mu_m)^+$ where m is an integer ≥ 3, odd or divisible by 4. We have one archimedean absolute value v_0 on $\mathbf{Q}$, the ordinary absolute value. The field K is totally real, meaning that any embedding of K into the complex numbers actually lies in the real numbers. We can find a cyclotomic number generating K, namely $K = \mathbf{Q}(\epsilon)$, where

$$\epsilon = (1 - \varsigma)(1 - \varsigma^{-1})$$

and $\varsigma = e^{2\pi i/m}$. The Galois group of K over $\mathbf{Q}$ is isomorphic to $(\mathbf{Z}/m\mathbf{Z})^*/\pm 1$, under the map $a \mapsto \sigma_a$.

We may define the **partial zeta function** relative to $K/\mathbf{Q}$:

$$\varsigma(\sigma_a, s) = \sum_{\substack{n=1 \\ n \equiv \pm a(m)}}^{\infty} n^{-s}.$$

Classically, we have $\varsigma(\sigma_a, 0) = 0$ and

$$\varsigma'(\sigma_a, 0) = -\frac{1}{2}\log|\sigma_a \epsilon|.$$

Of course, we could replace ϵ by ϵ^{-1} to get rid of the minus sign in that equation. Stark [**St IV**] interprets these relations as defining analytically a "unit" which becomes the generator of an abelian (ramified) extension of the rationals. Indeed, if m is composite, then ϵ is a unit, and if $m = p^\nu$ is a prime power, then ϵ is divisible only by the prime lying above p (p-unit).

Finally, Stark notes that $K(\epsilon^{1/2})$ is abelian over $\mathbf{Q}$. Indeed, if we write $K = K_m = \mathbf{Q}(\mu_m)$, then

$$K(\epsilon^{1/2}) = \begin{cases} K_{4m} & \text{if } m \text{ is odd,} \\ K_{2m} & \text{if } m \text{ is even.} \end{cases}$$

Stark has conjectured the existence of a similar situation, involving units, which can be used to generate "class fields" (ramified abelian extensions) of a number field. He has given proofs only in those cases when we know how to parametrize units as values of classical analytic functions of "exponential" type (parametrizing the exponential map on algebraic Lie groups), namely the cyclotomic case as above, and the case of modular units over imaginary quadratic fields, as mentioned in the last two sections. I shall briefly summarize the general situation envisioned by Stark. I base the exposition on lectures of Tate, and also on [**Ta 1**] and [**Ta 2**].

Let k be a number field and K an abelian extension with Galois group G. Denote by v an archimedean place of k. Then v denotes either an embedding of k in the real numbers, or in the complex numbers (up to complex conjugation). We let k_v denote the completion of k at v, which is isomorphic to either $\mathbf{R}$ or $\mathbf{C}$ as the case may be. Let v' be an archimedean place of K lying above v. Then the completion $K_{v'}$ is also equal to $\mathbf{R}$ or $\mathbf{C}$, and $K_{v'}$ is an extension of k_v of degree 1 or 2. If $K_{v'}$ is an extension of degree 2 of k_v, then the local Galois group G_v is cyclic of order 2, generated by complex conjugation. These are the archimedean analogues of the p-adic places which correspond to embeddings in the field $\mathbf{C}_p$, completion of the algebraic closure of $\mathbf{Q}_p$. Let $[K_{v'} : k_v] = n_v$ be the local degree. We define

$$||\alpha||_{v'} = |\alpha|_{v'}^{n_v}$$

where $|\alpha|_{v'}$ is the absolute value on K induced by the embedding v'.

As with prime ideals, we say that v splits completely in K if the number of distinct embeddings of K in $\mathbf{C}$ lying above v is equal to the degree $[K : k]$. Equivalently, we could say that for any extension v' of v to K, v' is real if and only if v is real. If v is real and v' is complex, then we say that v ramifies in K, or that v' is ramified over v.

Let R = set of primes of k ramified in K. Let

$$S \supset R \cup S^\infty(k)$$

be a finite set of places with at least two elements.

Let $\mathfrak{p}$ be a prime of k unramified in K, and let $\mathfrak{P}$ be a prime ideal of K lying above $\mathfrak{p}$. Let $G_{\mathfrak{P}}$ be the decomposition group of $\mathfrak{P}$, meaning the subgroups of elements $\sigma \in G$ such that $\sigma\mathfrak{P} = \mathfrak{P}$. Since we assumed that G is abelian, it follows that $G_{\mathfrak{P}}$ depends only on $\mathfrak{p}$, not on the choice of $\mathfrak{P}$, and $G_{\mathfrak{P}}$ is therefore denoted by $G_{\mathfrak{p}}$. Furthermore, $G_{\mathfrak{p}}$ acts on the residue class field $\mathfrak{o}_K/\mathfrak{P}$ over $\mathfrak{o}_k/\mathfrak{p}$, and is cyclic, with a canonical generator, the Frobenius element $\sigma_{\mathfrak{p}}$ such that

$$\sigma_{\mathfrak{p}}\, x \equiv x^{\mathbf{N}\mathfrak{p}} \bmod \mathfrak{P}.$$

The **partial zeta function** associated with a given $\sigma \in G$ is defined to be the partial sum

$$\zeta_S(\sigma, s) = \sum_{\substack{(\mathfrak{a},S)=1 \\ \sigma(\mathfrak{a})=\sigma}} \mathbf{N}\mathfrak{a}^{-s}\,.$$

Remark: If $S = \mathcal{R} \cup S^{\infty}(k)$, then we omit the subscript S, and we write $\zeta_S(\sigma, s) = \zeta(\sigma, s)$. *Let us assume*:

there is one archimedean place v_0 of k which splits completely in K.

Then basic analytic number theory shows that the partial zeta function vanishes of order ≥ 1 at $s = 0$. We are interested in the derivative of $\zeta_S(\sigma, s)$ at $s = 0$ following Stark, who observed that the formulas for this derivative at $s = 0$ are structurally more transparent than the analogous formulas for $\zeta_S(\sigma, s)$ at $s = 1$, considered more classically. (These formulas can be obtained from the functional equation of Hurwitz zeta functions.)

We shall say that we are in the **special case** if S has exactly two elements; then

$$\{S^{\infty}(k) - v_0\} \cup \mathcal{R}$$

has precisely one element (which may therefore be archimedean, or may be a single ramified prime) v_1, called **special**. By a **(v_0, S)-unit** we shall mean an element ϵ of K^* such that in non-special cases:

$$||\epsilon||_v = 1 \text{ for any place } v \text{ of } K, v \nmid v_0.$$

In the special case, then the above condition should hold for $v \nmid v_0, v \nmid v_1$, and then we require in addition that the values

$$||\epsilon||_v \text{ for } v \mid v_1$$

are equal to each other.

In particular, if we are not in the special case, then such ϵ are units; and in the special case, if the special element v_1 corresponds to a prime ideal, then such ϵ are $\mathfrak{p}$-adic units for all other primes $\mathfrak{p}$, i.e. they are what is usually called S-units. Note that the special case is rare, but that the classical cases $k = \mathbf{Q}$ or $k =$ imaginary quadratic field often give rise to special cases.

Conjecture 8.1. *Denote by v some extension of v_0 to K. Let $W = w_K$ be the number of roots of unity in K. Then there exists a (v_0, S)-unit ϵ in K, well-determined up to a root of unity in $\boldsymbol{\mu}_K$, such that*

(i) $\zeta'_S(\sigma, 0) = -\frac{1}{W} \log ||\sigma\epsilon||_v$ *for all* $\sigma \in G$.

(ii) $K(\epsilon^{1/W})$ *is abelian over* k.

In the special case when k is an imaginary quadratic field, the case of complex multiplication, Stark proved his conjecture by using the classical theory of L-series (Kronecker limit formula in this case), as well as the theory of modular units in a fairly precise form since we want ϵ itself, not just some power of ϵ, to fit the Stark formula giving the derivative of the partial zeta function at $s = 0$. At this point it becomes important to have as large a group of modular units as possible. Note that usually one expects to have $K = k(\epsilon)$, but one has to impose some conditions to insure this.

Despite attempts by Stark and by Shintani [**Sh**], even the case when k is a real quadratic field is still unknown. Stark's paper [**St IV**] contains numerical computations which confirm the conjecture strikingly in special fields, and some cubic extensions. Shintani [**Sh 2**] proves the conjecture in special, and non-trivial cases.

Gross [**Gro 1**] has formulated conjectures as above for p-adic L-functions. Also Tate pointed out that instead of taking an archimedean absolute value v_0, one could take a prime $\mathfrak{p}_0$ of k, also splitting completely in K, to obtain a new case for the complex zeta function, as follows [**Ta 2**].

Let $\mathfrak{p} \mid \mathfrak{p}_0$ in K. For simplicity, we assume that the set

$$S = \mathcal{R} \cup S^{\infty}(k) \cup \{\mathfrak{p}_0\}$$

has ≥ 3 elements, so $\mathcal{R} \cup S^{\infty}(k)$ has ≥ 2 elements. This assumption amounts to omitting the special case, which is so degenerate as to be of no interest here.

Conjecture 8.2. *Let $\mathfrak{p}$ be a prime of K lying above $\mathfrak{p}_0$. There exists a $\mathfrak{p}_0$-unit ϵ such that*

(i) $\zeta(\sigma, 0) = \frac{1}{W} \operatorname{ord}_{\mathfrak{p}} (\sigma\epsilon)$;

(ii) $K(\epsilon^{1/W})$ *is abelian over* k.

The absence of the derivative is due to a technical transformation as follows. Let

$$\zeta_{[\mathfrak{p}]}(\sigma, s) = (1 - \mathbf{N}\mathfrak{p}^{-s})\zeta(\sigma, s)$$

be the zeta function from which the $\mathfrak{p}$-Euler factor has been deleted. Then

$$\begin{aligned} \zeta'_{[\mathfrak{p}]}(\sigma, 0) &= (\log \mathbf{N}\mathfrak{p}_0)\zeta(\sigma, 0) \\ &= -\frac{1}{W}\log\|\sigma\epsilon\|_{\mathfrak{p}} \quad \text{(conjecturally)} \\ &= \frac{1}{W}\operatorname{ord}_{\mathfrak{p}}(\sigma\epsilon)(\log \mathbf{N}\mathfrak{p}), \end{aligned}$$

and $\mathbf{N}\mathfrak{p}_0 = \mathbf{N}\mathfrak{p}$ since we assumed that $\mathfrak{p}_0$ splits completely in K, so the formula for $\zeta(\sigma, 0)$ drops out.

Following Brumer, one defines the **Stickelberger element** associated with the extension K/k to be

$$\theta = \sum_{\sigma \in G} \zeta(\sigma, 0)\sigma^{-1}.$$

Siegel has shown that $\theta \in \mathbf{Q}[G]$, i.e. that $\varsigma(\sigma, 0)$ is a rational number. If there is more than one ramified prime, then one gets the formal factorization

$$(\epsilon) = \mathfrak{p}^{W\theta},$$

and if there is exactly one ramified prime, then a power of that prime may occur additonally in the factorization of ϵ.

As in the cyclotomic case, define the **integralizing ideal** I to be the ideal of $\mathbf{Z}[G]$ generated by W and elements of the form $\sigma_{\mathfrak{c}} - \mathbf{N}\mathfrak{c}$ for ideals $\mathfrak{c}$ prime to W and to the ramified primes. This is the annihilator in $\mathbf{Z}[G]$ of $\boldsymbol{\mu}(K)$. It is known by Deligne-Ribet [**De-R**] or Barsky-Cassou-Nogues [**CN**] that $I\theta \subset \mathbf{Z}[G]$. (This is a much more difficult result than in the cases we encountered previously.)

Conjecture 8.3. *The ideal $I\theta$ annihilates the class group* $\mathrm{Cl}(K)$. *If $x \in I$ and $\mathfrak{a}$ is an ideal of K, then*

$$\mathfrak{a}^{x\theta} = (\alpha_x)$$

where α_x satisfies $\|\alpha_x\|_v = 1$ for any archimedean absolute value v of K, and so α_x is well-defined up to a root of unity. Furthermore, if $\mathfrak{a}^{W\theta} = (\epsilon)$, then $K(\epsilon^{1/W})$ is abelian over k. Finally, if $(W\mathfrak{a}, \mathfrak{p}) = 1$, then

$$\mathfrak{a}^{(\sigma_{\mathfrak{p}} - \mathbf{N}\mathfrak{p})\theta} = (\alpha_{\mathfrak{p}}) \quad \textit{with } \alpha_{\mathfrak{p}} \equiv 1 \bmod \mathfrak{p}.$$

The annihilation of the ideal class group by the Stickelberger ideal $I\theta$ has been conjectured for some time by Brumer. The rest was formulated by Stark and Tate, who pointed out that Conjecture 8.2 implies Conjecture 8.3, which is the analogue of the Stickelberger theorem for cyclotomic fields. For related questions, see Coates-Sinnott [**Co-Si**] and [**Co 2**].

§9. Higher regulators: number fields

Stark [**St II**] has also formulated a non-abelian theory, with arbitrary representation of the Galois group G, starting from the Artin formalism of L-functions. This leads into higher dimensional regulators (determinants formed with the logarithms of units).

In this theory, we deal with two naturally defined $\mathbf{Q}$-vector spaces which become isomorphic over the complex numbers. The $\mathbf{Q}$-spaces give natural representations of a Galois group. After extending the scalars to $\mathbf{C}$, and choosing bases over $\mathbf{Q}$, the isomorphism between the spaces can be represented by a square matrix, whose determinant is well defined modulo a non-zero *rational number*, because changing bases over $\mathbf{Q}$ introduces only changes by rational matrices. A representation of G then leads to a determinant which conjecturally can be expressed as a transcendental number, times an algebraic number which transforms in a "functorial" way under the action of the Galois group. This algebraic number is the analogue of the Stark unit mentioned in §8, where the "determinant" was a one by one determinant.

The exposition in this section is based on lectures by Tate on Stark's conjectures, especially [**Ta 1**], [**Ta 2**].

Let K/k be a finite Galois extension of a number field, with $\mathrm{Gal}(K/k) = G$. Let V denote a $\mathbf{C}$-linear representation of G, or the finite dimensional vector space associated with it. Let:

$S^\infty(K) =$ set of archimedean places of K; elements v range over $S^\infty(K)$.

$U(K) =$ free abelian group on $S^\infty(K)$.

$U_0(K) =$ elements of $U(K)$ of degree 0 (the augmentation module).

$E(K) =$ group of units of $K = \mathfrak{o}_K^*$.

The groups $U(K), U_0(K), E(K)$ are G-modules. A theorem of Herbrand asserts that the units $E(K)$ contain a subgroup of finite index which is G-isomorphic to $U_0(K)$. Then the $\mathbf{Q}$-vector spaces are isomorphic:

$$\mathbf{Q} \otimes U_0(K) \approx \mathbf{Q} \otimes E(K).$$

To begin with some algebraic considerations, we first look at $\mathbf{C}U_0(K)$. We shall deal with the units afterward.

Let $U_0 = U_0(K)$, and let

$$\theta : \mathbf{C}U_0 \to \mathbf{C}U_0$$

be an automorphism (of $\mathbf{C}$-vector spaces). Then θ induces an endomorphism

$$\theta^* : \mathrm{Hom}_G(V^*, \mathbf{C}U_0) \to \mathrm{Hom}_G(V^*, \mathbf{C}U_0),$$

where V^* is the dual space of V. We define

$$\delta(V, \theta) = \det(\theta^*, \mathrm{Hom}_G(V^*, \mathbf{C}U_0)).$$

Then $V \mapsto \delta(V, \theta)$ satisfies the formalism of Artin, which we recall. Let H be a subgroup of G. Let Ind_G^H denote the induced representation, characterized by the formula

$$\mathrm{Hom}_G(\mathrm{Ind}_G^H(V), W) = \mathrm{Hom}_G(V, \mathrm{Res}_H^G(W)),$$

for any G-space W and H-space V, so induction is the adjoint of restriction. Let Inf denote inflation from G/H to G if H is normal. We have:

(1) $\delta(V_1 \bigoplus V_2, \theta) = \delta(V_1, \theta)\delta(V_2, \theta)$;

(2) $\delta(\mathrm{Ind}_G^H V, \theta) = \delta(V, \theta)$ if V is a representation of a subgroup H;

(3) $\delta(\mathrm{Inf}_G^{G/H} V, \theta) = \delta(V, \theta \mid (\mathbf{C}U_0)^H)$ if V is a representation of G/H;

(4) $\delta(V, \theta_1\theta_2) = \delta(V, \theta_1)\delta(V, \theta_2)$;

(5) $\delta(V, \theta)^\sigma = \delta(V^\sigma, \theta^\sigma)$ for any $\sigma \in \mathrm{Aut}(\mathbf{C})$.

The first four are easily proved. The fifth also once we make the following remarks. The number $\delta(V,\theta)$ depends only on the isomorphism class of V. An automorphism σ operates on V, say after a choice of basis so that σ operates on the coordinates of $\mathbf{C}^{\dim V}$. Thus V^σ is well defined up to an isomorphism. With this explanation, (5) is also easy.

Next, we need the regulator map

$$\lambda : \mathbf{C}E(K) \to \mathbf{C}U_0(K)$$

defined on units by

$$\lambda(\epsilon) = \sum_v \log \|\epsilon\|_v \cdot [v],$$

and extended by $\mathbf{C}$-linearity to a $\mathbf{C}$-isomorphism. If

$$\varphi : \mathbf{C}U_0 \to \mathbf{C}E(K)$$

is an isomorphism, then

$$\theta = \lambda \circ \varphi : \mathbf{C}U_0 \to \mathbf{C}U_0$$

is a possible automorphism.

If φ is the $\mathbf{C}$-linear extension of a G-embedding

$$U_0 \to E(K),$$

then one has the special value

$$\delta(1_k, \lambda \circ \varphi) = \pm(E(K) : \varphi(U_0(k)))R_k/w_k \tag{6}$$

where R_k is the regulator of k and w_k the number of roots of unity in k.

Finally, we need to recall the definition of the Artin L-function associated with the representation V of $G = \mathrm{Gal}(K/k)$. For each prime $\mathfrak{p}$ of k the conjugacy class of a Frobenius element $\sigma_\mathfrak{p}$ in G is well defined modulo the inertia group $I(\mathfrak{p})$, defined up to conjugacy. The **Artin L-function** is defined for $\mathrm{Re}(s) > 1$ by the product

$$L(s,V) = \prod_\mathfrak{p} \det(1 - \sigma_\mathfrak{p} \mathbf{N}\mathfrak{p}^{-s} \mid V^{I(\mathfrak{p})}),$$

where $V^{I(\mathfrak{p})}$ is the part of V fixed by $I(\mathfrak{p})$. We let $c(V)$ be the coefficient of the leading term at $s = 0$, that is

$$L(s,V) \sim c(V)s^{r(V)}, \quad \text{for } s \to 0,$$

where $r(V)$ is the order of the zero of $L(s,V)$ at $s = 0$.

Define

$$A(V,\varphi) = \frac{\delta(V, \lambda \circ \varphi)}{c(V)}.$$

Then $A(V,\varphi)$ depends only on the isomorphism class of the complex representation V, for fixed φ, and so may be written $A(\chi,\varphi)$. We may now state Stark's conjecture.

Conjecture 9.1. *For any $\sigma \in \mathrm{Aut}(\mathbf{C})$, we have*

$$A(V,\varphi)^\sigma = A(V^\sigma, \varphi^\sigma).$$

By (4) and (5), if the conjecture is true for some φ, then it is true for every φ.

Property (6) shows that the conjecture is true for trivial V. Stark in [St II] has shown that if the character of V is rational valued, then there is some positive integer m such that the conjecture is true for $m.V$, in other words,

$$(A(V,\varphi)^m)^\sigma = (A(V^\sigma, \varphi^\sigma))^m.$$

The proof is obtained by using the induction property, and the fact that any representation with rational character has an integral multiple which is a sum of induced representations of trivial characters on some subgroups of G. Then one uses the theorem for trivial representations, together with the Artin formalism (2) and (3). Tate has proved Conjecture 9.1 when the character of V is rational valued.

It is then possible to obtain units (conjecturally) in a manner similar to that of the preceding section, as follows.

Assume that V is irreducible and non-trivial, and that $r(V) = 1$, so the L-function vanishes of order 1 at the origin.
Then $\mathrm{Hom}_G(V, \mathbf{C}U_0)$ *is 1-dimensional, and there is an embedding*

$$V \to \mathbf{C}U_0$$

unique up to a scalar multiple.

Let φ satisfy Conjecture 9.1, *and assume in addition that φ is the* **C**-*linear extension of a G-embedding*

$$U_0 \to E(K).$$

If x is in the image of V under this embedding, then we have

$$\lambda \circ \varphi(x) = \delta(V, \lambda \circ \varphi)x.$$

Let F be the field generated over the rationals by the character values of G. Then $A(V,\varphi)$ lies in F.

By Frobenius reciprocity and the functional equation for the L-function, one sees that there is one archimedean place v of k such that

$$\dim V^{G_v} = 1, \text{ but } V^{G_{v'}} = 0 \quad \text{for } v' \in S^\infty(k), v' \neq v.$$

We denote by this same letter v an extension of the place to K. Let $\chi = \chi_V$ be the character of V. Then an element x can be taken to be

$$x = \sum_{\sigma \in G} \chi(\sigma^{-1})(\sigma[v] - [v]).$$

Let

$$\epsilon_{\sigma,\varphi} = \varphi(\sigma[v] - [v]).$$

If Y is any subset of G, let $\chi(Y) = \sum \chi(y)$, where the sum is taken for $y \in Y$.

Theorem 9.2. *Assume Conjecture 9.1. Suppose that V is irreducible, non-trivial; that $r(V) = 1$; and φ is the* **C**-*linear extension of a G-embedding $U_0 \to E(K)$. Then*

$$L'(0,V) = (A(V,\varphi)\chi(G_v))^{-1} \sum_{\sigma\in G} \chi(\sigma^{-1}) \log \|\epsilon_{\sigma,\varphi}\|.$$

The above theorem gives an expression in the non-abelian case analogous to that of the preceding section in the abelian case. Stark gives proofs of such a formula in special cases in [**St II**] and [**St B**]. The theorem is proved by juggling with the orthogonality relations of characters. It then gives rise to:

Theorem 9.3. *Assume in addition that V gives a faithful representation of G. Let*

$$c(\sigma) = \mathrm{Tr}_{F/\mathbf{Q}}\, \chi(\sigma)/A(V,\varphi),$$

and let m be a positive integer such that $mc(\sigma) \in \mathbf{Z}$ for all $\sigma \in G$. Let

$$\epsilon_m = \prod_{\sigma\in G} \epsilon_\sigma^{mc(\sigma^{-1})}.$$

Then K is the smallest Galois extension of k which contains ϵ_m.

These two statements summarize Stark's insight into the possibility of generating non-abelian extensions by units appearing in Artin L-functions derivatives at $s = 0$. Stark has proved special cases which could be reduced to the complex multiplication situation by induced characters. Chinburg (Thesis, Harvard, 1980) has made computations confirming the existence of the expected units to 13 decimal places in tetrahedral cases over the rationals.

§10. Higher regulators: curves (da capo)

We shall now describe results of Anderson [**An 1**], [**An 2**] giving a geometric context for the same formalism as in the preceding section.

Let X be a projective non-singular curve defined over a number field k. Then we have the De Rham cohomology group

$$H^1_{DR}(X,k)$$

of differential forms of second kind modulo exact forms. For every embedding $\tau : k \to \mathbf{C}$ of k into the complex numbers, the curve X^τ is defined over k^τ, and we have the corresponding space $H^1_{DR}(X^\tau, k^\tau)$, as well as $H^1_{DR}(X^\tau, \mathbf{C})$ obtained by extension of scalars from k^τ to $\mathbf{C}$. There is a natural isomorphism

$$\lambda = \lambda_\tau : H^1_{DR}(X^\tau, \mathbf{C}) \to H^1_{\mathrm{top}}(X^\tau, \mathbf{C})$$

which to each differential form ω associates the functional

$$\gamma \mapsto \int_\gamma \omega$$

for every cycle γ. This will play a role analogous to the regulator map of §9.

Now suppose G is a finite group of automorphisms of X, also defined over k. Then G gives rise to a finite covering

$$X \to X/G,$$

and G operates on $H^1_{DR}(X, k)$.

For each τ we let

$$\varphi_\tau : H^1_{\text{top}}(X^\tau, k^\tau) \to H^1_{DR}(X^\tau, k^\tau)$$

be an isomorphism of k^τ-vector spaces. Then $\lambda \circ \varphi_\tau$ is an automorphism of $H^1_{\text{top}}(X^\tau, \mathbf{C})$, after extending φ_τ by $\mathbf{C}$-linearity.

Let V be a representation of G over k, so V^τ is a representation of V over k^τ, whence over $\mathbf{C}$. Let θ_τ be any automorphism of $H^1_{\text{top}}(X^\tau, \mathbf{C})$. Then θ_τ induces an automorphism θ^*_τ of $\mathrm{Hom}_G(V^*, H^1_{\text{top}}(X^\tau, \mathbf{C}))$, where V^* is the dual space. Define

$$\delta(X^\tau, V^\tau, \theta_\tau) = \det(\theta^*_\tau : \mathrm{Hom}_G(V^*, H^1_{\text{top}}(X^\tau, \mathbf{C}))).$$

This symbol satisfies the Artin formalism, as listed in the preceding section. Cf. [**L 6**]. In practice, we shall take $\theta_\tau = \lambda \circ \varphi_\tau$.

On the other hand, Anderson defines a constant $c(X, V)$ with the gamma function as follows. Let $x \in X$ be a point of X in a fixed algebraic closure of k, and let $G(x)$ be the isotropy group of x in G; let $e(x)$ be the order of $G(x)$, namely the ramification index of x. Let $T(x)$ be the tangent space of x, as one-dimensional $G(x)$-module. Define

$$c(X, V) = (2\pi i)^{d(X,V)} \prod_x \prod_{i=1}^{e(x)} \Gamma\left(\frac{i}{e(x)}\right)^{d(i,x,V)}$$

where

$$d(X, V) = (g(X/G) - 1) \dim V + \dim V^G$$

and

$$d(i, x, V) = \dim \mathrm{Hom}_{G(x)}(T(x)^{\otimes i}, V).$$

Also define

$$A(X^\tau, V^\tau, \varphi_\tau) = \frac{\delta(X^\tau, V^\tau, \lambda \circ \varphi_\tau)}{c(X, V)}.$$

Then Anderson proves:

Theorem 10.1. *For any automorphism $\sigma \in \mathrm{Aut}(\mathbf{C})$, we have*

$$A(X^\tau, V^\tau, \varphi_\tau)^\sigma = A(X^{\tau\sigma}, V^{\tau\sigma}, \varphi_{\tau\sigma})\alpha(\sigma, \tau),$$

with some element $\alpha(\sigma, \tau) \in k^{\tau\sigma}$. [Notation: $k^{\tau\sigma} = (k^\tau)^\sigma$.] Furthermore, if $w = w(k)$ is the number of roots of unity in k, then

$$A(X^\tau, V^\tau, \varphi_\tau)^w \in k^\tau.$$

This last assertion shows that the number $A(X^\tau, V^\tau, \varphi_\tau)$ in fact generates a Kummer extension of k^τ, and by appropriate choice of φ_τ, Anderson can make the factor $\alpha(\sigma, \tau)$ equal to a root of unity. For this and further aspects of $\alpha(\sigma, \tau)$, cf. [**An 2**].

The expressions for the "periods" in Anderson's theory form a geometric counterpart to the arithmetic conjectures of Stark. Ultimately, these two extremes will be covered by the general theory of schemes of finite type over $\mathbf{Z}$.

Say R is a finitely generated subring of a finitely generated extension of the rational numbers. Assume that R is regular (all its local rings are regular) for simplicity. The prime ideals of R constitute the spectrum $X = \operatorname{spec} R$, and the maximal ideals P are called the closed points. Let $\mathbf{N}P$ denote the number of elements in the residue class field R/P, necessarily finite. One defines the **Hasse-Weil zeta function**

$$\varsigma(X, s) = \prod_P (1 - \mathbf{N}P^{-s}).$$

Such a function provides a way to reflect many arithmetic and geometric properties of X. if R is the ring of integers of a number field, then this function is the Dedekind zeta function. It can be defined for a scheme, covered by a finite number of such affine pieces.

Both in §9 and §10 the theorems or conjectures were formulated for number fields. Ultimately, they will be generalized to schemes of finite type. At the present time, only very special cases have been handled, mostly concerning elliptic curves with complex multiplication, modular curves, and Fermat type curves. However, one has already a good view of conjectural statements which started from the Birch-Swinnerton Dyer conjecture for elliptic curves, as given by Tate in two very valuable papers [**Ta 3**] and [**Ta 4**]. See also [**Ta 5**]. Just to make the link with these papers, I reproduce one of his conjectures as follows:

Conjecture 10.2. *If X is a regular ring R of finite type over $\mathbf{Z}$, then the order of $\varsigma(X, s)$ at the point $s = \dim X - 1$ is equal to*

$$\operatorname{rank} R^* - \operatorname{rank} \operatorname{Cl}(R),$$

where R^* is the group of units of R, and $\operatorname{Cl}(R)$ is the group of divisor classes.

One needs to go further and give a description of the coefficient of $(s - d + 1)^r$ in the expansion

$$\varsigma(X, s) = c(X)(s - d + 1)^r + \text{lower order terms}$$

at $s = d - 1$, where $d = \dim X$. Deligne [**De**] gives the value of the zeta function suitably normalized at the "critical point" in the context of "motives", when the zeta function does not vanish. To cover the cases considered by Birch-Swinnerton-Dyer-Stark-Tate with $r > 0$, there still remains to fit the considerations of §9, and the period considerations of §10 into this pattern, for regular schemes of finite type over $\mathbf{Z}$, especially the modular scheme (cf.

Beilenson [**Be**]). But I hope I have fulfilled my objective to lead the reader (who has come this far) into the unknown, concerning units and class groups in number theory and algebraic geometry.

Appendix: Distributions

Because of special interest shown in this topic, and some questions asking "why the Bernoulli polynomials and not others" in the theory described in §2 and §3, I have extracted here some general remarks from my AMS talk at the summer meeting.

Let

$$\varphi : \mathbf{Q}/\mathbf{Z} \to A$$

be a mapping into some abelian group which satisfies the relation for every positive integer N:

$$N^k \sum_{j=1}^{N-1} \varphi\left(x + \frac{j}{N}\right) = \varphi(Nx).$$

Here k is some positive integer. Then φ is said to be a **distribution of degree** k. Relations as above are satisfied by many functions in classical analysis and number theory, and I shall list a series of examples.

Note that if $\psi : \mathbf{Q}/\mathbf{Z} \to B$ is another distribution, then a homomorphism of φ to ψ is defined to be a homomorphism $h : A \to B$ making the following diagram commutative:

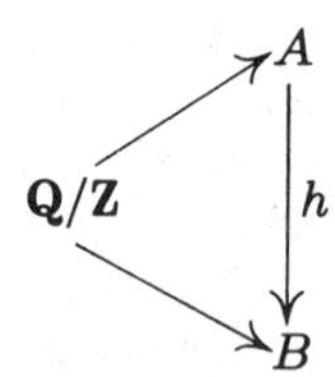

Some of the examples which we shall give will be homomorphic images of each other, but appearing under different disguises.

A distribution is called **odd** or **even** according as the function is odd or even. If multiplication by 2 is invertible on A, then any A-valued distribution can be uniquely decomposed as a direct sum of an even and an odd distribution. In the subsequent examples, each distribution is naturally equipped with a parity.

Bernoulli distribution. For each positive integer k, there exists a unique polynomial $\mathbf{B}_k$ with complex coefficients, leading coefficient 1, of degree k, such that the map

$$x \mapsto \frac{1}{k}\mathbf{B}_k(\langle x \rangle)$$

is a distribution of degree $k-1$. This polynomial is the Bernoulli polynomial, which has rational coefficients, and can be given by the generating series

$$\frac{te^{tX}}{e^t - 1} = \sum_{k=0}^{\infty} \mathbf{B}_k(X)\frac{t^k}{k!}.$$

If you check back to §2 or §3, you note that we discussed the Stickelberger element only at first level p, although we alluded to the necessity of considering all levels p^n. Let

$$G_n \approx (\mathbf{Z}/p^n\mathbf{Z})^*.$$

Then there is a natural homomorphism of group rings

$$\mathbf{Z}[G_{n+1}] \to \mathbf{Z}[G_n].$$

If one writes down the Stickelberger element at level $n+1$, then one expects its image under this homomorphism to be the Stickelberger element at level n. It is immediately seen that this compatibility condition amounts to the distribution relations for powers of p. This explains "why" the Bernoulli polynomials were forced in the context of the p-towers. Note that the Bernoulli distribution has parity $(-1)^k$.

The Fourier-Bernoulli distribution. Let

$$f_k(x) = \sum_{k=1}^{\infty} \frac{e^{2\pi i n x}}{n^k}.$$

Then it is immediately verified that f_k defines a distribution of degree $k-1$. In fact, the variable x can be taken to be in $\mathbf{R}/\mathbf{Z}$ rather than $\mathbf{Q}/\mathbf{Z}$. Furthermore, we have the following lemma of Rohrlich.

Lemma. *Let f be in $L^2(\mathbf{R}/\mathbf{Z})$, and assume that f satisfies the distribution relations of degree $k-1$. Let*

$$f(x) = \sum_{-\infty}^{\infty} c_n e^{2\pi i n x}$$

be the Fourier series of f. Then: $c_0 = 0$; $c_n = c_1/n^k$ for $n > 0$; $c_{-n} = c_{-1}/n^k$ for $n > 0$.

Proof. By definition, and using simple transformations, we have:

$$\begin{aligned}
c_n &= \int_0^1 f(x)e^{-2\pi i n x}\,dx \\
&= N^{k-1}\int_0^1 \sum_{j=0}^{N-1} f\left(\frac{x}{N}+\frac{j}{N}\right)e^{-2\pi i n x}\,dx \\
&= N^k \sum_{j=0}^{N-1} f\left(u+\frac{j}{N}\right)e^{-2\pi i n N u}\,du \\
&= N^k \int_0^1 f(u)e^{-2\pi i n N u} \\
&= N^k c_{nN}.
\end{aligned}$$

For $n = 0$, pick $N \neq 0$ to conclude $c_0 = 0$. Then take $n = 1$ or $n = -1$ to conclude the proof.

In view of the uniqueness theorem, the Fourier series for the Bernoulli distribution must be a linear combination of the above Fourier series and its conjugate. In fact, one knows classically that

$$\mathbf{B}_k(\langle x \rangle) = -\frac{k!}{(2\pi i)^k} \sum_{n \neq 0} \frac{e^{2\pi i n x}}{n^k}.$$

The Hurwitz zeta function. For $0 < u \leq 1$ let

$$\zeta(s, u) = \sum_{n=0}^{\infty} \frac{1}{(n+u)^s}.$$

This expression defines an analytic function of the complex variable s for $\operatorname{Re}(s) > 1$, and can be analytically continued into the whole plane, except for a simple pole at $s = 1$. For each real number t, let $\{t\}$ be the unique number congruent to $t \bmod \mathbf{Z}$, and such that

$$0 < \{t\} \leq 1.$$

Then one verifies at once that the map into the additive group of meromorphic functions given by

$$x \mapsto \zeta(s, \{x\})$$

is a distribution of degree $-s$. Here we can take s to be any complex number by analytic continuation. The Bernoulli distribution is a homomorphic image of the Hurwitz distribution, via the homomorphism evaluation at $s = 1 - k$, because of the classical Hurwitz relation:

$$\zeta(1-k, u) = -\frac{1}{k}\mathbf{B}_k(u).$$

The gamma distribution. Define

$$G(z) = \frac{1}{\sqrt{2\pi}}\Gamma(z).$$

We view G as defined on $\mathbf{Q}/\mathbf{Z}$ with the origin deleted, but then with values in the factor group

$$G : \mathbf{Q}/\mathbf{Z} \to \mathbf{C}^*/\mathbf{Q}_{\mathrm{a}}^*$$

of the multiplicative group of complex numbers modulo the group of nonzero algebraic numbers. The classical identity

$$\prod_{j=0}^{N-1} \frac{1}{\sqrt{2\pi}}\Gamma\left(z + \frac{j}{N}\right) = \frac{1}{\sqrt{2\pi}}\Gamma(Nz)N^{1/2-Nz}$$

shows that G defines a distribution.

Furthermore, this distribution also depends on the Hurwitz distribution. If we take the power series expansion of the Hurwitz zeta function at any complex number s_0, then the value defines a distribution. But we may also take the coefficients of higher powers $(s - s_0)^m$ for $m \geq 1$. Then we get the distribution relations, and in addition other terms coming from the overflow from lower terms. In particular, we have the classical expansion at $s = 0$, namely

$$\varsigma(s,u) = \frac{1}{2} - u + \log\left(\frac{1}{\sqrt{2\pi}}\Gamma(u)\right)s + O(s^2).$$

The distribution relations for $\log \frac{1}{\sqrt{2\pi}}\Gamma$ follows from the distribution relations for the Hurwitz zeta function. The extra term $N^{1/2-Nz}$ is explainable structurally from the constant term in the expansion, which is none other than $-\mathbf{B}_1(u)$, itself the first Bernoulli distribution. The next coefficient (that of s^2) would also satisfy the distribution relations, modulo the expressions due to the preceding terms.

We note that the gamma distribution is odd, because of the classical relation

$$\frac{1}{\sqrt{2\pi}}\Gamma(z)\frac{1}{\sqrt{2\pi}}\Gamma(1-z) = \frac{1/2}{\sin \pi z}.$$

Here of course we take z rational, so $\sin \pi z$ is algebraic.

Cyclotomic numbers. The function

$$x \mapsto e^{2\pi i x} - 1$$

defines a distribution into the multiplicative group of complex numbers (except at $x = 0$). This is immediate from the relation

$$\prod_{\varsigma^N=1} (1 - \varsigma X) = 1 - X^N.$$

If x ranges over rational numbers, then $e^{2\pi i x} - 1$ is just the numerator of the cyclotomic units, and may be called the cyclotomic numbers distribution. If the values are viewed as in $\mathbf{C}^*/\boldsymbol{\mu}$, then the distribution is even.

Modular units. In §6 we had defined the functions g_a with $a \in \mathbf{Q}^2$, $a \notin \mathbf{Z}^2$. It can easily be shown that if one changes a by a pair in $\mathbf{Z}^2$, then g_a changes by multiplication with a root of unity. We view the association

$$a \mapsto g_a$$

as a map of $\mathbf{Q}^2/\mathbf{Z}^2$ into the multiplicative group of meromorphic functions, modulo constants. Then this map satisfies the analogue of the distribution relations on $\mathbf{Q}^2/\mathbf{Z}^2$. It is an even distribution.

The Lobatchevski distribution. Define the Lobatchevski function

$$\lambda(x) = -\int_0^x \log\left|e^{2\pi it} - 1\right| dt.$$

Then λ is a distribution, being composed of the cyclotomic numbers distribution, the absolute value, the logarithm (which are homomorphisms), and the integral which is easily seen to preserve the distribution relations. Milnor has investigated this distribution in connection with hyperbolic geometry as follows. Let H be hyperbolic 3-space. This is the set of points

$$(x_1, x_2, y) \in \mathbf{R} \times \mathbf{R} \times \mathbf{R}^+$$

so (x_1, x_2) is an ordinary point in the plane, and $y > 0$. We endow H with the metric

$$\frac{dx_1^2 + dx_2^2 + dy^2}{y^2}.$$

Select four distinct points in the plane, and let T be the tetrahedron in H whose vertices are at these points. Then it can be shown that opposite dihedral angles are equal. (The dihedral angles are the angles between the faces of the tetrahedron.) Let α, β, γ be the dihedral angles. Then

$$\alpha + \beta + \gamma = \pi,$$

and the volume of the tetrahedron is precisely given in terms of the Lobatchevski function, by

$$\mathrm{Vol}(T) = \iiint_T \frac{dx_1\, dx_2\, dy}{y^3} = \lambda(\alpha) + \lambda(\beta) + \lambda(\gamma).$$

The search for relations among such volumes had led Milnor to consider the Lobatchevski function and its relations, now known as the distribution relations, and to show that it had the maximum rank (its values being viewed as generated a vector space over the rationals). We discuss this systematically below.

The Stickelberger distribution. Let $h : \mathbf{Q}/\mathbf{Z} \to \mathbf{C}$ be a distribution. Let $G(N) \approx (\mathbf{Z}/N\mathbf{Z})^*$ under an association $a \mapsto \sigma_a$, as in cyclotomic theory. Define

$$g_N(x) = \frac{1}{|G(N)|} \sum_{a \in (\mathbf{Z}/N\mathbf{Z})^*} h(xa)\sigma_a^{-1},$$

for $x \in \frac{1}{N}\mathbf{Z}/\mathbf{Z}$. Then g_N takes values in the group algebra $\mathbf{C}[G(N)]$, and if $M \mid N$, then the image of $g_N(x)$ under the canonical homomorphism $G(N) \to G(M)$ is equal to $g_M(x)$. Thus we may define

$$\mathrm{St}_h(x) = \lim g_N(x)$$

in the injective limit of the group algebras (as vector spaces over $\mathbf{C}$), ordered by divisibility, with the injection from one level to the next given by sending one group element to the sum of all the group elements lying above it under the canonical homomorphism. Then one sees that $x \mapsto \mathrm{St}_h(x)$ is a distribution, called the Stickelberger distribution St_h associated with h.

The universal distribution. Fix an integer $N > 1$, and consider the subgroup of $\mathbf{Q}/\mathbf{Z}$ consisting of those elements with order N, that is

$$\frac{1}{N}\mathbf{Z}/\mathbf{Z} = (\mathbf{Q}/\mathbf{Z})_N = Z_N \quad \text{(by definition).}$$

One can then form the universal distribution (restricted to Z_N) in the obvious way. We start with the free abelian group $Fr(Z_N)$, and factor out the distribution relations (say of degree 0 for simplicity) with level M dividing N. If we denote the subgroup of these relations by $DR(N)$, then the universal distribution is simply the factor group $\mathbf{U}(N) = Fr(Z_N)/DR(N)$, with the natural map of Z_N into this factor group. A theorem of Kubert [**Ku 2**] (see [**Ku-L 1**] or [**L 1**], Chapter 2 and Yamamoto [**Ya**]) asserts:

The universal distribution $U(N)$ is a free abelian group on $\phi(N)$ generators. Let

$$N = \prod p_i^{n_i}$$

be the factorization of N into prime powers. A free basis for $\mathbf{U}(N)$ is given by the elements

$$\sum \frac{a_i}{p_i^{n_i}}$$

with $a_i \in (\mathbf{Z}/p_i^{n_i}\mathbf{Z})^*$ (so a_i prime to p_i) and $a_i \neq 1$, or $a_i = 0$.

The general philosophy is that if a distribution arises naturally, and is not "obviously" special, then it is in fact universal, possibly with the parity odd or even. For example, the Bernoulli distribution is obviously not universal since it is rational valued, but it can be shown that its associated Stickelberger distribution is universal of the appropriate parity (for values into abelian groups where 2 is invertible). The cyclotomic number distribution is universal even (with values as above). This is a reformulation of a theorem of H. Bass. The modular units give a universal even distribution, cf. [**Ku-L 1**]. Rohrlich conjectured that the gamma distribution is universal odd. Since the values are in the group $\mathbf{C}^*$ modulo non-zero algebraic numbers, this would amount to a theorem in the theory of transcendental numbers. Similarly, Milnor conjectured that the Lobatchevski distribution is universal odd (values in the additive group of complex numbers). For a general discussion of distributions, cf. Chapter 1 of [**Ku-L 1**], and the bibliography contained in that book, as well as [**L 1**], Chapter 2, containing proofs for all the theorems mentioned here. Distributions on projective systems arose in number theory through the work of Iwasawa, reformulated by Mazur. The point of view on injective systems taken here stems from the Kubert-Lang series of papers. The subtleties involving 2-torsion in the universal distribution have been dealt with systematically by Kubert, cf. [**Ku 3**] and the bibliography at the end of [**Ku-L 1**]. They involve the cohomology of ± 1 in the universal distribution, and have applications to modular functions, and possibly to algebraic topology.

In this Appendix, I did not want to go systematically into the study of distributions. I merely wanted to point out the general pattern underlying much of the formalism which arises in connection with §2, §3, §4, §8 and in other parts of mathematics.

Note: The author wishes to point out to the reader that some errors may have been introduced in this paper during the corrections process, after the material had been proofread.

BIBLIOGRAPHY

[An 1] G. ANDERSON, Logarithmic derivatives of Dirichlet L-functions and the periods of abelian varieties, to appear.

[An 2] G. ANDERSON, to appear.

[Be] N. BEILINSON, Higher regulators and values of L-functions of curves, Functional Analysis 14 No. 2 (1980), pp. 46–47.

[Be] R. BERGELSON, The index of the Stickelberger ideal of order k on $C^k(N)$, to appear, Annals of Math.

[Bl] S. BLOCH, Algebraic K-theory and class field theory for arithmetic surfaces, to appear.

[Bo] F. A. BOGOMOLOV, On the algebraicity of l-adic representations, CR Acad. Sci. Paris 290 No. 15 (1980) pp. 701–703.

[CN] P. CASSOU-NOGUÈS, Valeurs aux entiers négatifs des fonctions zeta et fonctions zeta p-adiques, Invent. Math. 51 (1979) pp. 29–59.

[Co 1] J. COATES, p-adic L-functions and Iwasawa's theory, Durham Conference on algebraic number theory and class field theory, 1976.

[Co 2] J. COATES, Fonctions zeta partielles d'un corps de nombres totalement réel, Seminaire Delange-Pisot-Poitou, 1974–1975.

[Co-L] J. COATES and S. LICHTENBAUM, On l-adic zeta functions, Ann. of Math. 98 (1973) pp. 498–550.

[Co-Si 1] J. COATES and W. SINNOTT, On p-adic L-functions over real quadratic fields, Invent. Math 25 (1974) pp. 253–279.

[Co-Si 2] J. COATES and W. SINNOTT, Integrality properties of the values of partial zeta functions, Proc. London Math. Soc. (1977) pp. 365–384.

[Co-Wi 1] J. COATES and A. WILES, On the conjecture of Birch and Swinnerton-Dyer, Invent. Math. 39 (1977) pp. 223–251.

[Co-Wi 2] J. COATES and A. WILES, Kummer's criterion for Hurwitz numbers, Kyoto Conference on Algebraic Number Theory, 1977.

[Co-Wi 3] J. COATES and A. WILES, On p-adic L-functions and elliptic units, J. Austr. Math. Soc. 26 (1978) pp. 1–25.

[Da-H] H. DAVENPORT and H. HEILBRONN, On the density of discriminants of cubic fields II, Proc. Royal Soc. 322 (1971) pp. 405–420.

[De] P. DELIGNE, Valeurs de fonctions L et periodes d'integrales, Proc. Symp. Pure Math. Vol. 33 (1979) pp. 313–346.

[De-Ra] P. DELIGNE and M. RAPOPORT, Schémas de modules des courbes élliptiques, Springer Lecture Notes 349 (1973).

[Dr] V. G. DRINFELD, Two theorems on modular curves, Functional analysis and its applications, Vol. 7 No. 2 translated from the Russian April–June 1973 pp. 155–156.

[Fe-W] B. FERRERO and L. WASHINGTON, The Iwasawa invariant μ_p vanishes for abelian number fields, Ann. Math. 109 (1979) pp. 377–395.

[Gra] G. GRAS, Classes d'idéaux des corps abéliens et nombres de Bernoulli généralisés, Annales Institut Fourier Université de Grenoble, Tome XXVII, Fasc. 1 (1977) pp. 1–66.

[Gre 1] R. GREENBERG, On p-adic L-functions and cyclotomic fields I, Nagoya Math. J. No. 56 (1974) pp. 61–77.

[Gre 2] R. GREENBERG, On p-adic L-functions and cyclotomic fields II, Nagoya Math. J. No. 67 (1977) pp. 139–158.

[Gro 1] B. GROSS, p-adic L-series at $s = 0$, to appear.

[Gro 2] B. GROSS, On the periods of Abelian integrals and a formula of Chowla and Selberg, Invent. Math. 45 (1978) pp. 193–211.

[Iw 1] K. IWASAWA, On Γ-extensions of algebraic number fields, Bull. AMS 65 (1959) pp. 183–192.

[Iw 2] K. IWASAWA, On p-adic L-functions, Ann. of Math. 89 (1969) pp. 198–205.

[Iw 3] K. IWASAWA, A class number formula for cyclotomic fields, Ann. of Math. 76 (1962) pp. 171–179.

(For a more complete list of eighteen papers by Iwasawa on the subjects of concern here, cf. the bibliography at the end of my *Cyclotomic Fields* Vol. 2.)

[Kato] K. KATO, Higher Local Class Field Theory, Proc. Japan Acad. 53 (1977) pp. 140–143 and 54 (1978) pp. 250–255.

[Ka-L] N. KATZ and S. LANG, Finiteness Theorems in Geometric Class Field Theory, to appear in l'Enseignement Mathematique, 1982.

[Ka-M] N. KATZ and B. MAZUR, to appear.

[Ke 1] D. KERSEY, Modular units inside cyclotomic units, Ann. of Math. 112 (1980) pp. 361–380.

[Ke 2] D. KERSEY, The index of modular units, to appear.

[Kl] F. KLEIN, Uber die elliptischen Normalkurven der n-ten Ordnung, Abh. math.-phys. Klasse Sächsischen Kgl. Gesellschaft Wiss. Bd 13, Nr. IV (1885) pp. 198–254.

[Ku 1] D. KUBERT, The 2-primary component of the ideal class group in cyclotomic fields, to appear.

[Ku 2] D. KUBERT, The universal ordinary distribution, Bull. Soc. Math. France **107** (1979), 179–202.

[Ku 3] D. KUBERT, The $\mathbf{Z}/2\mathbf{Z}$ cohomology of the universal ordinary distribution, Bull. Soc. Math. France **107** (1979), 203–224.

[Ku-L 1] D. KUBERT and S. LANG, Modular Units, Springer Verlag, 1981.

[Ku-L 2] D. KUBERT and S. LANG, Modular units inside cyclotomic units, Bull. Soc. Math. France 107 (1979) pp. 161–178.

[Kum 1] E. KUMMER, Mémoire sur la théorie des nombres complexes composeés de racines de l'unité et de nombres entiers, J. Math. Pure et Appliquees, XVI (1851) pp. 377–498 (=Collected Works I, especially p. 452).

[Kum 2] E. KUMMER, Theorie der idealen Primfaktoren der complexen Zahlen, welche aus den Wurzeln der Gleichung $\omega^n = 1$ gebildet sind, wenn n eine zusammengesetzte Zahl ist, Math. Abh. Konig. Akad. Wiss. Berlin (1856) pp. 1–47, (Collected Works I, especially p. 583). *Note*: In CW I, p. 626, Kummer gives what is known as the Stickelberger congruence for Gauss sums in terms of factorials.

[L 1] S. LANG, Cyclotomic Fields Vols. 1 and 2, Springer Verlag 1979 and 1980.

[L 2] S. LANG, Elliptic functions, Addison Wesley, 1974.

[L 3] S. LANG, Division points on curves, Ann. Mat. Pura Appl. IV, Tomo LXX (1965) pp. 229–234.

[L 4] S. LANG, Unramified class field theory over function fields in several variables, Ann. of Math. 64 (1956) pp. 286–325.

[L 5] S. LANG, Sur les séries L d'une variété algébrique, Bull. Soc. Math. France 84 (1956) pp. 385–407.

[L 6] S. LANG, L-series of a covering, Proc. Nat. Acad. Sci. USA (1956).

[Lgds] R. LANGLANDS, Modular forms and l-adic representations, Springer Lecture Notes 349 (1973) pp. 361–500.

[Le 1] H. W. LEOPOLDT, Eine Verallgemeinerung der Bernoullische Zahlen, Abh. Math. Sem. Hamburg (1958) pp. 131–140.

[Le 2] H. W. LEOPOLDT, Uber Einheitengruppe und Klassenzahl reeler abelscher Zahlkörper, Abh. Deutsche Akad. Wiss. Berlin Math. 2 (1954) Akademie Verlag.

[Le 3] H. W. LEOPOLDT, Uber die Arithmetik algebraischen Zahlkörper, J. reine angew. Math 209 (1962) pp. 54–71.

(For a more complete list of eleven papers of Leopoldt on cyclotomic fields and p-adic L-functions, cf. the bibliography at the end of *Cyclotomic Fields* Vol. 2.)

[Ma 1] B. MAZUR, Modular curves and the Eisenstein ideal, Pub. IHES No. 47 (1977) pp. 33–186.

[Ma 2] B. MAZUR, Rational isogenies of prime degree, Invent. Math. 44 (1978) pp. 129–162.

[Ma-W] B. MAZUR and A. WILES, to appear.

[Mi 1] J. MILGRAM, Odd index subgroups of units in cyclotomic fields and applications, Springer Lecture Notes no. 854 (1981).

[Mi 2] J. MILGRAM, Patching techniques in surgery and the solution of the compact space form problem, to appear.

[O-T] F. OORT and J. TATE, Group schemes of prime order, Ann. Scient. Ec. Norm. Sup. serie 4, 3 (1970) pp. 1–21.

[Pa] A. M. PARSHIN, Class field theory for arithmetical schemes, preprint, and also Uspekhi Matem. Nauk 39 (1975) p. 253 and Izvestija Acad. Nauk. SSSR Ser. Matem. 40 (1976) pp. 736–773.

[Ra] K. RAMACHANDRA, Some applications of Kronecker's limit formula, Ann. of Math. 80 (1964) pp. 104–148.

[Ray 1] M. RAYNAUD, Schémas en groupes de type $(p, \ldots, p)$, Bull. Soc. Math. France 102 (1974) pp. 241–280.

[Ray 2] M. RAYNAUD, Faisceaux amples sur les schémas en groupes et les éspaces homogènes, Springer Lecture Notes 119 (1970).

[Ri] K. RIBET, A modular construction of unramified p-extensions of $\mathbf{Q}(\mu_p)$, Invent. Math. 34 (1976) pp. 151–162.

[Ro 1] G. ROBERT, Unités élliptiques, Bull. Soc. Math. France Supplément, Décembre 1973 No. 36.

[Ro 2] G. ROBERT, Nombres de Hurwitz et unités élliptiques, Ann. Scient. Ec. Norm. Sup. 4e serie t. 11 (1978) pp. 297–389.

[Se 1] J. P. SERRE, Propriétés Galoisiènnes des points d'ordre fini des courbes élliptiques, Invent. Math. 15 (1972) pp. 259–331.

[Se 2] J. P. SERRE, Classes des corps cyclotomiques d'après Iwasawa, Bourbaki Seminar 1958.

[Shim] G. SHIMURA, Introduction to the arithmetic theory of automorphic functions, Iwanami Shoten and Princeton University Press 1971.

[Shin 1] T. SHINTANI, On a Kronecker limit formula for real quadratic fields, J. Fac. Sci. Univ. Tokyo Sec. IA, 24 (1977) pp. 167–199.

[Shin 2] T. SHINTANI, On certain ray class invariants of real quadratic fields, J. Math. Soc. Japan Vol. 30 No 1 (1978) pp. 139–167.

[Sie] C. L. SIEGEL, Lectures on advanced analytic number theory, Tata Institute Lecture Notes 1961.

[Sin 1] W. SINNOTT, On the Stickelberger ideal and the circular units of a cyclotomic field, Ann. of Math. 108 (1978) pp. 107–134.

[Sin 2] W. SINNOTT, On the Stickelberger ideal and the circular units of an abelian field, Invent. Math. 62 (1980) pp. 181–234.

[St] H. STARK, L-functions at $s = 1$:

I: Advances in Math. 7 (1971) pp. 301–343,

II: Ibid. 17 (1975) pp. 60–92,

III: Ibid 22 (1976) pp. 64–84,

IV: Ibid 35 (1980) pp. 197–235.

[St B] H. STARK, Class fields and modular forms of weight 1, Springer Lecture Notes 601, 1976; (Bonn conference on modular forms in one variable).

[Ta 1] J. TATE, On Stark's conjectures on the behavior of $L(s, \chi)$ at $s = 0$, to appear, Shintani Memorial Volume, J. Fac. Sci. Tokyo 1982.

[Ta 2] J. TATE, Les conjectures de Stark sur les fonctions L d'Artin en $s = 0$, notes d'un cours à Orsay rédigées par D. Bernardi et N. Schappacher, Lecture Notes to appear in the Birkhauser Boston Series.

[Ta 3] J. TATE, Algebraic cycles and poles of zeta functions, in Arithmetical Algebraic Geometry, Conference held at Purdue University, 1963, Harper and Row, New York 1965.

[Ta 4] J. TATE, The conjecture of Birch and Swinnerton-Dyer and a geometric analogue, Dix exposés sur la cohomologie des schémas, North Holland, 1968 (=Seminaire Bourbaki 352, 1966).

[Ta 5] J. TATE, Arithmetic of Elliptic Curves, Invent. Math. 23 (1974) pp. 179–206.

[Va 1] H. S. VANDIVER, Fermat's last theorem and the second factor in the cyclotomic class number, Bull. AMS 40 (1934) pp. 118–126.

[Va 2] H. S. VANDIVER, Fermat's last theorem, Am. Math. Monthly 53 (1946) pp. 555–576.

[Wall] C. T. C. WALL, Classification of hermitian forms VI, Ann. of Math. 103 (1976) pp. 1–80.

[Wash] L. WASHINGTON, The non-p-part of the class number in a cyclotomic $\mathbf{Z}_p$-extension, Invent. Math. 49 (1978) pp. 87–97.

[Wi] A. WILES, Modular curves and the class group of $\mathbf{Q}(\zeta_p)$, Invent. Math. 58 (1980) pp. 1–35.

[Ya 1] K. YAMAMOTO, The Gap group of multiplicative relationships of Gaussian sums, Symposia Mathematica No. 15 (1975) pp. 427–440.

[Ya 2] K. YAMAMOTO, On a conjecture of Hasse concerning multiplicative relations, of Gaussian sums, J. Combin. Theory 1 (1966) pp. 476–489.

[Yu] JING YU, A cuspidal class number formula for the modular curves $X_1(N)$, Math. Ann. 252 (1980) pp. 197–216.

[Zi] H. ZIMMER, Lecture Notes in Math., vol. 262, Springer-Verlag, New York, 1972.

Current address: Department of Mathematics, Box 2155 Yale Station, Yale University, New Haven, Connecticut 06520

Volume dedicated to Shafarevic
(Birkhäuser, 1984–1985): 155–171

Conjectured Diophantine Estimates on Elliptic Curves

Serge Lang

To I.R. Shafaravich

Let A be an elliptic curve defined over the rational numbers $\mathbf{Q}$. Mordell's theorem asserts that the group of points $A(\mathbf{Q})$ is finitely generated. Say $\{P_1, \ldots, P_r\}$ is a basis of $A(\mathbf{Q})$ modulo torsion. Explicit upper bounds for the heights of elements in such a basis are not known. The purpose of this note is to conjecture such bounds for a suitable basis. Indeed, $\mathbf{R} \otimes A(\mathbf{Q})$ is a vector space over $\mathbf{R}$ with a positive definite quadratic form given by the Néron-Tate height: if A is defined by the equation

$$y^2 = x^3 + ax + b, \quad a, b \in \mathbf{Z},$$

and $P = (x, y)$ is a rational point with $x = c/d$ written as a fraction in lowest form, then one defines the x-height

$$h_x = \log \max(|c|, |d|).$$

There is a unique positive definite quadratic form h such that

$$h(P) = \frac{1}{2} h_x(P) + O_A(1) \quad \text{for } P \in A(\mathbf{Q}).$$

Then $A(\mathbf{Q}) \bmod A(\mathbf{Q})_{\mathrm{tor}}$ can be viewed as a lattice in the vector space $\mathbf{R} \otimes A(\mathbf{Q})$ endowed with the quadratic form h.

I assume that the reader is acquainted with the Birch-Swinnerton Dyer conjecture. For an elegant self contained presentation in the form which I shall use, see Tate [Ta 1]. Starting with the Birch-Swinnerton Dyer formula, and several conjectures from the analytic number theory associated with the L-function of the elliptic curve, I shall give arguments showing how one can get bounds for the heights of a suitably selected basis $\{P_1, \ldots, P_r\}$ of the lattice. A classical theorem of Hermite gives bounds for an almost-orthogonalized basis of a lattice in terms of the volume of the fundamental domain, so the problem is to estimate this volume. Hermite's theorem will be recalled for the convenience of the reader in the last section.

The conjectured estimates will also tie in naturally with a conjecture of Marshall Hall for integral points, which we discuss in §2. I shall give some numerical examples in §3.

Manin [Ma] gave a general discussion showing how the Birch-Swinnerton Dyer conjecture and the Taniyama-Weil conjecture that all elliptic curves over $\mathbf{Q}$ are modular give effective means to find a basis for the Mordell-Weil group $A(\mathbf{Q})$ and an estimate for the Shafarevich-Tate group. The Taniyama-Weil conjecture will play no role in the arguments of this paper, and I shall propose a much more precise way of estimating the regulator and the heights in a basis as in Hermite's theorem. Manin also attributes a "gloomy joke" to Shafarevich in the context of his Theorem 11.1. It is therefore a pleasure to dedicate the conjectures and this paper to Shafarevich.

1. Rational Points

We let $\langle P, Q\rangle$ be the bilinear symmetric form associated with the Néron-Tate height, namely

$$\langle P, Q\rangle = h(P+Q) - h(P) - h(Q).$$

Then

$$|P| = \sqrt{\langle P, P\rangle} \quad \text{and} \quad h(P) = \frac{1}{2}|P|^2.$$

If r is the rank of $A(\mathbf{Q})$, that is the dimension over $\mathbf{R}$ of $\mathbf{R} \otimes A(\mathbf{Q})$, then Birch-Swinnerton Dyer conjecture that the L-function $L_A(s) = L(s)$ has a zero of order r at $s = 1$, and that the coefficient of $(s-1)^r$ in the Taylor expansion is given by the formula

$$\frac{1}{r!}L^{(r)}(1) = |\text{III}|\frac{2^{-r}\,|\det\langle P_i, P_j\rangle|}{|A(\mathbf{Q})_{\text{tor}}|^2}\pi_\infty \prod_{p|\Delta} \pi_p.$$

On the right-hand side, $|\text{III}|$ is a positive integer, the order of the Shafarevich-Tate group.

The determinant is taken with respect to any basis $\{P_1, \ldots, P_r\}$. Its absolute value is called the **regulator**,

$$R_A = R = |\det\langle P_i, P_j\rangle|,$$

and is the square of the volume of a fundamental domain of the lattice $A(\mathbf{Q})/A(\mathbf{Q})_{\mathrm{tor}}$ in $\mathbf{R} \otimes A(\mathbf{Q})$.

As usual, the discriminant is given by

$$\Delta = -16(4a^3 + 27b^2).$$

The terms π_p for $p|\Delta$ are given by the integral of $|\omega|$ over $A(\mathbf{Q}_p)$, where the differential form ω is the one associated with the global minimum model. As Tate pointed out, π_p is an integer for each p. This property is all we need to know about π_p for our purposes. Essentially.

$$\omega = dx/y,$$

except that the equation $y^2 = x^3 + ax + b$ is not the equation of a minimal model, and one has to introduce extra coefficients, cf. [Ta 1] where the matter is explained in detail. For simplicity of notation, I shall continue to write the equation in standard Weierstrass form as above.

The period π_∞ is given by the integral

$$\pi_\infty = \int_{A(\mathbf{R})} |\omega|.$$

Finally $|A(\mathbf{Q})_{\mathrm{tor}}|$ is the order of the (finite) torsion group.

We are now interested in estimating all the terms to get a bound for $|Ш|R$, which will yield a bound for R since $|Ш|$ is an integer.

Since the π_p are integers, they work for us in getting an upper bound for R.

Mazur [Ma] has shown that $|A(\mathbf{Q})_{\mathrm{tor}}|$ is bounded by 16, and hence this torsion number has a limited effect on the desired estimates.

Concerning the r-th derivative of the L-function. I am indebted to H. Montgomery and David Rohrlich for instructive discussions on the analytic number theory of more classical zeta functions and L-series. In particular, for the Riemann zeta function $\zeta_{\mathbf{Q}}(s)$, the Riemann hypothesis implies that

$$\zeta_{\mathbf{Q}}\left(\frac{1}{2} + it\right) \ll t^{\epsilon(t)},$$

where $\epsilon(t)$ tends to 0 as $t \to \infty$. More precisely, Montgomery conjectures [Mo], formula (10) that one can take

$$\epsilon(t) = c(\log t \log\log t)^{-1/2}$$

with some constant c. Without the $\log\log t$, Titchmarsh already observed that one cannot improve the exponent $(\log t)^{-1/2}$, see [Ti], Theorem 8.12. The Riemann hypothesis implies by Theorem 14.14 that

$$\zeta_{\mathbf{Q}}\left(\frac{1}{2}+it\right) \ll t^{c/\log\log t},$$

but is not known to imply Montgomery's conjecture, or even the weaker formulation with the Titchmarsh exponent without the $\log\log t$.

For Dirichlet L-series over the rationals, with conductor q, Montgomery also conjectures that one has

$$\frac{1}{r!}L_q^{(r)}\left(\frac{1}{2}\right) \ll q^{\epsilon(q)}c_2^r(\log q)^r$$

with a universal constant c_2.

It is known in the case of elliptic curves with complex multiplication, that $L_A(s)$ is a Hecke L-series, which is so normalized that $L_A^{(r)}(1)$ corresponds to the r-th derivative of the Hecke L-series in its usual form at $s = 1/2$. So to get bounds for $L_A^{(r)}(1)$, on needs bounds for the Hecke L-series at $s = 1/2$. I once gave a general principle which allows one to prove (or conjecture) what happens for more general L-series, on the basis of what happens for the Riemann zeta function. To quote [L 2]: whenever you see a t in an estimate, with a logarithm, then replace it by $d_\chi t^n$, where $d_\chi = d_K \mathbf{N}\mathfrak{f}_\chi$, d_K is the absolute value of the discriminant of a number field K, and $n = [K : \mathbf{Q}]$. In line with this principle, I expect

$$\frac{1}{r!}L_A^{(r)}(1) \ll N^{\epsilon(N)}c_3^r(\log N)^r$$

where N is the conductor of the curve, cf. [Ta 1], end of §6; and $\epsilon(N)$ tends to 0 as $N \to \infty$. Furthermore, $\epsilon(N)$ may have the same shape as in Montgomery's conjecture, namely

$$\epsilon(N) = c(\log N \log\log N)^{-1/2}.$$

Note that the conductor is divisible by the same primes as the descriminant Δ, and $\mathrm{ord}_p N \leq \mathrm{ord}_p \Delta$ so $N \leq |\Delta|$. For the precise order of N at a prime p, see [Ta 1], end of §6, and [Ta 2] for a systematic way of computing N, which is an isogeny invariant. Thus the above inequalities for N in terms of Δ hold when Δ is replaced by a minimal discriminant $\Delta_{\min}$.

This leaves the real period π_∞ to be estimated from below. The Néron differential of a minimal model will not differ too much from dx/y. Let us compute roughly the integral

$$\int_{-\infty}^{\infty} \frac{dx}{\sqrt{x^3 + ax + b}}$$

where intervals of x such that $x^3 + ax + b$ is negative are to be disregarded in this integral. If $b^2 \geq |a|^3$ we change variables and let $x = |b|^{1/3}u$, $dx = |b|^{1/3}du$. If $b^2 < |a|^{1/2}u$.

For definiteness, say $b^2 \geq |a|^3$. Then the integral becomes

$$\frac{1}{|b|^{1/6}} \int_{-\infty}^{\infty} \frac{du}{\sqrt{u^3 + cu + 1}}$$

where $-1 \leq c \leq 1$. The integral for $|u| \geq 2$ is bounded from below by a universal constant. The integral

$$\int_{-2}^{2} \frac{du}{\sqrt{u^3 + cu + 1}}$$

is also bounded from below because $|u^3 + cu + 1|$ is bounded from above for c, u lying in bounded intervals. A similar anslysis if $b^2 \leq |a|^3$ yields the lower bound

$$|\pi_\infty| \gg 1/H^{1/12} \quad \text{where } H = \max(|a|^3, |b|^2).$$

Putting all this together, we arrive at the following conjecture.

Conjecture 1. *Let $H(A) = \max(|a|^3, |b|^2)$. For all elliptic curves $y^2 = x^3 + ax + b$ with $a, b \in \mathbf{Z}$ we have*

$$|\text{III}|R \ll H(A)^{1/12} N^{\epsilon(N)} c^r (\log N)^r$$

with some universal constant c, and $\epsilon(N) \to 0$ as $N \to \infty$. In fact, $\epsilon(N)$ may have the explicit form

$$\epsilon(N) = c'(\log N \log\log N)^{-1/2}.$$

Note that if either a or $b = 0$, then $H(A)^{1/12}$ can be replaced by $|\Delta|^{1/12}$. In light of the Hall conjecture below, it should always be the case that

$$H(A) \ll |\Delta|^{6+\epsilon}.$$

In any case, $H(A)$ is absolutely homogeneous of degree 12, in the sense that $H(cA) = |c|^{12}H(A)$, where cA denotes the elliptic curve obtained by multiplying all the quantities a, b, x, y with the appropriate power of c, corresponding to their weight.

The shape of the inequality in Conjecture 1 is analogous to the upper bound for the regulator of a number field K by easy estimates of the residue of the Dedekind zeta function at $s = 1$, apparently first noted by Landau, of the form

$$h_K R_K \leq c^{[K:\mathbf{Q}]} d_K^{1/2} (\log d_K)^{[K:\mathbf{Q}]-1}$$

where d_K is the absolute value of the discriminant. Siegel [Si] more carefully determined precise and very good constants in this estimate. In [L 1] I made the following conjecture:

Conjecture 2. *For all elliptic curves over the integers as above, one has the lower bound for the canonical height:*

$$h(P) \gg \log|\Delta_{\min}|$$

for any rational point P which is not a torsion point.

This conjecture was proved by Silverman [Si] when the j-invariant is an integer. Silverman [Sil 3] has also shown that the conjecture is best possible. Taking into account the Hermite theorem, we may now give an upper bound for an appropriate basis of the Mordell-Weil group. I leave out the integral factor $|Ш|$ on the left for simplicity.

Conjecture 3. *There exists a basis $\{P_1, \ldots, P_r\}$ for $A(\mathbf{Q})$ modulo torsion, ordered by ascending height, such that:*

$$h(P_1) \ll H(A)^{1/12r} N^{\epsilon(N)/r} \log N \left(\tfrac{2}{\sqrt{3}}\right)^{(r-1)/2}$$
$$h(P_r) \ll H(A)^{1/12} N^{\epsilon(N)} (\log N) c^{r(r-1)/2}$$

Indeed, we have $h(P) = \frac{1}{2}|P|^2$, so the upper bound for $h(P_1)$ is precisely the Hermite bound in Theorem 4.2. As to the highest point, we use

Theorem 4.2 (1) and the *lower bound* of Conjecture 2, which allows us to divide by $(\log|\Delta_{\min}|)^{r-1}$ on both sides so that we end up with only one factor of $\log N$ on the right hand side.

The above conjectures are phrased for the canonical height h. To apply them to the ordinary height h_x of the x-coordinate, we need an upper bound on $|h - \frac{1}{2}h_x|$ which will be discussed in the next section.

On the other hand, observe that the lower bound of Conjecture 2 allows us to estimate $|Ш|$ from above. Indeed, referring to Hermite's theorem in §4, we have

$$\begin{aligned} R = \det(L)^2 &= |u_1|^2 \cdots |u_r|^2 \\ &\geq \left(\frac{3}{4}\right)^{(r-1)} (2h(P_1))^r \\ &\gg c_0^{r^2} (\log|\Delta_{\min}|)^r \end{aligned}$$

by Conjecture 2. Hence the upper bound for $|Ш|R$ in Conjecture 1 yields an upper bound for $|Ш|$ of the same nature, except that we can cancel $(\log|\Delta_{\min}|)^r$.

The above arguments do not seem to give further insight in the question raised in [L 1], p. 92, concerning the ratios of successive minima, or equivalently the lengths of elements in an almost orthogonalized basis.

It is a good question to determine to what extent the factor $c^r(\log N)^r$ in Conjecture 1 must really be there. After extending $\mathbf{Q}$ to the field of 2-torsion points the standard proof of the Mordell-Weil theorem of course bounds r in terms of the number of prime factors of N but also in terms of the 2-rank of the class number or the p-rank of the class number of p-torsion points, for any prime p. The question is whether this second part of the bound is, in fact, significant over the rational numbers.

Supposing r is like the number of distinct prime factors of N, one knows that if N is a product of the first r primes, then r is asymptotic to $\log N/\log\log N$. On the other hand, the average value of r is $\log\log N$.

2. Integral Points

We begin by recalling an estimate for $|h - \frac{1}{2}h_x|$, see Demjanenko [De], Zimmer [Zi] or [L 1], p. 99, where I piece together the local estimates to give a global one as follows.

Theorem 2.1. *Let $y^2 = x^3 + ax + b$ be an elliptic curve with $a, b \in \mathbf{Z}$. Let h be the canonical height. Then*

$$\left|h - \frac{1}{2}h_x\right| \leq \frac{1}{6}\log|\Delta| + \frac{1}{6}\log\max(1, |j|) + O(1).$$

In particular,

$$\left|h - \frac{1}{2}h_x\right| \leq \frac{1}{6}\log H(A) + O(1).$$

Proof. Let h_v be the local canonical height, or in other words the Néron function at v, normalized as in Tate, cf. [L 1] Chapter I, §7 and §8, and Chapter III, §4. Let $h_{x,v}$ be the local height of the x-coordinate, that is

$$h_{x,v}(P) = \log\max\{1, |x(P)|_v\}.$$

We now distiguish the two cases when v is archimedean or not.

First let v be the archimedean absolute value on $\mathbf{Q}$. Then in [L 1], Theorem 8.4 of Chapter I, I proved

$$\left|h_v - \frac{1}{2}h_{x,v}\right| \leq \begin{cases} \frac{1}{12}\log|\Delta| + O(1) & \text{if } |j| \leq C_0 \\ \frac{1}{12}\log|\Delta| + \frac{1}{6}|v(j)| + O(1) & \text{if } |j| \geq C_0 \end{cases}$$

where C_0 is some appropriate constant. The two cases correspond to whether j is small or large at the archimedian absolute value. In the present case, Δ is an integer so $\log|\Delta| \geq 0$. Furthermore $j = -2^{12}3^3a^3/\Delta$, so we get $\log|j| = \log|a^3| - \log|\Delta| + O(1)$, (To get all bounds in terms of Δ, see Conjecture 4.)

On the other hand, if v is non-archimedean, Tate has shown that

$$\min\left\{0, \frac{1}{24}v(j)\right\} \leq h_v - \frac{1}{2}h_{x,v} \leq \frac{1}{12}v(\Delta).$$

See Theorem 4.5 of Chapter III. Recall that $v(z) = -\log|z|_p$ if $v = v_p$ is the p-adic absolute value, whence in this case

$$\left|h_v - \frac{1}{2}h_{x,v}\right| \leq \frac{1}{12}v_p(\Delta)$$

because $j = -2^{12}3^3a^3/\Delta$ and a, Δ are p-integral. Taking the sum we then obtain a bound

$$\left|h - \frac{1}{2}h_x\right| = \left|\sum\nolimits_v \left(h_v - \frac{1}{2}h_{x,v}\right)\right| \le \sum\nolimits_v \left|h_v - \frac{1}{2}h_{x,v}\right|.$$

The $O(1)$ comes only from the archimedean v, there is no such constant for $v = v_p$ with a prime number p. The theorem follows at once. (*Note:* one can get something slightly better, see the Remark in [L 1], p. 32.)

Theorem 2.1 suffices amply to apply the bound of Conjecture 3 to the ordinary height h_x. However, I want the estimate for $|h - \frac{1}{2}h_x|$ to be entirely in terms of Δ. For this purpose, among many others, one can use a conjecture of Marshall Hall [Hall].

Conjecture 4. *If (x, y) is an integral point on the elliptic curve $y^2 = x^3 + b$ with $b \in \mathbf{Z}$, then $|x| \ll b^2$.*

Actually, Stark and Trotter after some theoretical probabilistic considerations have suggested that the conjecture should be with an exponent $2 + \epsilon$, or at least some appropriate powers of $\log|b|$ occurring as a factor of b^2. I shall refer to the Hall conjecture so modified. Hall also expressed his conjecture as an inequality

$$|t^2 - u^3| \gg \max\{|u|^{1/2}, |t|^{1/3}\}$$

whenever t, u are integers. Silverman has pointed out to me that [B-C-H-S] yields parametrizations of curves with integral points having the order of magnitude $|\Delta|^5$, by polynomials $g^2 - f^3$ with integer coefficients, having degree $1 + \frac{1}{2}\deg f$. Presumably $|\Delta|^6$ is the best possible power of Δ. Davenport [Da] has proved the Hall conjecture for polynomials in its original form, without the ϵ.

If we now use the Hall conjecture on the equation $4a^3 + 27b^2 = \Delta_0$ (see the remark below), we can apply Theorem 2.1 to get

$$\left|h - \frac{1}{2}h_x\right| \le (1 + \epsilon)\log|\Delta| + O(1).$$

This is of special interest in that $\log|\Delta|$ occurs on the right hand side with multiplicity $1 + \epsilon$. As already mentioned, in some cases one might get an even better estimate.

Remark. In using the Hall conjecture, the "variables" a, b have fixed coefficients, namely 4 and 27. Trivial changes of coordinates (and of Δ_0 by a bounded small factor) reduce such equations to the case when the coefficients are equal to 1. For example we multiply the equation by $4^2 3^3$ and let $u = -12a$, $t = 2^2 c^3 b$ to get $t^2 - u^3 = 4^2 3^3 \Delta_0$.

It is not completely clear how the Hall conjecture should extend to an arbitrary equation

$$y^2 = x^3 + ax + b$$

with $a, b \in \mathbf{Z}$. It seems reasonable to expect:

Conjecture 5. *If (x, y) is an integral point on this elliptic curve, then*

$$|x| \ll \max(|a|^3, |b|^2)^k = H(A)^k$$

for some fixed positive number k independent of $a, b \in \mathbf{Z}$.

The analogous statement has been proved in the function field case by Schmidt [Sch]. It is not clear to me what the exponent k should be. Added in proofs: Stark has tentatively suggested $k = 5/3 + \epsilon$, but he has counterexamples showing that probabilistic arguments can be very misleading, and that one cannot make the conjecture until it is checked in the function field case. For instance, he gives the example

$$y^2 = x^3 + x + \frac{1}{4}t^2,$$

which has the solution $x = t^6 + 2t^2$ and $y = t^9 + 3t^5 + \frac{3}{2}t$.

3. Some Numerical Examples

There exist some tables in the literature giving bases for the Mordell-Weil group, mostly when the rank is 1. Selmer [Se 1], [Se 2] found some solutions which at first sight appear quite large. He works with the curve in Fermat form

$$X^3 + Y^3 = DZ^3, \quad 0 < D \leq 500.$$

I shall give here his two largest solutions. I am indebted to J. Brette for using the computer to put Selmer's solutions in the Weierstrass form

$$y^2 = x^3 + b, \quad \text{with } b = -2^4 3^3 D^2.$$

Since we are interested in the height, it is best to clear denominators and consider the homogeneous form

$$w^2 z = u^3 - 2^4 3^3 D^2 z^3 = u^3 + bz^3$$

to be solved in relatively prime integers (u, w, z). Then

$$x = u/z \quad \text{and} \quad y = w/z.$$

The transformation is given by the formulas

$$u' = 2^2 3DZ, \quad w' = 2^2 3^2 D(Y - Z), \quad z' = X + Y.$$

One still has to divide u', z' by their g.c.d. to get x in lowest form. The rank of $A(\mathbf{Q})$ is 1 in each of the following two cases, and the given values for (u, w, z) yield a generator of $A(\mathbf{Q})$ mod torsion. Note how the height h_x drops to about two-thirds of the height of $h_{X/Z}$, as it should for theoretical reasons since X/Z has degree 3 and x has degree 2. In this special case, one might apply Theorem 2.1 to h_x and an analogue to $h_{X/Z}$ to see this. The behavior is remarkably uniform for this particular algebraic family depending on D.

$D = 346$

$$u = 35,208,298,044,859,842,638,816,896,575,916$$
$$z = 94,295,149,257,506,211,891,484,409,025$$
$$(u', z') = 956,232,024,756,251,670$$

$$\Delta_0 = 3^3(2^4 3^3 D^2)^2 \sim 7.2 \times 10^{16}$$
$$\log \Delta_0 \sim 38.8$$

For the conductor, we use the formula in Stephens [St], p. 125, which for $N \not\equiv \pm 2 \bmod 9$ gives the conductor of the Hecke character as

$$\mathfrak{f}_\chi = 3 \prod_{p|D} p,$$

and therefore

$$N = d_{\mathbf{Q}(\sqrt{-3})}\mathrm{N}\mathfrak{f}_\chi = 3\mathrm{N}\mathfrak{f}_\chi.$$

Since $D = 2 \times 173 \equiv 4 \bmod 9$ we find:

$$\begin{aligned} N &= 2^2 3^3 173^2 = 3,232,332 \\ \log N &\sim 15 \end{aligned}$$

Using $\epsilon(N) = (\log N \log\log N)^{-1/2}$, we also find:

$$\begin{aligned} H(A)^{1/12} N^{\epsilon(N)} \log N &\sim 3.04 \times 10^3 \\ h_x &\sim 72.5 \end{aligned}$$

This puts the bound at about 40 times h_x.

$D = 382$

$$\begin{aligned} u &= 96,793,912,150,542,047,971,667,215,388,941,033 \\ z &= 195,583,944,227,823,667,629,245,665,478,169 \\ (u', z') &= 384,647,097,245,468,469,552 \\ \Delta_0 &= 27(2^4 3^3 D^2)^2 \sim 1.07 \times 10^{17} \\ \log \Delta_0 &\sim 38.9 \\ N &= 2^2 3^3 191^2 = 3,939,948 \\ \log N &\sim 15.2 \\ H(A)^{1/12} N^{\epsilon(N)} \log N &\sim 4.9 \times 10^3 \\ h_x &\sim 80.4 \end{aligned}$$

The estimate is again roughly the same as in the preceding example.

I owe the next example to Andrew Bremner, who considers the curve

$$y^2 = x^3 + 317x.$$

This is the "other" case with complex multiplication. Bremner parametrizes a generator of the Mordell-Weil group in the form

$$x = (a/b)^2, \quad y = ac/b^3$$

where

$$\begin{aligned} a^2 &= 317v^4 - 4u^4 \\ b &= 2uv \\ c &= 317v^4 + 4u^4 \\ u &= 48,869 \text{ and } v = 73,265. \end{aligned}$$

We have

$$\begin{aligned} \Delta_0 &\sim 1.3 \times 10^8 \\ \log \Delta_0 &\sim 18 \quad . \end{aligned}$$

Now we use the formula in Birch-Stephens [B-S], p. 297, giving the conductor of χ_D for $y^2 = x^3 - Dx$, namely

$$\mathfrak{f}_\chi = 4 \prod_{p|D} p$$

If $D \equiv 3 \bmod 4$. Here $D = -317 \equiv 3 \bmod 4$, so $N = d_{\mathbf{Q}(i)} \mathrm{N}\mathfrak{f}_\chi$, and we find:

$$\begin{aligned} N &= 2^6 317^2 = 6,431,296 \\ \log N &\sim 15.7 \\ H(A)^{1/12} N^{\epsilon(N)} \log N &\sim 718 \\ h_x &\sim 81 \end{aligned}$$

So h_x is about one tenth of its presumed upper bound.

It should be noted that most generators in Selmer's tables, for instance, are considerably smaller than the upper bound given by the conjecture. Statistically, it would seem that they follow quite a different distribution, namely

$$R \ll \log|\Delta|.$$

For the exponential height, $H_x = \exp h_x$, this would mean that on the average, one has

$$H_x \ll |\Delta|^k$$

for some constant k. Even the two largest Selmer points have heights compatible with such a better bound taking $k = 2$; while the Bremner point fits with $k = 3$ or 4. Most other points in Selmer's tables, and all the points in Cassels' early table [Ca] or Podsypanin [Po] fit with $k = 1$. Within the range of such tables, it seems impossible to distinguish a polynomial estimate as above from an exponential estimate (for the exponential height) as in Conjecture 3. It would be interesting to have a more precise statistical analysis of the possibility of a polynomial estimate.

We note that if $f(x, y)$ is an irreducible homogeneous polynomial of degree ≥ 3 over $\mathbf{Z}$ then there exists a constant k such that for any *integral* solution (x, y) of the equation

$$f(x,y) = m \in \mathbf{Z}, \quad m \neq 0$$

one has

$$\max(|x|, |y|) \ll |m|^k,$$

so the estimate is known to be polynomial in this case. This was proved by Feldman [Fe], following Baker's method. On the other hand, for the standard Weierstrass equation $y^2 = x^3 + b$, the best known result is exponential and due to Stark [St], using similar methods, namely

$$\max(|x|, |y|) \leq \exp(C_\epsilon |b|^{1+\epsilon}).$$

As to the constants which appear throughout in the estimates, I don't know any explicit bounds for them, although in number theory, one does not expect such constants to be "large" (whatever that means). However, Silverman analyzed the constant in Conjecture 2, and found that in the inequality

$$h(P) \geq c_1 \log|\Delta_{\min}|$$

one had $c_1 < 5 \times 10^{-4}$ in some examples [Si].

The estimate of $O(1)$ in Theorem 2.1, which comes from the archimedean solution value, could be given explicitly, but I have not done so. Zimmer [Zi] gives a good explicit form as $2 \log 2$ in his version.

The constants which come up in Conjecture 1 are of a much more complicated nature, what with Riemann hypothesis estimates coming up. For Hecke L-series, if one is satisfied with the $\epsilon(N)$ which is implied by the Riemann hypothesis, then the constants could be estimated following the general approach of [L 2]. Similar analytic techniques might work more generally for the L-function of an elliptic curve, say assuming the functional equation and Riemann hypothesis. Cf. the remarks at the end of [L 2]. Of course, the Montgomery conjecture (or a weaker form along Titchmarsh) seems to lie outside the range of such techniques. These considerations anyhow lie at the periphery of this kind of analytic number theory.

4. The Hermite Theorem

This section is essentially an appendix, to state the Hermite theorem concerning lattices in euclidean space.

Theorem 4.1. *Let L be a lattice in a vector space V of dimension r over $\mathbf{R}$, with a positive definite quadratic form. Then there exists an orthogonal bases $\{u_1, \ldots, u_r\}$ of V, and a basis $\{e_1, \ldots, e_r\}$ of L having the following properties.*

(i) $e_1 = u_1$ *is a vector of minimal length in* L.

(ii) $e_i = b_{i,1}u_1 + \cdots + b_{i,i-1}u_{i-1} + u_i$
with $b_{i,j} \in \mathbf{R}$ *for* $i = 1, \ldots, r$ *and* $|b_{i,j}| \leq 1/2$ *for all* i, j.

(iii) $|u_i| \leq |e_i| \leq \left(\frac{2}{\sqrt{3}}\right)^{i+1} |u_i|$.

(iv) $|u_i| \leq \left(\frac{2}{\sqrt{3}}\right)|u_{i+1}|$.

The theorem is proved by induction, essentially by Gram-Schmidt orthogonalization, and taking minimal vectors successively.

A basis of the lattice as in Theorem 4.1 is called **almost orthogonalized**. One gets the following bounds for the lengths of such basis elements.

Theorem 4.2. *Let $\{e_1, \ldots, e_r\}$ be an almost orthogonalized basis of L. Then:*

(1) $|e_1| \cdots |e_r| \leq \left(\frac{2}{\sqrt{3}}\right)^{r(r-1)/2} \det(L)$

where $\det(L)$ *is the volume of a fundamental domain.*

(2) $|e_1| \leq \left(\frac{2}{\sqrt{3}}\right)^{(r-1)/2} \det(L)^{1/r}$.

Proof. The basis $\{e_1, \ldots, e_r\}$ of L is obtained from the orthogonal basis V by a triangular matrix all of whose diagonal elements are equal to 1. The volume of the rectangular box spanned by $u_1, \ldots, u_r$ is equal to the product of the lengths of the sides, so

$$\det(L) = |u_1| \cdots |u_r|.$$

The first assertion follows by using (ii) in Theorem 4.1. The second assertion follows immediately.

Note. The appearance of $r(r-1)/2$ (so essentially r^2) surprised me at first, but I saw no way of reducing it to the first power of r.

References

[B-C-H-S] B. BIRCH, S. CHOWLA, M. HALL, A. SCHINZEL, *On the difference* $x^3 - y^2$, Norske Vid. Selsk. Forrh. 38 (1965), pp. 65–69.

[B-S] B. BIRCH and N. STEPHENS, *The parity of the rank of the Mordell-Weil group*, Topology 5 (1966) pp. 295–299.

[Ca] J. W. CASSELS, *The rational solutions of the Diophantine equation* $Y^2 = X^3 - D$, Acta Math. 82 (1950) pp. 243–273. *Addenda and corrigenda to the above*, Acta Math. 84 (1951), p. 299.

[Da] H. DAVENPORT, *On* $f^3(t) - g^2(t)$, Kon. Norsk Vid. Selsk. For. Bd. 38 Nr. 20 (1965) pp. 86–87.

[De] V. A. DEMJANENKO, *Estimate of the remainder term in Tate's formula*, Mat. Zam. 3 (1968), pp. 271–278.

[Fe] N. I. FELDMAN, *An effective refinement of the exponent in Liouville's theorem*, Izv. Akad. Nauk 35 (1971) pp. 973–990, AMS Transl. (1971), pp. 985–1002.

[H] M. HALL, *The diophantine equation* $x^3 - y^2 = k$, *Computers in Number Theory*, Academic Press (1971), pp. 173–198.

[L 1] S. LANG, *Elliptic curves: diophantine analysis*, Springer Verlag, 1978.

[L 2] S. LANG, *On the zeta function of number fields*, Invent. Math. 12 (1971), pp. 337–345.

[M] J. MANIN, *Cyclotomic fields and modular curves*, Russian Mathematical Surveys Vol. 26 No. 6, Published by the London Mathematical Society, Macmillan Journals Ltd, 1971.

[Mo] H. MONTGOMERY, *Extreme values of the Riemann zeta function*, Comment. Math. Helvetici 52 (1977), pp. 511–518.

[Po] V. D. PODSYPANIN, *On the equation* $x^3 = y^2 + Az^6$, Math. Sbornik 24 (1949), pp. 391–403 (See also Cassels' corrections in [Ca]).

[Sch] W. SCHMIDT, *Thue's equation over function fields*, J. Australian Math. Soc. (A) 25 (1978) pp. 385–422.

[Se 1] E. SELMER, *The diophantine equation $ax^3 + by^3 + cz^3$*, Acta Math. (1951), pp. 203–362.

[Se 2] E. SELMER, *Ditto, Completion of the tables*, Acta. Math. (1954), pp. 191–197.

[Sie] C. L. SIEGEL, *Abschätzung von Einheiten*, Nachr. Wiss. Göttingen (1969), pp. 71–86.

[Sil 1] J. SILVERMAN, *Lower bound for the canonical height on elliptic curves*, Duke Math. J. Vol. 48 No. 3 (1981), pp. 633–648.

[Sil 2] J. SILVERMAN, *Integer points and the rank of Thue Elliptic curves*, Invent. Math. 66 (1962), pp. 395–404.

[Sil 3] J. SILVERMAN, *Heights and the Specialization map for families of abelian varieties*, to appear.

[St] H. STARK, *Effective estimates of solutions of some diophantine Euations*, Acta. Arith. 24 (1973), pp. 251–259.

[Ste] N. STEPHENS, *The diophantine equation $X^3+Y^3 = DZ^3$ and the conjectures of Birch-Swinnerton Dyer*, J. reine angew. Math. 231 (1968), pp. 121–162.

[Ta 1] J. TATE, *The arithmetic of elliptic curves*, Invent. Math. (1974), pp. 179–206.

[Ta 2] J. TATE, *Algorithm for determining the Type of a Singular Fiber in an Elliptic Pencil*, Modular Functions of One Variable IV, Springer Lecture Notes 476 (Antwerp Conference) (1972), pp. 33–52.

[Ti] E. C. TITCHMARSH, *The theory of the Riemann zeta function*, Oxford Clarendon Press, 1951.

[Zi] H. ZIMMER, *On the difference of the Weil height and the Néron-Tate height*, Math. Z. 147 (1976), pp. 35–51.

Received August 24, 1982

Professor Serge Lang
Department of Mathematics
Yale University
Box 2155 Yale Station
New Haven, Connecticut 06520

Séminaire de théories des nombres
(Paris: Birkhäuser Boston, 1983): 407-419

VOJTA'S CONJECTURE

Serge Lang
Department of Mathematics
Yale University
New Haven, CT 06520
U.S.A.

§ 1. Nevanlinna theory

Let $f : \mathbb{C}^d \longrightarrow X$ be a holomorphic map, where X is a complex non-singular variety of dimension d. Let D be an effective divisor on X, with associated invertible sheaf $\mathcal{L}$. Let s be a meromorphic section of $\mathcal{L}$, with divisor $(s) = D$. We suppose that f is non-degenerate, in the sense that its Jacobian is not zero somewhere. For positive real r we define

$$m(D,r,f) = \int_{\mathrm{Bd}\,B(r)} -\log|f^*s|^2 \sigma .$$

where σ is the natural normalized differential form invariant under rotations giving spheres area 1. When $d = 1$, then $\sigma = d\theta/2\pi$. Actually, $m(D,r,f)$ should be written $m(s,r,f)$, but two sections with the same divisor differ by multiplication with a constant, so $m(s,r,f)$ is determined modulo an additive constant. One can select this constant such that $m(s,r,f) \geqq 0$, so by abuse of notation, we shall also write $m(D,r,f) \geqq 0$.

We also define

$N(D,r,d)$ = normalized measure of the analytic divisor in the ball of radius r whose image under f is contained in D; (Cf. Griffiths [Gr] for the normalization.)

$$N(D,r,f) = \int_0^r [N(D,r,f)-N(D,0,f)]\,\frac{dr}{r} + N(D,0,f)\log r .$$

$$T(D,r,f) = m(D,r,f) + N(D,r,f) .$$

Remark. If $d = 1$ then $N(D,r,f) = n(D,r,f)$ is the number of points in the disc of radius r whose image under f lies in D.

One formulation of the FIRST MAIN THEOREM (FMT) of Nevanlinna theory runs as follows. *The function* $T(D,r,f)$ *depends only on the linear equivalence class of* D, *modulo bounded functions* $O(1)$.

The first main theorem is relatively easy to prove. More important is the SECOND MAIN THEOREM (SMT), which we state in the following form:

Let D *be a divisor on* X *with simple normal crossings (SNC, meaning that the irreducible components of* D *are non-singular, and intersect transversally). Let* E *be an ample divisor, and* K *the canonical class. Given* ε, *there exists a set of finite measure* $Z(\varepsilon)$ *such that for* r *not in this set,*

$$m(D,r,f) + T(K,r,f) \leqq \varepsilon T(E,r,f) .$$

This is an improved formulation of the statement as it is given for instance in Griffiths [Gr], p. 68, formula 3.5.

§ 2. Weil functions

Let X be a projective variety defined over $\mathbb{C}$ or $\mathbb{C}_p$ (p-adic complex numbers = completion of the algebraic closure of $\mathbb{Q}_p$). Let $\mathcal{L}$ be an invertible sheaf on X and let ρ be a smooth metric on $\mathcal{L}$. If s is a meromorphic section of $\mathcal{L}$ with divisor D, we define the associated Weil function (also called Green's function)

$$\lambda(P) = -\log|s(P)| \quad \text{for} \quad P \notin \operatorname{supp}(D) .$$

If we change the metric or s with the same divisor, λ changes by a bounded smooth function, so is determined mod $O(1)$. We denote such a function by λ_D. It has the following properties:

The association $D \longmapsto \lambda_D$ is a homomorphism mod $O(1)$.

If $D = (f)$ on an open set U (Zariski) then there exists a smooth function α on U such that

$$\lambda_D(P) = -\log |f(P)| + \alpha .$$

If D is effective, then $\lambda_D \geqq -O(1)$ (agreeing that values of λ_D on D are then ∞).

If v denotes the absolute value on $\mathbb{C}_v$ then we write

$$v(a) = -\log|a|_v$$

for any element $a \in \mathbb{C}_v$, so we can write

$$\lambda_D = v\circ f + \alpha .$$

In the sequel, metrics will not be used as such; only the associated Weil functions and the above properties will play a role. Note that these Weil-Green functions need not be harmonic. In some cases, they may be, for instance in the case of divisors of degree 0 on a curve. But if the divisor has non-zero degree, then the Green function is not harmonic.

In the sequel, we shall deal with global objects, and then the Weil functions and others must be indexed by v , such as $\lambda_{D,v}$, α_v, etc.

§ 3. Heights (Cf.[La])

Let K be a number field, and let $\{v\}$ be its set of absolute values extending either the ordinary absolute value on $\mathbb{Q}$, or the p-adic absolute values such that $|p|_v = 1/p$. We let K_v be the completion, and K_v^a its algebraic closure. Then we have the product formula

$$\sum_v d_v v(a) = 0$$

where $d_v = \dfrac{[K_v : \mathbb{Q}_v]}{[K : \mathbb{Q}]}$ and $a \in K$, $a \neq 0$. We let $\|a\|_v = |a|_v^{d_v}$

Let $(x_o,\ldots,x_n) \in \mathbb{P}^n(K)$ be a point in projective space over K . We define its height

$$h(P) = \sum_v \log \max_i \|x_i\|_v$$

If $K = \mathbb{Q}$ and $x_o, \ldots, x_n \in \mathbb{Z}$ are relatively prime, then

$$h(P) = \log \max |x_i|$$

where the absolute value is the ordinary one. From this it is immediate that there is only a finite number of points of bounded height and bounded degree.

Let

$$\varphi : X \longrightarrow \mathbb{P}^n$$

be a morphism of a projective non-singular variety into projective space. We define

$$h_\varphi(P) = h(\varphi(P)) \quad \text{for} \quad P \in V(K^a) .$$

The basic theorem about heights states:

There exists a unique homomorphism $c \longmapsto h_c$

$$\text{Pic}(X) \longrightarrow \text{functions from } X(K^a) \text{ to } \mathbb{R} \text{ modulo bounded functions}$$

such that if D *is very ample, and* $\mathcal{O}(D) = \varphi^*\mathcal{O}_{\mathbb{P}}(1)$, *then*

<u>H 1</u>. $$h_c = h_\varphi + O(1) .$$

In the above statement, we denote by h_c any one of the functions in its class mod bounded functions. Similarly, if D lies in c , we also write h_D instead of h_c . This height function also satisfies the following properties:

<u>H 2</u>. If D is effective, then $h_D \geq -O(1)$.

H 3. If E is ample and D any divisor, then

$$h_D = O(h_E) .$$

In particular, if E_1, E_2 are ample, then

$$h_{E_1} \gg \ll h_{E_2} .$$

We are using standard notation concerning orders of magnitude. Since according to our conventions, a given height h_E is defined only mod bounded functions, the notation $h_D = O(h_E)$ or $h_D \ll h_E$ means that there exists a constant C such that for all points P with $h_E(P)$ sufficiently large, we have $|h_D(P)| \leqq Ch_E(P)$.

Essential to the existence and uniqueness of such height functions h_c is the property of elementary algebraic geometry that given any divisor D , if E is ample, then $D + mE$ is very ample for all $m \geqq m_o$.

A fundamental result also states that one can choose metrics ρ_v "uniformly" such that

$$h_D = \sum_v d_v \lambda_{D,v} + O(1)$$

The right hand side depends on Green-Weil functions $\lambda_{D,v}$, and so is a priori defined only for P outside the support of D . Since h_D depends only on the linear equivalence class of D mod $O(1)$, we can change D by a linear equivalence so as to make the right hand side defined at a given point.

Now let S be a finite set of absolute values on K . We define, relative to a given choice of Weil-Green functions and heights:

$$m(D,S) = \sum_{v\in S} d_v \lambda_{D,v}$$

$$N(D,S) = \sum_{v\notin S} d_v \lambda_{D,v}$$

Then

$$h_D = m(D,S) + N(D,S) ,$$

and one basic property of heights says that h_D depends only on the linear equivalence class of D. This is Vojta's translation of FMT into the number theoretic context, with the height h_D corresponding to the function $T(D)$ of Nevanlinnna theory.

Remark. The properties of heights listed above also hold for T, as well as others listed for instance in [La], e.g. if D is algebraically equivalent to 0, then $T(D) = O(T(E))$ for E ample. As far as I can tell, in the analytic context, there has been no such systematic listing of the properties of T, similar to the listing of the properties of heights as in number theory.

Vojta's translation of SMT yields his conjecture:

Let X be a projective non-singular variety defined over a number field K. Let S be a finite set of absolute values on K. Let D be a divisor on X rational over K and with simple normal crossings. Let E be ample on X. Give ε. Then there exists a proper Zariski closed subset $Z(S,D,E,\varepsilon) = Z(\varepsilon)$ such that

$$m(D,S,P) + h_{\mathcal{K}}(P) \leq \varepsilon h_E(P) \quad \textit{for} \quad P \in X(K) - Z(\varepsilon) .$$

Or in other words,

$$\sum_{v \in S} d_v \lambda_{D,v} + h_{\mathcal{K}} \leq \varepsilon h_E \quad \text{on} \quad X(K) - Z(\varepsilon)$$

where $\mathcal{K}$ is the canonical class.

EXAMPLES

Example 1. Let $X = \mathbb{P}^1$, $K = \mathbb{Q}$, $E = (\infty)$. Let α be algebraic, and let

$$f(t) = \prod_{\sigma} (\sigma\alpha - t)$$

where the product is taken over all conjugates $\sigma\alpha$ of α over $\mathbb{Q}$. Let D be the divisor of zeros of f. The canonical class K is just $-2(\infty)$. A rational point P corresponds to a rational value $t = p/q$ with $p,q \in \mathbb{Z}$, $q > 0$, and p,q relatively prime. We let S consist of the absolute value at infinity. If $|f(p/q)|$ is small, then p/q is close to some root of f. If p/q is close to α, then it has to be far away from the other conjugates of α. Consequently Vojta's inequality yields from the definitions:

$$-\log|\alpha - p/q| - 2h_\infty(p/q) \leqq \varepsilon h_\infty(p/q)$$

with a finite number of exceptional fractions. Exponentiating, this reads

$$\left|\alpha - \frac{p}{q}\right| \geqq \frac{1}{q^{2+\varepsilon}},$$

which is Roth's theorem.

Remark. Some time ago, I conjectured that instead of the q^ε in Roth's theorem, one could take a power of $\log q$ (even possibly $(\log q)^{1+\varepsilon}$). Similarly, in Vojta's conjecture, the right hand side should be replaced conjecturally by $O(\log h_E)$. If one looks back at the Nevanlinna theory, one then sees that the analogous statement is true, and relies on an extra analytic argument which is called the lemma on logarithmic derivatives. Cf. Griffiths [Gr].

Example 2. Let $X = \mathbb{P}^n$ and let $D = L_o+\dots+L_n$ be the formal sum of the hyperplane coordinate sections, with L_o at infinity, and $E = L_o$. Let φ_i be a rational function such that

$$(\varphi_i) = L_i - L_o .$$

Let S be a finite set of absolute values. Note that in the case of $\mathbb{P}^n$, the canonical class K contains $-(n+1)L_o$. Consequently, Vojta's inequality in this case yields

$$\prod_i \prod_{v\in S} \| \varphi_i(P) \|_v \geqq \frac{1}{H(P)^{n+1+\varepsilon}}$$

for all P outside the closed set $Z(\varepsilon)$. This is Schmidt's theorem, except that Schmidt arrives at the conclusion that the exceptional set is a finite union of hyperplanes. In order to make Vojta's conjecture imply Schmidt strictly, one would have to refine it so as to give a bound on the degrees of the components of the exceptional set, which should turn out to be 1 if the original data is linear.

Example 3. Let X be a curve of genus $\geqq 2$. Take S empty. The canonical class has degree $2g-2$ where g is the genus, and so is ample. Then Vojta's inequality now reads

$$h_K \leqq \varepsilon h_E \quad \text{on} \quad X(K) ,$$

except for a finite set of points. Since K is ample, such an inequality holds only if $X(K)$ is finite, which is Falting's theorem.

Example 4. This is a higher dimensional version of the preceding example. Instead of assuming that X is a curve, we let X have any dimension, but assume that the canonical class is ample. The same inequality shows that the set of rational points is not Zariski dense.

This goes toward an old conjecture of mine, that if a variety is hyperbolic, then it has only a finite number of rational points. The effect of hyperbolicity should be to eliminate the exceptional Zariski set in Vojta's conjecture. For progress concerning this conjecture in the function field case, cf. Noguchi [No], under the related assumption that the cotangent bundle is ample, and that the rational points are Zariski dense.

To apply the argument of Vojta's inequality it is not necessary to assume that the canonical invertible sheaf is ample, it suffices to be in a situation when for any ample divisor E, $h_E = O(h_K)$. This is the case for varieties of general type, which means that the rational map of X defined by a sufficiently high multiple of the canonical class gives a rational map of dimension $d = \dim X$. Then we have $h^0(mK) \gg m^d$ for m sufficiently large, and we use the following lemma.

Lemma. *Let X be a non-singular variety. Let E be very ample on X, and let D be a divisor on X such that $h^0(mD) \gg m^d$ for $m \geqq m_0$. Then there exists m_1 such that $h^0(mD-E) \gg m^d$, and in particular, $mD-E$ is linearly equivalent to an effective divisor, for all $m \geqq m_1$.*

Proof. First a remark for any divisor D. Let E' be ample, and such that $D + E'$ is ample. Then we have an inclusion

$$H^0(mD) \subset H^0(mD + mE') ,$$

which shows that $h^0(mD) \leqq h^0(mD + mE') = \chi(m(D + E'))$ for m large because the higher cohomology groups vanish for m large, so $h^0(mD) \gg m^d$.

Now for the lemma, without loss of generality we can replace E

by any divisor in its class, and thus without loss of generality we may assume that E is an irreducible non-singular subvariety of X. We have the exact sequence

$$0 \longrightarrow \mathcal{O}(mD-E) \longrightarrow \mathcal{O}(mD) \longrightarrow \mathcal{O}(mD)|E \longrightarrow 0 ,$$

whence the exact cohomology sequence

$$0 \longrightarrow H^0(X,mD-E) \longrightarrow H^0(X,mD) \longrightarrow H^0(E,(\mathcal{O}(D)|E)^{\otimes m})$$

noting that $\mathcal{O}(mD)|E = (\mathcal{O}(D)|E)^{\otimes m}$. Applying the first remark to this invertible sheaf on E we conclude that the dimension of the term on the right is $\ll m^{d-1}$, so $h^0(X,mD-E) \gg m^d$ for m large, and in particular is positive for m large, whence the lemma follows.

For $mD - E$ effective, we get $h_E \leqq h_{mD} + O(1)$ as desired.

Example 5. Let A be an abelian variety, and let D be a very ample divisor with SNC. Let S be a finite set of absolute values of K containing the archimedean ones. Let $\varphi_1,\ldots,\varphi_n$ be a set of generators for the space of sections of $\mathcal{O}(D)$. Let $\mathfrak{o}_S$ be the ring of S-integers in K (elements of K which are integral at all $v \notin S$). A point $P \in A(K)$ is said to be S-integral relative to these generators if $\varphi_i(P) \in \mathfrak{o}_S$ for $i = 1,\ldots,n$. On the set of such S-integral points, we have

$$\sum_{v\in S} d_v \lambda_{D,v} = h_D + O(1)$$

immediately from the definitions. The canonical class is 0. Then again Vojta's inequality shows that the set of S-integral points as above is not Zariski dense.

This is in the direction of my old conjecture that on any affine open subset of an abelian variety, the set of S-integral points is finite. However, in this stronger conjecture, we again see the difference between finiteness and the property of not being Zariski dense.

Example 6. Hall's conjecture Marshall Hall conjectured that if x, y are integers, and $x^3-y^2 \neq 0$ then

$$|x^3-y^2| \geqq \max(|x^3|, |y^2|)^{\frac{1}{6}-\varepsilon}$$

with a finite number of exceptions. Actually, Hall omitted the ε, but Stark and Trotter for probabilistic reasons have pointed out that it is almost certainly needed, so we put it in.

Vojta has shown that his conjecture implies Hall's. We sketch the argument. Let

$$f : \mathbb{P}_1^2 \longrightarrow \mathbb{P}_2^2$$

be the rational map defined on projective coordinates by

$$f(x,y,z) = (x^3, y^2 z, z^3) .$$

Then f is a morphism except at $(0,1,0)$. We have indexed projective 2-space by indices 1 and 2 to distinguish the space of departure and the space of arrival. We let $L = L_1$ be the hyperplane at infinity on $\mathbb{P}_1^2$, and L_2 the hyperplane at infinity on $\mathbb{P}_2^2$.

Let C be the curve in $\mathbb{P}_1^2$ defined by $x^3-y^2 = 0$. Let φ be the rational function defined by

$$\varphi(x,y) = x^3-y^2 .$$

Then the divisor of φ is given by

$$(\varphi) = C - 3L .$$

In terms of heights, Hall's conjecture can be formulated in the form

$$\log |\varphi(x,y)| \geq \frac{1}{6} h_{L_2} \circ f(x,y) + \text{error term},$$

or if v denotes the ordinary absolute value on $\mathbb{Q}$,

$$(1) \qquad v \circ \varphi(x,y) \leq -\frac{1}{6} h_{L_2} \circ f(x,y) + \text{error term}.$$

Note that $v \circ \varphi = \lambda_{(\varphi)}$ is a Weil function associated with the divisor (φ). Thus Hall's conjecture amounts to an inequality on Weil functions. By blowing up the point of indeterminacy of f and the singularity of C at $(0,0)$, one obtains a variety X and a corresponding morphism $f_1 : X \longrightarrow \mathbb{P}^2$ making the following diagram commutative:

$$\begin{array}{ccc} X & & \\ \downarrow & \overset{f_1}{\searrow} & \\ \mathbb{P}^2_1 & \longrightarrow & \mathbb{P}^2_2 . \end{array}$$

The blow ups are chosen so that the exceptional divisor and C have simple normal crossings. By taking D to be their sum together with the hyperplane at infinity, Vojta shows that his conjecture implies Hall's. By a similar technique, Vojta shows that his conjecture implies several other classical diophantine conjectures. I refer the reader to his forthcoming paper on the subject.

BIBLIOGRAPHY

[Gr] P. Griffiths, *Entire holomorphic mappings in one and several variables:* Hermann Weyl Lectures, Institute for Advanced Study, Institute for Advanced Study, Princeton Univ. Press, Princeton NJ, 1976.

[La] S. Lang, *Fundamentals of Diophantine Geometry*, Springer Verlag, 1984.

[No] J. Noguchi, *A higher dimensional analogue of Mordell's conjecture over function fields*, Math. Ann. 258 (1981) pp. 207-212.

[Vo] P. Vojta, *Integral points on varieties*, Thesis, Harvard, 1983.

Reprinted from

Séminaire de Théorie des Nombres,
Paris 1984–85
Edited by Catherine Goldstein

Progress in Mathematics, Volume 63

Printed in the U.S.A.

Birkhäuser
Boston · Basel · Stuttgart

VARIÉTÉS HYPERBOLIQUES ET ANALYSE DIOPHANTIENNE

Serge Lang

Dans [La 2] je traite en détail un certain nombre de relations entre les notions suivantes :

- notions de distance et de mesure;
- notions de géométrie différentielle et de courbure;
- notions de géométrie algébrique (amplitude);
- notions diophantiennes.

En accord avec les organisateurs du Séminaire de Théorie des Nombres pour lequel j'ai résumé un certain ensemble de conjectures, il a paru utile de publier mon exposé séparément, ce que je fais ici. Je commence par donner toutes les définitions qui s'imposent, puis je donne certains énoncés, constituant un programme. Je terminera par de courtes indications bibliographiques.

Hyperbolicité.

Pour les besoins de la cause, toutes les variétés seront supposées projectives, définies sur $\mathbb{C}$. On peut donner a priori deux définitions de l'hyperbolicité (complexe). La première, plus simple, à la Brody est que toute application holomorphe de $\mathbb{C}$ dans la variété est constante. La seconde (démontrée équivalente à la première par Brody [Br]) est dûe à Kobayashi, et nous la rappelons.

Soit X une variété et $x, y \in X$ deux points de X. On va définir une pseudo distance d_X dite de Kobayashi. On joint x, y par une suite de disques, c'est-à-dire qu'on considère une suite finie

$$f_i : \mathbb{D} \longrightarrow X \qquad (i = 1,\ldots,m)$$

d'applications holomorphes du disque unité dans X, avec des points $x_1 = x, x_2,\ldots,x_m = y$ dans X et des points $p_i, q_i \in \mathbb{D}$ tels que

$$f_1(p_1) = x, \qquad f_m(q_m) = y \qquad \text{et} \qquad f_i(q_i) = f_{i+1}(p_{i+1}).$$

Dans le disque unité $\mathbb{D}$ on a la métrique hyperbolique (de Poincaré), où en un point z la longueur d'un vecteur tangent $v \in T_z(\mathbb{D})$ est donnée par la formule

$$|v|_z^2 = \frac{|v|^2}{(1-|z|^2)^2} .$$

On a donc la distance hyperbolique $d(p_i,q_i)$ dans D, et l'on définit

$$d_X(x,y) = \inf \sum_{i=1}^{m} d_{hyp}(p_i,q_i),$$

où le inf est pris pour toutes les suites de disques $(f_1,\ldots,f_m)$ et les choix des p_i,q_i comme ci-dessus.

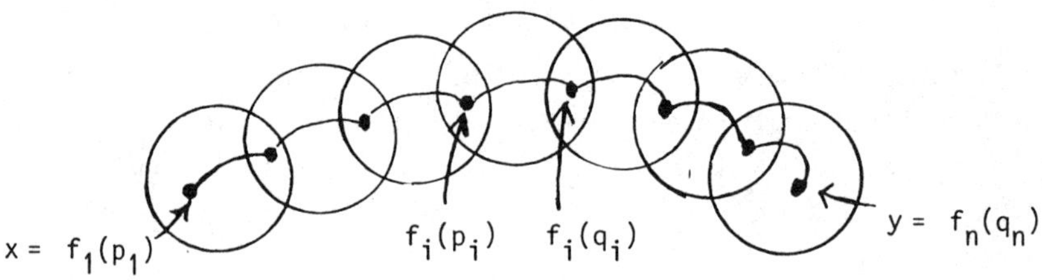

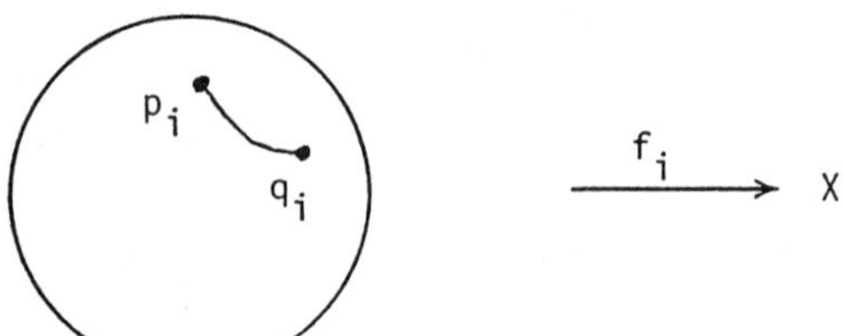

Si $X = \mathbb{C}$ alors on a $d_X(x,y) = 0$ pour tout $x, y \in \mathbb{C}$, car x, y sont contenus dans des disques de rayons arbitrairement grands. Par dilatation, on peut trouver une application holomorphe de $\mathbb{D}$ dans $\mathbb{C}$ telle que x, y soient les images de deux points p, q tout près de l'origine, et la distance hyperbolique est voisine de la distance euclidienne au voisinage de l'origine, donc la distance de Kobayashi entre x, y est arbitrairement petite, donc égale à 0.

On dit que X est <u>Kobayashi hyperbolique</u> si d_X est une distance, c'est-à-dire que $x \neq y$ implique $d_X(x,y) > 0$.

Au lieu de disques et de distances, on peut aussi envoyer des polydisques dans X, de la même dimension que X. Soit $n = \dim X$. On définit alors la <u>mesure de Kobayashi</u> μ_X par le même procédé. Soit B Borel-mesurable dans X, et soit $\{U_i\}$ une suite dénombrable d'ouverts dans $\mathbb{D}^n$, avec des applications holomorphes

$$f_i : \mathbb{D}^n \longrightarrow X$$

telles que la réunion des $f_i(U_i)$ recouvre B. On définit

$$\mu_X(B) = \inf \Sigma \, \mu_{hyp}(U_i)$$

où μ_{hyp} est la mesure hyperbolique dans $\mathbb{D}^n$, et le inf est pris sur tous les choix de f_i et U_i comme ci-dessus. On dit que X est <u>mesure hyperbolique</u> si pour tout ouvert V non vide dans X on a

$$\mu_X(V) > 0.$$

Etant donné une certaine propriété, on aura à traiter de sa <u>pseudofication</u>, c'est-à-dire de la même propriété mais seulement valable pour un ouvert non vide de Zariski. On peut pseudofier l'hyperbolicité de Brody et celle de Kobayashi. Etant donné une variété X, je définis son <u>ensemble exceptionnel</u>.

$\mathrm{Exc}(X) =$ clôture Zariskienne de la réunion de toutes les images d'applications holomorphes non constantes

$$f : \mathbb{C} \longrightarrow X.$$

On dira que X est <u>Brody pseudo hyperbolique</u> si $\mathrm{Exc}(X)$ est un sous-ensemble propre, c'est-à-dire $\mathrm{Exc}(X) \neq X$.

On dira que X est <u>Kobayashi pseudo hyperbolique</u> s'il existe un sous-ensemble algébrique $Y \neq X$ tel que si $x, y \in X$ et $d_X(x,y) = 0$, alors $x = y$ ou bien $x, y \in Y$. Je conjecture que ces deux définitions sont équivalentes, et qu'on peut prendre $Y = \mathrm{Exc}(X)$. En raison de cette conjecture, j'omettrai les préfixes Kobayashi ou Brody devant pseudo hyperbolique.

On remarquera que la notion de mesure hyperbolique est déjà pseudofiée.

Exemples de variétés hyperboliques.

(i) Quotients compacts d'un domaine borné par un groupe discret opérant librement. Toute application holomorphe de $\mathbb{C}$ dans ce quotient se remonte au revêtement universel, et est constante par le théorème de Liouville.

(ii) Sous variétésde variétés abéliennes ne contenant pas de sous tores translatés de dimension >0 [Gr 2].

(iii) Brody-Green [Br-Gr] montrent que

$$z_0^d + z_1^d + z_2^d + z_3^d + (tz_0z_1)^{d/2} + (tz_0z_2)^{d/2} = 0$$

est hyperbolique pour $t \neq 0$ (perturbation de l'hypersurface de Fermat), et d pair ≥ 50. Pour t petit, c'est une variété non-singulière, et on rappelle qu'en dimension ≥ 2 une hypersurface non-singulière est simplement connexe. Brody [Br] a également montré que l'hyperbolicité est une condition ouverte, et l'exemple de Brody-Green montre qu'elle n'est pas fermée : pour tout $t = 0$, il y a des droites sur Fermat.

Courbure et forme de Ricci.

Supposons X non singulière. Par une forme volume Ψ sur X on entend une forme qui s'exprime en coordonnées complexes locales sous la forme

$$\Psi(z) = h(z)\ \Phi(z),$$

où

$$\Phi(z) = \prod_{i=1}^{n} \frac{\sqrt{-1}}{2\pi}\, dz_i \wedge d\bar{z}_i$$

est la forme euclidienne et $h > 0$ est C^∞. On définit la forme de Ricci de Ψ localement par

$$\mathrm{Ric}(\Psi) = dd^c \log h,$$

qui est indépendant de la représentation en coordonnées locales, puisque $dd^c \log(g\bar{g}) = 0$ pour g holomorphe. Rappelons que d est l'opérateur différentiel ordinaire, $d = \partial + \bar{\partial}$, et

$$d^c = \frac{1}{2\pi\sqrt{-1}}\,\frac{1}{2}\,(\partial - \bar{\partial}), \qquad \text{d'où} \qquad dd^c = \frac{\sqrt{-1}}{2\pi}\,\partial\bar{\partial}.$$

On aura également besoin d'une généralisation d'une forme volume, qui admet des zéros sur un sous-ensemble analytique propre. Donc on dit que Ψ est une forme de pseudo-volume si

$$\Psi(z) = |g(z)|^{2q} h(z)\Phi(z),$$

ou Φ, h sont comme ci-dessus, g est holomorphe, et $q \in \mathbb{Q}$, $q > 0$. On définit Ric d'une forme de pseudo-volume par la même formule.

Etant donnée une forme volume Ψ, il existe une fonction G_Ψ, que j'appelle la fonction de Griffiths, telle que si $n = \dim X$ on a

$$\frac{1}{n!} \operatorname{Ric}(\Psi)^n = G_\Psi \Psi .$$

On peut aussi écrire

$$G_\Psi = \frac{1}{n!} \operatorname{Ric}(\Psi)^n/\Psi.$$

Si $n = 1$, alors G = -(courbure de Gauss), et en coordonnées locales,

$$G_\Psi(z) = \frac{1}{h} \frac{\partial^2 \log h}{\partial z \partial \bar{z}}$$

Passons maintenant à des notions algébriques.

Mordellicité.

La variété X est définie par des équations polynomiales, et possède un corps de définition de type fini sur $\mathbb{Q}$. On dira que X, ou un ouvert de Zariski, est mordellique, s'il ne possède qu'un nombre fini de points rationnels dans tout corps de type fini sur $\mathbb{Q}$ contenant un corps de définition pour X.

On dira que X est pseudo mordellique s'il existe un sous-ensemble algébrique $Y \neq X$ tel que $X - Y$ soit mordellique.

Amplitude et classe canonique.

Supposons X non-singulière, et soit K la classe canonique. On rappelle que K est ample s'il existe un produit tensoriel $K^{\otimes m} = K^m$ tel qu'une base $(s_0,\ldots,s_N)$ des sections dans $H^0(X,K^m)$ engendre K et donne un plongement projectif

$$\varphi_m : X \longrightarrow \mathbb{P}^N .$$

On dit que K est <u>très ample</u> si on peut prendre $m = 1$.

On dira que K est <u>pseudo ample</u> s'il existe m tel que l'application rationnelle φ_m donne un plongement projectif d'un ouvert de Zariski non vide (pour les géomètres algébristes, ceci équivaut à la condition que la dimension de Kodaira soit maximale). On dira que X est <u>très canonique</u> (resp. <u>canonique</u>, resp. <u>pseudo canonique</u>) selon que sa classe canonique est <u>très ample</u> (resp. <u>ample</u>, resp. <u>pseudo ample</u>), ce qui rend la terminologie fonctorielle par rapport aux idées.

Si X est singulière, on dira que X est pseudo canonique selon qu'une désingularisation (et donc toute désingularisation) est pseudo canonique.

Ceci termine la liste des définitions des notions dont nous avons besoin pour formuler le programme suivant.

Définissons l'<u>ensemble exceptionnel algébrique</u>.

$\mathrm{Exc}_{\mathrm{alg}}(X) =$ réunion des images de toutes les applications rationnelles non constantes de $\mathbb{P}^1$ et des variétés abéliennes dans X.

Alors je conjecture que les deux ensembles exceptionnels (analytique comme ci-dessus, et algébrique) sont égaux. En outre, $X - \mathrm{Exc}(X)$ est mordellique.

Tant que l'égalité entre les deux ensembles exceptionnels n'est pas démontrée, cette conjecture donne lieu à plusieurs variantes, y compris que $\mathrm{Exc}(X)$ n'est pas mordellique.

On voudrait savoir quand le sous-ensemble exceptionnel est propre. L'équivalence entre les conditions suivantes fournirait une réponse.

P 1. X est pseudo canonique.

P 2. X est pseudo mordellique.

P 3. X est pseudo hyperbolique.

P 3'. Il existe un sous-ensemble algébrique propre contenant toutes les images d'applications rationnelles non-constantes de $\mathbb{P}^1$ et de variétés abéliennes dans X.

P 4. Si X est non-singulière, il existe une forme de pseudo volume Ψ telle que $\mathrm{Ric}(\Psi)$ soit positive.

P 5. X est mesure hyperbolique.

Enfin, sans la pseudofication, l'équivalence des conditions suivantes donnerait une bonne caractérisation que X soit mordellique.

1. Toutes les sous-variétés de X, y compris X, sont pseudo canoniques.

2. X est mordellique.

3. X est hyperbolique.

3'. Toute application rationnelle de $\mathbb{P}^1$ et d'une variété abélienne dans X est constante.

4. Si X est non-singulière, il existe une forme ω positive de type (1,1), et $B > 0$ tels que pour toute immersion $Y \subset X$ d'une variété complexe non-singulière de dimension 1, on ait

$$G_{\omega|Y} \geq B.$$

5. Toutes les sous-variétés de X, y compris X, sont mesures hyperboliques.

On remarquera que $\dim Y = 1$ implique que la restriction $\omega|Y$ est une forme volume sur Y, donc que la fonction de Griffiths est définie pour $\omega|Y$. Cette quatrième condition est une condition de courbure.

Aucune équivalence entre les conditions dans chaque groupe n'est connue aujourd'hui. Certaines implications sont connues, par exemple

$$\mathbf{P}\,1 \Longrightarrow \mathbf{P}\,4 \Longrightarrow \mathbf{P}\,5$$

sont dues à Kodaira et Kobayashi-Ochiai [K-O 1], et dans la deuxième liste, $4 \Longrightarrow 3$ est dû à Grauert et Reckziegel [G-R] et Kobayashi, qui a aussi posé la question de savoir si la réciproque est vraie, voir [Ko 2], [Ko 3]. Mis à part $3 \Rightarrow 4$, je conjecture l'équivalence des autres conditions.

Dans [La 1] j'avais conjecturé qu'hyperbolique implique mordellique. Ici, je complète cette conjecture comme ci-dessus. J'ai essayé de tirer au clair les conditions variées qui sont apparues dans la littérature pour le cas des variétés de dimension supérieures, définies sur des corps de fonctions, c'est-à-dire le cas des familles algébriques où les points rationnels correspondent à des sections. Des résultats diophantiens on été obtenus dans ce contexte (à la suite de Manin et Grauert) par Kobayashi-Ochai [K-O 2], Noguchi [No], où les hypothèses suivantes apparaissaient :

Fibré cotangent ample;
classe canonique ample;
classe canonique pseudo ample.

Chaque hypothèse implique la suivante sans que la réciproque soit vraie en général. Kobayashi [Ko 4] a démontré que

$$\text{Fibré cotangent ample} \Longrightarrow \text{hyperbolique},$$

et conjecturalement,

$$\text{hyperbolique} \Longrightarrow \text{class canonique ample}.$$

Riebesehl [Ri] considère le cas des familles algébriques sous l'hypothèse de courbure (de Gauss) négative.

Des contributions substantielles à la clarification entre la géométrie différentielle et la géométrie algébrique sont aussi dûes à Griffiths [Gri], Green-Griffiths [Gr-Gr] qui, par exemple, ont conjecturé que si la classe canonique est ample, alors toute application holomorphe $\mathbb{C} \longrightarrow X$ n'est pas Zariski dense. Enfin, Vojta [Vo] a puissamment contribué au sujet en établissant une correspondance conjecturale entre le Deuxième Théorème Principal de la Théorie de Nevanlinna et la théorie des hauteurs pour les variétés sur les corps de nombres. Les conjectures de Vojta sont quantitatives, alors que les conjectures du présent article sont qualitatives.

Remarquons qu'une des particularités du sujet est la difficulté extrème de démontrer dans des cas particuliers qu'une variété est mordellique. Jusqu'au Théorème de Faltings, on ne connaissait aucun exemple de courbe définie sur un corps de nombre pour laquelle on sache démontrer que la courbe est mordellique.

BIBLIOGRAPHIE

[Br] R. Brody.- Compact manifolds and hyperbolicity, Trans. Amer. Math. Soc. 235 (1978), 213-219.

[Br-Gr] R. Brody et M. Green.- A family of smooth hyperbolic hypersurfaces in P_3, Duke Math. J. (1977), 873-874.

[G-R] H. Grauert et H. Reckziegel.- Hermitesche Metriken und normale Familien Holomorpher Abbildungen, Math. Z. 89 (1965), 108-125.

[Gr 1] M. Green.- Some Picard theorems for holomorphic maps to algebraic varieties, Am. J. Math. 97 (1975), 43-75.

[Gr 2] M. Green.- Holomorphic maps to complex tori, Am. J. Math. (1978), 615-620.

[Gr-Gr] M. Green et P. Griffiths.- Two applications of algebraic geometry to entire holomorphic mappings, Chern Symposium, Berkeley, June 1979, Springer-Verlag 1980, 41-74.

[Gri] P. Griffiths.- Holomorphic mapping into canonical algebraic varieties, Ann. Math. 98 (1971), 439-458.

[Ko 1] S. Kobayashi.- Volume elements, holomorphic mappings and Schwarz lemma, Proc. Symposia in Pure Math. 11 (1968), AMS Publ., 253-260.

[Ko 2] S. Kobayashi.- Hyperbolic Manifolds and holomorphic mappings, Marcel Dekker, New York, 1970.

[Ko 3] S. Kobayashi.- Intrinsic distances, measures and geometric function theory, Bull. AMS 82 (1976), 357-416.

[Ko 4] S. Kobayashi.- Negative vector bundles and complex Finsler structures, Nagoya Math. J. 57 (1975), 153-166.

[K-O 1] S. Kobayashi et T. Ochiai.- Mappings into compact complex manifolds with negative first Chern class, J. Math. Soc. Japan 23 (1971), 137-148.

[K-O 2] S. Kobayashi et T. Ochiai.- Meromorphic mappings into compact complex spaces of general type, Invent. Math. 31 (1975), 7-16.

[La 1] S. Lang.- Higher dimensional diophantine problems, Bull. Amer. Soc. 80 (1974), 779-787.

[La 2] S. Lang.- Hyperbolic and diophantine analysis, à paraître, Bull. AMS, 1986.

[No] J. Noguchi.- A higher dimensional analogue of Mordell's conjecture over function fields, Math. Ann. 258 (1981), 207-212.

[Ri] D. Riebesehl.- Hyperbolische Komplex Raüme und die Vermutung von Mordell, Math. Ann. 257 (1981), 99-110.

[Sw-D] P. Swinnerton-Dyer.- A solution of $A^5+B^5+C^5=D^5+E^5+H^5$, Proc. Cam. Phil. Soc. 48 (1952), 516-518.

[Vo] P. Vojta.- à paraître, voir aussi sa thèse, Harvard, 1984.

Serge LANG
Department of Mathematics
University of California
Berkeley, Cal. 94720
U. S. A.

BULLETIN (New Series) OF THE
AMERICAN MATHEMATICAL SOCIETY
Volume 14, Number 2, April 1986

HYPERBOLIC AND DIOPHANTINE ANALYSIS

BY SERGE LANG

Dedicated to F. Hirzebruch

In this survey we consider Kobayashi hyperbolicity, in which there is an interplay between five notions:

analytic notions of distance and measure;
complex analytic notions;
differential geometric notions of curvature (Chern and Ricci form);
algebraic notions of "general type" (pseudo ampleness);
arithmetic notions of rational points (existence of sections).

I am especially interested in the relations of the first four notions with diophantine geometry, which historically has intermingled with complex differential geometry. One of the main points of this survey is to arrive at a certain number of conjectures in an attempt to describe at least some of these relations coherently.

Throughout this article, by an **algebraic set** we mean the set of zeros of a finite number of homogeneous polynomials

$$P_j(x_0,\ldots,x_N)=0, \qquad j=1,\ldots,m$$

in projective space $\mathbf{P}^N$ over $\mathbf{C}$, so $x_0,\ldots,x_N$ are the projective coordinates. An algebraic set will be called a **variety** if it is irreducible—that is, the polynomials can be chosen so that they generate a prime ideal. If X has no singularities, then X is also a complex manifold. In any case, X has a complex structure. From now on, we let X denote a variety.

Among the possible complex analytic properties of X we shall emphasize that of being **hyperbolic**. There are several equivalent definitions of this notion, and one of them, due to Brody, is that every holomorphic map of $\mathbf{C}$ into X is constant. At the beginning of this article, we shall give three possible characterizations, including Kobayashi's original definition, and prove the equivalence between them, following Brody.

On the other hand, X also has an algebraic structure. For one thing, taking the algebraic subsets of X as closed subsets defines the Zariski topology. Thus the **Zariski closure** of a set S is the smallest algebraic set containing S, and is equal to the intersection of all algebraic sets containing S. Furthermore, the polynomials P_j have coefficients in some field F_0, finitely generated over the rational numbers, and this gives rise to diophantine properties as follows.

Received by the editors March 13, 1985 and, in revised form, July 3, 1985.
1980 *Mathematics Subject Classification* (1985 *Revision*). Primary 14J99, 14G95, 32H20, 11D41.

Let F be a field containing F_0. A point with projective coordinates $(x_0, \ldots, x_n)$ is said to be **rational** over F if the affine coordinates

$$x_0/x_j, \ldots, x_n/x_j$$

lie in F for some index j such that $x_j \neq 0$, and hence for every such index. The set of F-rational points of X is denoted by $X(F)$. When we speak of **rational points**, we shall always mean rational points in some field of finite type over the rationals, that is, finitely generated over the rationals. No other qualification will be made.

We shall say that X (or a Zariski open subset) is **mordellic** if it has only a finite number of rational points in every finitely generated field over $\mathbf{Q}$ according to our convention. I once conjectured that a hyperbolic variety is mordellic. We shall put this conjecture in a much broader context.

Let X be any variety. We define the **analytic exceptional set** $\mathrm{Exc}(X)$ of X to be the Zariski closure of the union of all images of non-constant holomorphic maps $f: \mathbf{C} \to X$. Thus X is hyperbolic if and only if this exceptional set is empty. In general the exceptional set may be the whole variety, X itself. I conjecture that $X - \mathrm{Exc}(X)$ is mordellic.

It is also a problem to give an algebraic description of the exceptional set, giving rise to the converse problem of showing that the exceptional set is not mordellic, and in fact always has infinitely many rational points in a finite extension of a field of definition for X. We define the **algebraic exceptional set** $\mathrm{Exc}_{\mathrm{alg}}(X)$ to be the union of all non-constant rational images of $\mathbf{P}^1$ and abelian varieties in X. (An abelian variety is a variety which is complex analytically isomorphic to a complex torus.) I would conjecture that the analytic exceptional set is equal to the algebraic one. A subsidiary conjecture is therefore that $\mathrm{Exc}_{\mathrm{alg}}(X)$ is closed, and that X is hyperbolic if and only if every rational map of $\mathbf{P}^1$ or an abelian variety into X is constant. Similarly, until the equality between the two exceptional sets is proved, one has the corresponding conjecture that the complement of the algebraic exceptional set is mordellic.

Observe that the equality

$$\mathrm{Exc}(X) = \mathrm{Exc}_{\mathrm{alg}}(X)$$

would give an algebraic characterization of hyperbolicity. Such a characterization implies for instance that if a variety is defined over a field F_0 as above, and is hyperbolic in one imbedding of F_0 in $\mathbf{C}$, then it is hyperbolic in every imbedding of F_0 in $\mathbf{C}$, something which is by no means obvious, given the analytic definitions of "hyperbolic" in §§1 and 2.

We wish to characterize those varieties such that the exceptional set is a proper subset. We shall give conjecturally a number of equivalent conditions, which lead us into complex differential geometry, and algebraic geometry as well as measure theory.

The properties having to do with hyperbolicity from the point of view of differential geometry have been studied especially by Grauert-Reckziegel, Green, Griffiths, and Kobayashi. Such properties have to do with "curvature". In §4 we reproduce the proof that under certain "curvature" conditions, the variety is hyperbolic. Also in §4 we consider a weakening of this notion, namely **measure hyperbolic**, due to Kobayashi. The difference lies in looking at positive $(1,1)$-forms or volume (n,n)-forms on a variety of dimension n. One key aspect is that of the Ahlfors-Schwarz lemma, which states that a holomorphic map is measure decreasing under certain conditions.

In §3 we describe how to associate a $(1,1)$-form, the **Ricci form** $\mathrm{Ric}(\Psi)$, to a volume form Ψ. The positivity of the Ricci form is related in a fundamental way to hyperbolicity. In the differential geometric part of this article, the Schwarz lemma and the formalism attached to the positivity of the Chern form

$$dd^c \log|\sigma|^2,$$

where $|\sigma|$ is a length function of some sort, provide a uniform thread weaving through all the questions.

We shall also be led to weaken certain properties. Roughly speaking, given a property, the weakening of this property obtained by requiring that it holds only outside a proper algebraic subset may be called its **pseudofication**. Thus we shall deal with **pseudo volume forms**, for which we allow zeros on a proper analytic subset. The precise definition is given in §4. We can also say that X is **pseudo hyperbolic** if the exceptional set is a proper subset. We say that X is **pseudo mordellic** if there exists a proper algebraic subset Y such that $X - Y$ is mordellic.

In the direction of algebraic geometry, let $T = T_X$ be the complex tangent bundle, and $T^\vee$ the cotangent bundle dual of T, assuming now that X is non-singular. Let

$$K = K_X = \bigwedge^{\mathrm{top}} T^\vee$$

be the canonical bundle of differential forms of top degree. We say that X is **canonical** if there exists a positive integer m such that a basis $(s_0, \ldots, s_N)$ for the space of sections $H^0(X, K^{\otimes m})$ gives an imbedding

$$x \mapsto (s_0(x), \ldots, s_N(x))$$

of X into $\mathbf{P}^N$. We say that X is **pseudo canonical** (or of **general type**) if there exists some m such that the map is defined outside a proper algebraic subset, and gives a projective imbedding of the complement. A singular variety is called **pseudo canonical** if a desingularization is pseudo canonical.

Then the problem is to determine whether the following conditions are equivalent:

P1. X is pseudo canonical.
P2. X is pseudo mordellic.
P3. X is pseudo hyperbolic.
P3′. The algebraic exceptional set is a proper subset.
P4. If X is non-singular, there exists a pseudo volume form Ψ such that $\mathrm{Ric}(\Psi) > 0$.
P5. X is measure hyperbolic.

Without the pseudo, the problem is whether the following conditions are equivalent:

1. Every subvariety of X (including X itself) is pseudo canonical.
2. X is mordellic.
3. X is hyperbolic.
3′. Every rational map of $\mathbf{P}^1$ or an abelian variety into X is constant.
4. If X is non-singular, there exist a positive $(1,1)$-form ω and a number $B > 0$ such that for every complex one-dimensional immersed submanifold Y in X,

$$\operatorname{Ric}(\omega|Y) \geqq B\omega|Y.$$

5. Every subvariety of X (including X itself) is measure hyperbolic.

No two properties in each column are known to be equivalent. Some implications are known, and will be discussed at length. In order to make this article as accessible as possible to a broad audience, I reproduce complete proofs for several basic results of Brody, Grauert-Reckziegel, Griffiths, and Kobayashi-Ochiai complemented by Kodaira, proving some of the above implications. But I also survey other matters, with the selections made for their application to certain diophantine questions. Kobayashi's survey [**Ko 3**] takes other directions. As usual, I would like to remind the reader that a choice made here is not meant to exclude other points of view, and is only dictated by logistical boundary conditions.

Evidence for the diophantine conjectures comes from their self-coherence, rather than special cases. I remind the reader that until Faltings' theorem, there was not known a single example of a curve (variety of dimension 1) which was proved to be mordellic.

Some of the listed conditions, like **P1**, **P2**, **P3′** and **1, 2, 3′** are algebraic. The others are analytic. Different readers can thus be interested in different combinations of them.

In this article, I concentrate on the finiteness of rational points, and I omit the whole area of intermediate hyperbolicity involving holomorphic maps of balls of intermediate dimension—between 1 and the dimension of the manifold—originally defined by Eisenman in his thesis (1969), see [**E**]. Results analogous to those mentioned here have been obtained, and similar conjectures can be made, especially concerning the dimension and parametrization of the exceptional set discussed in §8. For an extensive bibliography of papers and a good survey of results on the analytic aspects of this intermediate-dimensional situation, see a forthcoming paper of Graham-Wu [**G-W**].

The conjectures of this paper concerning mordellic properties are qualitative. I refer the reader to Vojta [**Vo**] for the extraordinary connection which he established conjecturally between diophantine analysis and the Second Main Theorem of Nevanlinna theory. Vojta's conjectures amount to quantitative estimates for the heights of rational points in the higher-dimensional case, and involve diophantine approximations in an arithmetic-geometric setting. Some of the conjectures of this article can be viewed as providing more information on the sets where Vojta's conjectured inequalities do not hold.

In addition, parallel to the absolute diophantine questions there has also been progress in higher-dimensional results in the context of sections of algebraic families, e.g. by Kobayashi-Ochiai [**K-O** 2], Noguchi [**No**], and Riebesehl [**Ri**], following Manin and Grauert. I shall mention this work more precisely in an appendix.

I am very grateful to Artin, Friedman, Green, Griffiths, Harris, Siu, Sommese, and Vojta for useful conversations which have helped me sort out the literature and clarify my ideas. I especially thank Green and Griffiths for communicating to me their conjecture in §8, and explaining to me various matters pertaining to their paper [**Gr-Gr**].

Contents

§1. Kobayashi hyperbolicity

Let **D** be the open unit disc. The **Poincaré metric**, also called the **hyperbolic metric**, on **D** is defined by the form

$$\frac{dz\, d\bar{z}}{\left(1-|z|^2\right)^2}.$$

The tangent space $T_z(\mathbf{D})$ at a given point z can be identified with **C**, and if $v \in T_z(\mathbf{D}) = \mathbf{C}_z$ under this identification, then the hyperbolic norm of v under the metric is

$$|v|_{\mathrm{hyp},z} = \frac{|v|_{\mathrm{euc}}}{1-|z|^2},$$

where $|v|_{\mathrm{euc}}$ is the ordinary absolute value on **C**. Instead of **C**, we write $\mathbf{C}_z$ to specify the Poincaré metric on **C** at the point z of **D**. Note that for $z = 0$, the Poincaré metric is the euclidean metric.

Similarly, for any positive number r, we let $\mathbf{D}_r$ be the disc of radius r, with the Poincaré metric corresponding to the form

$$\frac{dz\, d\bar{z}}{\left(1-|z|^2/r^2\right)^2}.$$

The hyperbolic norm of a tangent vector v at z is then given by the similar formula as above, replacing $|z|^2$ by $|z|^2/r^2$.

We shall investigate the behavior of the Poincaré metric under holomorphic maps. We begin with a classical result, at the level of elementary complex variables.

Proposition 1.1 (Schwarz-Pick Lemma). *Let $f: \mathbf{D} \to \mathbf{D}$ be a holomorphic map of the disc into itself. Then*

$$\frac{|f'(z)|}{1 - |f(z)|^2} \leqq \frac{1}{1 - |z|^2}.$$

Proof. Fix $a \in \mathbf{D}$. Let

$$g(z) = \frac{z + a}{1 + \bar{a}z} \quad \text{and} \quad h(z) = \frac{z - f(a)}{1 - \overline{f(a)}\, z}.$$

Thus g and h are automorphisms of the disc which map 0 on a and $f(a)$ on 0 respectively. We let

$$F = h \circ f \circ g,$$

so that $F: \mathbf{D} \to \mathbf{D}$ is holomorphic and $F(0) = 0$. Then by the chain rule,

$$\begin{aligned} F'(0) &= h'(f(a))f'(a)g'(0) \\ &= \frac{1 - |a|^2}{1 - |f(a)|^2} f'(a) \end{aligned}$$

by a direct computation. By the ordinary Schwarz lemma, we have

$$|F'(0)| \leqq 1,$$

with equality if and only if F is an automorphism, so f is an automorphism. We also get the reformulation

$$\frac{|f'(a)|}{1 - |f(a)|^2} \leqq \frac{1}{1 - |a|^2},$$

which proves the proposition.

As already remarked in the proof of the proposition, we have equality at one point if and only if f is an automorphism. In particular, we can express the lemma invariantly in terms of the differential of f as follows.

Let $f: \mathbf{D} \to X$ be a holomorphic mapping into a complex hermitian manifold. Then we have an induced tangent linear map for each $z \in \mathbf{D}$:

$$df(z): T_z(\mathbf{D}) = \mathbf{C}_z \to T_{f(z)}(X).$$

Each complex tangent space has its norm: $T_{f(z)}$ has the hermitian norm, and $T_z(\mathbf{D}) = \mathbf{C}_z$ has the hyperbolic norm. We can define the **norm of the linear map** $df(z)$ as usual:

$$|df(z)| = \sup_v |df(z)v|/|v|$$

for $v \in T_z(\mathbf{D})$, $v \neq 0$. Then the Schwarz-Pick Lemma can be stated in the form:

Corollary 1.2. (i) *A holomorphic map* $f\colon \mathbf{D} \to \mathbf{D}$ *is distance decreasing for the hyperbolic norm.*

(ii) *An automorphism of* $\mathbf{D}$ *is an isometry.*

Unless otherwise specified, the norm on the tangent space $\mathbf{C}_z$ will be the hyperbolic norm.

Let X be a complex space. Let x, $y \in X$. We consider sequences of holomorphic maps

$$f_i\colon \mathbf{D} \to X, \qquad i = 1, \ldots, m$$

and points p_i, $q_i \in \mathbf{D}$ such that $f_1(p_1) = x$, $f_m(q_m) = y$, and

$$f_i(q_i) = f_{i+1}(p_{i+1}).$$

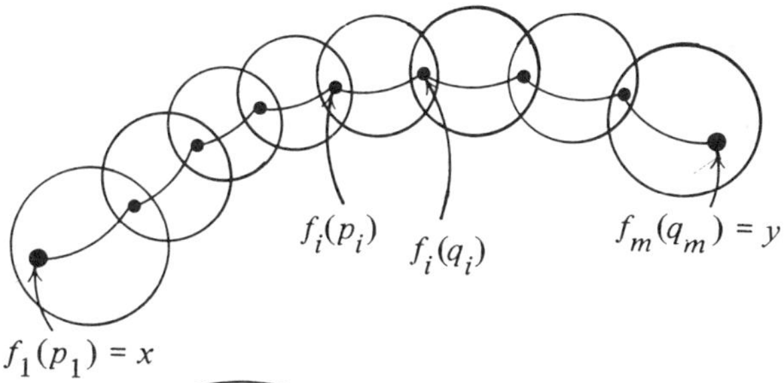

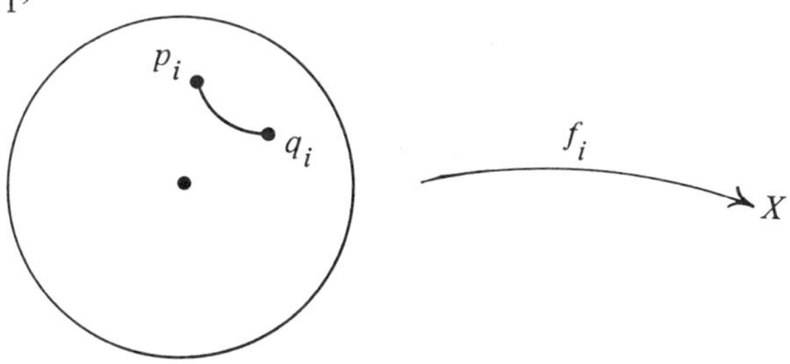

In other words, we join x to y by a chain of discs. We add the hyperbolic distances between p_i and q_i, and take the inf over all such choices of f_i, p_i, q_i to define the **Kobayashi semi distance**

$$d_X(x, y) = \inf \sum_{i=1}^{m} d_{\mathrm{hyp}}(p_i, q_i).$$

Then d_X satisfies the properties of a distance, except that $d_X(x, y)$ may be 0 if $x \neq y$, so d_X is a semi distance.

Example 1. If $X = \mathbf{D}$ is the hyperbolic disc, then d_X on X coincides with the hyperbolic distance by Corollary 1.2 (the fact that a holomorphic map of $\mathbf{D}$ into itself is distance decreasing).

Example 2. Let $X = \mathbf{C}$ with the euclidean metric. Then $d_X(x, y) = 0$ for all $x, y \in \mathbf{C}$. Indeed, given $x \neq y$ there exists a disc of arbitrarily large radius imbedded in $\mathbf{C}$ such that 0 maps to x and $y - x$ maps to y. By a dilation, we can map $f\colon \mathbf{D} \to \mathbf{C}$ such that $f(0) = x$ and $f(q) = y$ where q is very close to 0. But the hyperbolic metric is very close to the euclidean metric near 0, so $d_X(x, y)$ is arbitrarily small, so equal to 0.

It will be useful to consider the following generalization. Let X be a subset of a complex hermitian manifold Z. We can define d_X on X (rather than on Z) by taking the maps f_i to lie in X, and to be holomorphic as maps into Z. Then we obtain a semi distance on X. We say that X is **Kobayashi hyperbolic in** Z if this semi distance is a distance, that is if $x \neq y$ implies $d_X(x, y) \neq 0$. For simplicity, we shall say **hyperbolic** instead of Kobayashi hyperbolic. We need Z only to be able to define a holomorphic map of $\mathbf{D}$ into something. From now on, unless otherwise specified, we suppose that we deal with this situation. When we speak of a holomorphic map $f\colon \mathbf{D} \to X$ we mean a holomorphic map $f\colon \mathbf{D} \to Z$ whose image is contained in X.

The foundations of complex analytic spaces actually allow one to define analytic maps from one such space to another. Furthermore, if X is an analytic space imbedded in a complex manifold Z, then a map into X is analytic if and only if it is analytic viewed as a map into Z. Therefore the definition of the Kobayashi semi distance on X is intrinsic, independent of the imbedding of X into a manifold.

We shall also consider hermitian metrics, and for this it is useful to have the analytic space imbedded in a manifold so that we can use norms on a tangent space (which does not exist when X has singularities). For instance in Theorem 2.2 below, we give Brody's criterion for a compact subspace of a complex hermitian manifold to be hyperbolic. Without mentioning the norms, this result will imply the following statement.

Let X be a compact analytic space. Then X is hyperbolic if and only if every holomorphic map $f\colon \mathbf{C} \to X$ is constant.

In this context, it is useful to define an analytic space X to be **Brody hyperbolic** if every holomorphic map of $\mathbf{C}$ into X is constant. The above statement then can be rephrased to say that *for a compact analytic space, Brody hyperbolic is equivalent with Kobayashi hyperbolic.* Let $f\colon \mathbf{D} \to X$ be a holomorphic map. In line with Royden's approach [**Ro**] we define

$$c(f) = \sup_z |df(z)|,$$

where the sup is taken over all $z \in \mathbf{D}$. We also define

$$c(X) = \sup_f c(f),$$

where the sup is taken over holomorphic maps $f\colon \mathbf{D} \to X$ (so, according to our conventions, holomorphic maps $f\colon \mathbf{D} \to Z$ whose image is contained in X). We do not exclude that $c(f)$ and $c(X)$ are equal to ∞. Since, however, automorphisms of the disc operate transitively on the disc, it follows that we have

$$c(X) = \sup_f |df(0)|,$$

where the sup is taken as before. Obviously $c(f) \leqq c(X)$.

Lemma 1.3. *Let $f\colon \mathbf{D} \to X$ be holomorphic. Let $z_1, z_2 \in \mathbf{D}$. Then*

$$d_{\text{her}}(f(z_1), f(z_2)) \leqq c(f) d_{\text{hyp}}(z_1, z_2) \leqq c(X) d_{\text{hyp}}(z_1, z_2).$$

Proof. Let $\gamma\colon [0,1] \to \mathbf{D}$ be a geodesic between z_1 and z_2 in $\mathbf{D}$. Then $f \circ \gamma$ is a curve joining $f(z_1)$ and $f(z_2)$ and its length is

$$\int_0^1 |df(\gamma(t))\gamma'(t)|_{\text{her}}\, dt \leqq c(f) \int_0^1 |\gamma'(t)|_{\text{hyp}}\, dt,$$

so length of $f \circ \gamma \leqq c(f)$ length of γ, whence the lemma follows.

Theorem 1.4 (Brody [Br]). *If $c(X)$ is finite then X is hyperbolic. Conversely, if X is compact and hyperbolic then $c(X)$ is finite.*

Proof. Suppose $c(X)$ finite. By Lemma 1.3 for $x \neq y$ in X we obtain

$$d_X(x, y) \geqq (1/c) d_{\text{herm}}(x, y) > 0,$$

whence the first part of the theorem follows.

Conversely, suppose $c(X) = \infty$ and X compact. There exists a sequence of holomorphic maps

$$f_n\colon \mathbf{D} \to X$$

with $|df_n(0)|$ increasing to ∞. By compactness, say $\lim f_n(0) = x$. Take a chart at x

$$U \subset U^{\text{cl}} \subset V,$$

where U, V are open balls, say, centered at 0, and U^{cl} is the closure of U. Let S be the boundary of U and let s be the radius of S. Let $r < 1$ be a positive number such that $f_n(\mathbf{D}_r^{\text{cl}}) \subset U$. Then

$$f_n'(0) = \frac{1}{2\pi i} \int_{C_r} \frac{f_n(w)}{w^2}\, dw \quad \text{so} \quad |f_n'(0)| \leqq s/r.$$

Since the hyperbolic metric at $z = 0$ is the same as the euclidean metric, it follows that if $|f_n'(0)|$ is large, then $r = r(n)$ has to tend to 0. Therefore, given a positive integer m, there exists n such that $f_n(\mathbf{D}_{1/m})$ is not contained in U and therefore intersects the boundary S. Hence we can find a point x_m on S such that $x_m = f_n(p_m)$ with $p_m \in \mathbf{D}_{1/m}$. Then

$$d_X(x, x_m) = d_X(f_n(0), f_n(p_m)) \leqq d_{\text{hyp}}(0, p_m) \to 0$$

as $m \to \infty$. But a subsequence of x_m converges to a point $y \in S$ so $d_X(x, y) = 0$ and X is not hyperbolic. This proves the theorem.

In the compact case, Brody's criterion can be taken as a definition. Similarly, Theorem 2.2 in the next section will give another important equivalent condition. Thus from the start, we have three valuable characterizations of hyperbolic varieties, which exist on an equal footing. The Kobayashi condition and Theorem 1.4 put special emphasis on mapping discs into the manifold.

Remark 1. This context also allows transposing certain questions from the theory of transcendental numbers. For instance if $f\colon \mathbf{D} \to X$ is a holomorphic immersion into a hyperbolic variety X defined over a number field, if $df(0)$ is algebraic, and $f(0)$ is an algebraic point, then is $f(z_0)$ transcendental for z_0 algebraic $\neq 0$? Is the radius of the disc transcendental? See [**La 3**], and results of Wolfart [**Wo**], Wolfart-Wustholz [**Wo-Wu**] for special Riemann surfaces, relating the radius to special values of the gamma (beta) function which also arise as periods.

Remark 2. Instead of using the unit disk and having $|df(0)| \to \infty$, in the next section we shall use $|df(0)| = 1$, say, with discs of increasing radius. If the variety is hyperbolic, then this radius is bounded. For Riemann surfaces, this is the classical Landau-Schottky theorem, and a variation is due to Bloch. In this article dealing with the higher-dimensional case, I give one formulation of the generalizations in terms of the Schwarz lemma. For the other possible formulations and variants, explaining the historical motivation, see Griffiths [**Gri**]. One can again ask for the arithmetic nature (including transcendence) of the sup of radii r such that there is a holomorphic map

$$f\colon \mathbf{D}_r \to X$$

into a hyperbolic variety with $|df(0)| = 1$. This sup could be called the **Landau-Schottky constant** associated with the metric.

Similarly, one can ask for the arithmetic nature of the constant $c(X)$, which also depends on the metric.

Remark 3. Consider a holomorphic curve $f\colon \mathbf{D} \to X$ into a hyperbolic variety. The finiteness of rational points in a field F amounts to the property that if $\mathbf{D}_r$ is a closed subdisc, then there are only a finite number of values $z \in \mathbf{D}_r$ such that $f(z) \in X(F)$. In the theory of transcendental numbers, one considers such parametrizations, proving the transcendence of a uniformization map f at certain points. But Siegel in some cases also proved that the images of certain points if algebraic must have a high degree [**Si**]. I have had the impression for a long time that the methods using approximating functions should apply to this case to yield the finiteness.

Remark 4. Using $\mathbf{C}_p$ instead of $\mathbf{C}$ and the theory of rigid analytic spaces, it should be possible to translate p-adically the various characterizations of hyperbolicity. It then becomes a problem to determine the relations between the various possible definitions of p-adic hyperbolicity, and to relate these analytic properties with the algebraic characterization of Conjecture 5.6 and the diophantine properties listed in §5.

§2. Brody's criterion for hyperbolicity

If X is hyperbolic, then directly from the definition of the Kobayashi pseudo metric and Example 2 of §1, there cannot be a non-constant holomorphic map $f\colon \mathbf{C} \to X$. The converse is due to Brody [**Br**], and is the main goal of this section. We start with:

Lemma 2.1 (Brody's reparametrization lemma). *Let X be a subset of a complex hermitian manifold. Let $f\colon \mathbf{D}_r \to X$ be holomorphic. Let $c > 0$ and for $0 \leqq t \leqq 1$ let $f_t(z) = f(tz)$.*

(i) *If $|df(0)| > c$ then there exists $t < 1$ and an automorphism h of $\mathbf{D}$ such that if we let*

$$g = f_t \circ h,$$

then

$$\sup_{z \in \mathbf{D}_r} |dg(z)| = |dg(0)| = c.$$

(ii) *If $|df(0)| = c$, then we get the same conclusion allowing $t \leqq 1$.*

Proof. Let $m_t\colon \mathbf{D}_r \to \mathbf{D}_r$ be multiplication by t, so that f_t can be factored:

$$\mathbf{D}_r \xrightarrow{m_t} \mathbf{D}_r \xrightarrow{f} X.$$

Then $dm_t(z)v = tv$, so

$$|df_t(z)| = |df(tz)|\, t \frac{1 - |z|^2/r^2}{1 - |tz|^2/r^2}.$$

Let

$$s(t) = \sup_{z \in \mathbf{D}_r} |df_t(z)|.$$

Then $s(t)$ is increasing for $0 \leqq t \leqq 1$. Indeed, if $t_1 < t_2 < 1$ and z_1 is such that

$$s(t_1) = |df_{t_1}(z_1)|,$$

then we let $z_2 = t_1 z_1 / t_2$ so that $t_2 z_2 = t_1 z_1$. A simple comparison of the extra factor then shows that

$$|df_{t_1}(z_1)| < |df_{t_2}(z_2)| \leqq s(t_2).$$

If $t < 1$ then the extra factor

$$t\frac{r^2 - |z|^2}{r^2 - |tz|^2} \quad \text{is less than } 1,$$

and if $t = 1$ then this extra factor is equal to 1, so $s(t)$ is also increasing up to $t = 1$. Also we can write $tz = w$. Taking the sup for $z \in \mathbf{D}_r$ amounts to taking the sup for $w \in t\mathbf{D}_r$. If $t < 1$, we can even take the sup over the closure $t\mathbf{D}_r^{\mathrm{cl}}$. It follows that $s(t)$ is continuous for $0 \leqq t < 1$. Also $s(t) \to s(1)$ as $t \to 1$, even if $s(1) = \infty$. By assumption in the first part, $|df(0)| > c$, and hence $s(1) > c$.

Hence there exists $0 \leqq t < 1$ such that $s(t) = c$. Hence there is some $z_0 \in t\mathbf{D}_r^{\text{cl}}$ such that $|df_t(z_0)| = c$. Now let h: $\mathbf{D}_r \to \mathbf{D}_r$ be the automorphism such that $h(0) = z_0$ and let $g = f_t \circ h$. Then

$$|dg(0)| = |df_t(z_0)|\,|dh(0)| = |df_t(z_0)| = c,$$

thus proving the first part. The second part is proved similarly, allowing $t = 1$.

Theorem 2.2 [Br]. *Let X be a relatively compact complex subspace of a complex hermitian manifold, and suppose X is not hyperbolic. Then there exists a non-constant holomorphic map f: $\mathbf{C} \to X^{\text{cl}}$ such that*

$$|df(z)|_{\text{euc}} \leqq 1 \quad \textit{for all } z \in \mathbf{C} \textit{ and } |df(0)| = 1.$$

In particular, the hermitian area of $f(\mathbf{D}_r)$ is $\leqq \pi r^2$.

Remark. Since f: $\mathbf{C} \to X$ has domain $\mathbf{C}$, we are using the euclidean hermitian metric on $\mathbf{C}$, so the norm of $df(z)$ is now measured as going from the euclidean norm to the hermitian norm in X. This is the reason for putting the subscript on this norm in the statement of the theorem.

Proof. By Theorem 1.4, there exists a sequence of maps

$$f_n\colon \mathbf{D} \to X \quad \text{such that} \quad |df_n(0)| \to \infty.$$

Without loss of generality, making a dilation, we may consider a sequence of maps

$$f_n\colon \mathbf{D}_{r_n} \to X \quad \text{such that} \quad |df_n(0)| = 1$$

and the radii r_n increase to infinity. By Lemma 2.1 there exist holomorphic maps

$$g_n\colon \mathbf{D}_{r_n} \to X \quad \text{such that} \quad |dg_n(0)| = 1 = \sup_{z \in \mathbf{D}_{r_n}} |dg_n(z)|.$$

We want to show that given a compact subset K of $\mathbf{C}$ there exists a subsequence of $\{g_n\}$ which converges uniformly on K. This is a simple matter, based on the following arguments.

Let $z_0 \in \mathbf{C}$. By compactness it suffices to prove that there is a neighborhood of z_0 on which a subsequence converges uniformly. After passing to a subsequence, we may assume without loss of generality that $\{g_n(z_0)\}$ converges to a point x_0 of X. Now pick a chart at x_0, say an open set V and a holomorphic isomorphism

$$\varphi\colon V \to \mathbf{B}_s,$$

where $\mathbf{B}_s$ is the open ball of radius s centered at 0. We may assume that the closure of V is contained in another chart, isomorphic to a ball of bigger radius. Then by hypothesis, there is a positive number C such that

$$|(\varphi \circ g_n)'(z)| \leqq C \quad \text{for all } z \in g_n^{-1}(V^{\text{cl}}) \text{ and all } n.$$

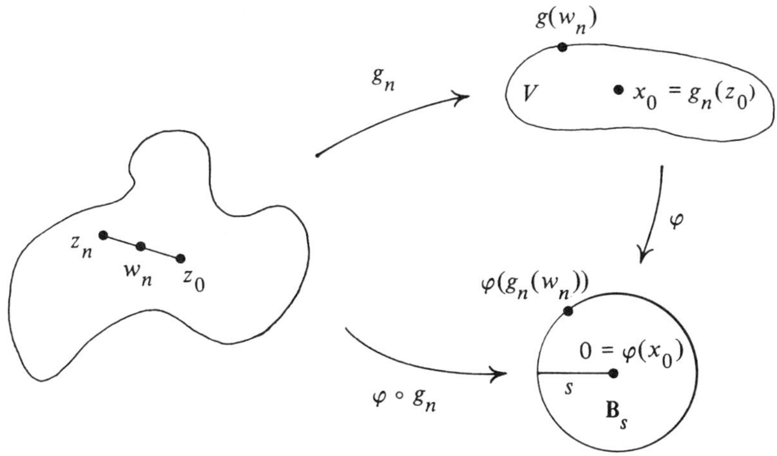

The arguments which follow are concerned with this simple situation: We have a sequence g_n: $U \to X$ of holomorphic maps of some open U in $\mathbf{C}$ into X, a point $z_0 \in U$ such that $g_n(z_0)$ converges to x_0, and charts as above, such that the derivative $(\varphi \circ g_n)'$ is bounded on $g_n^{-1}(V^{\mathrm{cl}})$. We now prove equicontinuity statements for this situation.

1. There exists a disc $\mathbf{D}_a(z_0)$ such that $g_n(\mathbf{D}_a) \subset V$ for all n, that is $\mathbf{D}_a(z_0) \subset g_n^{-1}(V)$ for all n.

Proof. Otherwise, there exists a sequence $\{z_n\}$ tending to z_0 such that $g_n(z_n) \notin V$ for arbitrarily large n. On the segment from z_0 to z_n, let w_n be the first point such that $g_n(w_n)$ lies on the boundary of V, so $\varphi \circ g_n(w_n)$ lies on the sphere bounding $\mathbf{B}_s$. Then

$$(\varphi \circ g_n)(w_n) = \int_{z_0}^{w_n} (\varphi \circ g_n)'(\zeta)\, d\zeta,$$

where the integral is taken over the line segment between z_0 and w_n. Thus we obtain the bound

$$|\varphi \circ g_n(w_n)| \leqq |w_n - z_0| C.$$

As $n \to \infty$, the left side approaches s and the right side approaches 0, a contradiction.

2. The sequence $\{g_n\}$ is equicontinuous at z_0. In fact for $z, z' \in \mathbf{D}_a(z_0)$ (where $\mathbf{D}_a(z_0)$ is the disc as in the previous statement), we have for all n:

$$|\varphi \circ g_n(z) - \varphi \circ g_n(z')| \leqq |z - z'| C.$$

Proof. Immediate from the integral formula, which can now be applied uniformly for all n.

This proves that the sequence $\{g_n\}$ converges uniformly on compact sets, by Ascoli's theorem.

Going back to Theorem 2.2 proper, we get a subsequence of $\{g_n\}$ converging uniformly on $\mathbf{D}_1$ to a map f. A further subsequence converges uniformly on $\mathbf{D}_2$, and so on. We can then extend f analytically to all of $\mathbf{C}$. Furthermore f is not constant since

$$|df(0)| = \lim_{n\to\infty} |df_n(0)| = 1.$$

Finally, consider the euclidean metric on $\mathbf{D}_r \subset \mathbf{C}$, together with the holomorphic map $f\colon \mathbf{C} \to X$ just obtained. The tangent map

$$df(z)\colon T_z(\mathbf{C}) = \mathbf{C} \to T_{f(z)}$$

is a linear map with euclidean norm on $T_z(\mathbf{C})$ and hermitian norm on $T_{f(z)}$ (of the ambient manifold). Let $z_0 \in \mathbf{C}$. Then

$$|df(z_0)|_{\text{euc}} = \lim_{n\to\infty} |dg_n(z_0)|_{\text{euc}} \leqq \lim_{n\to\infty} \left(1 - |z_0|^2/r_n^2\right)^{-1} = 1.$$

Finally, the area of $f(\mathbf{D}_r)$ is bounded by the euclidean area of $\mathbf{D}_r$, which is πr^2. This concludes the proof of the theorem.

I once made the conjecture [**La 1**]:

Conjecture 2.3. *Let X be a projective variety. If X is hyperbolic, then X is mordellic.*

To make the conjecture imply that a subvariety of an abelian variety has only a finite number of rational points unless these are contained in translated abelian subvarieties, I conjectured that a subvariety of an abelian variety which does not contain translations of abelian subvarieties $\neq 0$ is hyperbolic. This conjecture was proved by Mark Green [**Gr**] as follows.

Theorem 2.4. *Let X be a closed subset of a complex torus $\mathbf{T}$, with standard hermitian metric induced from a representation $\mathbf{T} = \mathbf{C}^d/\Lambda$, where Λ is a lattice. Assume also that X is an analytic subvariety, that is X is equal to its complex analytic closure. Then X is hyperbolic if and only if X does not contain a translated complex subtorus $\neq 0$.*

Proof. A complex subtorus in X would give a complex line $\mathbf{C} \to X$ which would contradict Brody's Theorem 2.2.

Conversely, if X is not hyperbolic, then Brody's theorem yields a holomorphic map $f\colon \mathbf{C} \to X$ such that $|df(z)| \leqq 1$ and $|df(0)| = 1$ (euclidean norms). Then f lifts to a holomorphic map into the universal covering space of $\mathbf{T}$, making the following diagram commutative:

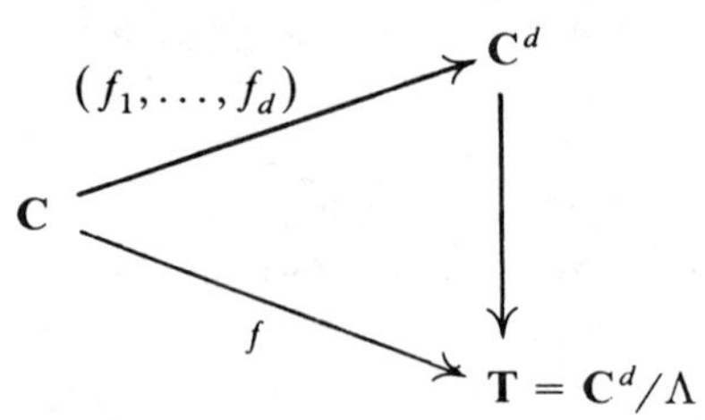

and we have $\sum_{i=1}^{d}|f_i'|^2 \leqq 1$. Hence f_i' is constant for all i by Liouville's theorem, so f_i is linear. After a translation of X we may assume without loss of generality that $f(0) = 0$, and f is a one-parameter subgroup. Since X is assumed to be a complex analytic variety, the complex analytic closure of the image of f is itself a complex analytic group, which is therefore a complex subtorus contained in X. This concludes the proof.

Remark. We have used an elementary lemma:

Let G be a group with a topology (not necessarily Hausdorff, since in the applications it may be a Zariski topology). Assume that given $x \in G$, translation by x is bicontinuous, and also that $x \mapsto x^{-1}$ is bicontinuous. Let H be an abstract subgroup of G. Then the closure $\overline{H}$ is a subgroup.

Proof. The closure of a subset S is the intersection of all closed subsets of G containing S. Since $x \mapsto x^{-1}$ is bicontinuous, it follows that $\overline{H}^{-1}$ is a closed set containing H^{-1}, so

$$\overline{H} = \text{closure}(H) = \text{closure}(H^{-1}) \subset \overline{H}^{-1}.$$

Applying the inverse map to this inclusion yields $\overline{H}^{-1} \subset \overline{H}$, so

$$\overline{H} = \overline{H}^{-1}.$$

Next let $h \in H$. Then $hH \subset \overline{H}$, so $H \subset h^{-1}\overline{H}$, which is closed, so $\overline{H} \subset h^{-1}\overline{H}$, whence finally $h\overline{H} \subset \overline{H}$. Therefore $H\overline{H} \subset \overline{H}$. Similarly, $\overline{H}H \subset \overline{H}$. Finally, let $h \in \overline{H}$. Then $hH \subset \overline{H}$ by what we have just proved, and hence $H \subset h^{-1}\overline{H}$ so $\overline{H} \subset h^{-1}\overline{H}$ so finally $h\overline{H} \subset \overline{H}$. So we have proved

$$\overline{H}\,\overline{H} \subset \overline{H}.$$

This concludes the proof of the lemma.

Note that we did not assume G to be a topological group because, in applications, the composition map $G \times G \to G$ is not continuous. Only translation by single elements is bicontinuous.

We give another application also due to Green [**Gr**].

Lemma 2.5. *Let* $\mathbf{T}$ *be a complex torus, and let X be an effective divisor on* $\mathbf{T}$, *not containing non-trivial translated complex subtoruses. If $T - X$ is not hyperbolic, then there exists a translation of a one-parameter group*

$$g\colon \mathbf{C} \to \mathbf{T} - X.$$

Proof. As in Theorem 2.2, there exists a sequence of holomorphic maps

$$g_n\colon \mathbf{D}_{r_n} \to \mathbf{T} - X$$

with $r_n \to \infty$ and such that

$$|dg_n(0)| = 1 = \sup_{z \in \mathbf{D}_{r_n}} |dg_n(z)|.$$

By a diagonal selection, after picking subsequences we may assume without loss of generality that $\{g_n\}$ converges uniformly on compact subsets of $\mathbf{C}$ (each compact set is contained in $\mathbf{D}_{r_n}$ for n sufficiently large), and the convergence is to an analytic function

$$g: \mathbf{C} \to \mathbf{T}.$$

The problem is now whether the image of g can intersect X.

If $g(\mathbf{C})$ is contained in X then g is constant by Theorem 2.4 and Theorem 2.2, but this is not possible since $|dg(0)| = \lim |dg_n(0)|$. Hence $g(\mathbf{C})$ is not contained in X. Suppose that $g(\mathbf{C})$ intersects X. We first prove that $g^{-1}(X)$ consists of isolated points in $\mathbf{C}$. Suppose X is defined locally on an open set V by a function $\varphi = 0$. Let $x \in X \cap V$ and let U be a connected component of $g^{-1}V$. Then $\varphi \circ g$ is a holomorphic function of U into $\mathbf{C}$ and hence its zeros are isolated or this function is the constant $\varphi(x) = 0$. If $\varphi \circ g(U) = 0$ this means that $g(U) \subset X$. Suppose this is the case. Let U' be a small disc such that $g(U')$ is contained in a chart where X is defined by one equation $\varphi' = 0$, and such that $U \cap U'$ is not empty. Then $\varphi \circ g(U \cap U') = 0$, and hence by analytic continuation, $g(U')$ is contained in X. In this way, we can see that every compact disc in $\mathbf{C}$ is mapped into X by g, and so $g(\mathbf{C}) \subset X$, which is a contradiction.

It then follows that $g^{-1}(X)$ consists of isolated points in $\mathbf{C}$. We shall now prove that $g(\mathbf{C})$ does not intersect X. Suppose there is some $z_0 \in \mathbf{C}$ such that $g(z_0) \in X$. Let S be a small circle centered at z_0 such that $g(S) \subset \mathbf{T} - X$. Such a circle exists since $g^{-1}(X)$ consists of isolated points in $\mathbf{C}$. We pick S small enough that $g(S) \subset V$, where V is a small open set containing $g(z_0)$, in which x is defined by the equation $\varphi = 0$. Let the winding number

$$W(\varphi \circ g(S), 0) = \text{number of zeros of } \varphi \circ g \text{ inside } S$$

and

$$W(\varphi \circ g_n(S), 0) = 0 \text{ because } g_n(\mathbf{D}_{r_n}) \text{ does not intersect } X.$$

Then

$$0 = \lim_{n \to \infty} W(\varphi \circ g_n(S), 0) = W(\varphi \circ g(S), 0),$$

so $\varphi \circ g$ has no zeros inside S, whence $g^{-1}(X) \cap \operatorname{Int}(S)$ is empty, contradicting the hypothesis that $g(z_0) \in X$. This proves that

$$g(\mathbf{C}) \subset \mathbf{T} - X.$$

By Theorem 2.2 we have $|dg(z)| \leqq 1$ for all $z \in \mathbf{C}$ and, as in Theorem 2.4, we conclude that g is the translation of a one-parameter subgroup in $\mathbf{T}$. This proves the theorem.

We recall that for an algebraic complex torus, that is, an abelian variety, if X does not contain a translated abelian subvariety of dimension > 0 then X is ample. Cf. my *Abelian Varieties*, Chapter IV, §2. The converse is false. For instance, the Jacobian of a curve which is a quadratic extension of an elliptic curve, of genus 3, ramified at infinity (in the Weierstrass form) has a theta divisor which contains an isogeneous image of the elliptic curve by pull-back.

Sometimes (see the end of Weil's thesis), people are not aware of this possibility. When it happens, the theta divisor contains infinitely many rational points in some finite extension of a field of definition.

Theorem 2.6. *Let* **T** *be a complex torus imbedded in projective space. Let X be an effective divisor which does not contain a translated abelian subvariety of dimension* > 0. *Then* $\mathbf{T} - X$ *is hyperbolic.*

Proof. Suppose $\mathbf{T} - X$ is not hyperbolic. After replacing X by a suitable multiple, we may assume without loss of generality that X is very ample, so is a hyperplane section in some projective imbedding. In Lemma 2.5, let $\mathbf{T}'$ be the Zariski closure of $g(\mathbf{C})$. Then $\mathbf{T}'$ is a complex subtorus (abelian subvariety), and $X \cap \mathbf{T}' = X'$ is a hyperplane section of $\mathbf{T}'$ in this imbedding. In particular, $X \cap \mathbf{T}'$ is not empty. The fact that $g(\mathbf{C})$ is contained in $\mathbf{T}' - X'$ now contradicts Theorem 3 of [**Ax**], which implies that if a holomorphic curve in a projective variety is Zariski dense, then the intersection with a hyperplane is not empty, and is infinite if and only if the curve is not algebraic. This concludes the proof.

Remark 1. Green's theorem gives an example when both the compact X and the non-compact $\mathbf{T} - X$ are hyperbolic. Note that $\mathbf{T} - X$ may have infinitely many rational points, while Conjecture 2.3 says that X cannot. Green's original proof of Theorem 2.6 did not rely on Ax's theorem, and proved somewhat more concerning the hyperbolic imbedding of $\mathbf{T} - X$ in $\mathbf{T}$. The variation of the proof given here is also due to Green (oral communication).

Observe that the argument used to prove Lemma 2.5 for the most part has nothing to do with complex toruses, and in fact proves the following statement:

Let Z be a variety and let X be a proper algebraic subset. Assume that X is Brody hyperbolic. If $Z - X$ is Brody hyperbolic, then $Z - X$ is Kobayashi hyperbolic.

Remark 2. I had conjectured Ax's result via the theory of transcendental numbers, in the context of analytic parametrizations. This shows another interrelation between number theory and complex differential-algebraic geometry.

The last two theorems of Green give examples of hyperbolic manifolds, including a non-compact case. Brody and Green give other examples of non-singular hypersurfaces which are deformations of the Fermat surface [**Br-Gr**]. The Fermat surface is not hyperbolic since it contains lines. The Brody-Green example is the variety X_t given by the equation

$$z_0^d + z_1^d + z_2^d + z_3^d + (tz_0z_1)^{d/2} + (tz_0z_2)^{d/2} = 0$$

where d is even $\geqq 50$. Thus X_0 is the Fermat surface, which is non-singular, and X_t is a non-singular hypersurface for small t. Recall that all non-singular hypersurfaces of the same degree are C^∞-isomorphic, but of course not complex analytically isomorphic. Furthermore, such hypersurfaces are simply connected in dimension $\geqslant 2$. Brody and Green show that X_t is hyperbolic for all small $t \neq 0$. Thus we have three classes of hyperbolic varieties:

(i) Compact quotients of bounded domains by a discrete group operating freely.
(ii) Subvarieties of abelian varieties not containing translated subtoruses of dimension > 0.
(iii) The Brody-Green family of perturbations of Fermat hypersurfaces.

In this connection, it is worth while to note that Brody [**Br**] showed that the space of hyperbolic complex structures on a given C^∞ manifold is open for the ordinary topology. I do not understand clearly the extent to which hyperbolicity is open for the Zariski topology. However, since a neighborhood of a point for the ordinary topology contains a generic point over some field of definition F_0, if one had an algebraic characterization of hyperbolicity, it would follow that when a special member of a family is hyperbolic, then the generic member of that family is also hyperbolic.

Pseudo-hyperbolicity.

As mentioned in the introduction, we can pseudofy hyperbolicity. Let X be a variety. Kiernan and Kobayashi [**K-K**] discuss the notion of X being hyperbolic **modulo a subset** Y, meaning that the Kobayashi semi distance in X satisfies $d_X(x, y) \neq 0$ unless $x = y$ or $x, y \in Y$. In the present system, I would define X to be **pseudo Brody hyperbolic** if the exceptional set $\mathrm{Exc}(X)$ is a proper subset; and X is **pseudo Kobayashi hyperbolic** if there exists a proper algebraic subset Y such that X is hyperbolic modulo Y. It is not known if the two definitions are equivalent, although I would certainly conjecture that they are, with $Y = \mathrm{Exc}(X)$. In fact, it is likely that the following should be true:

Let $x \neq y$ be distinct points such that $d_X(x, y) = 0$. Then there exists a non-constant holomorphic map $f\colon \mathbf{C} \to X$ such that y lies in the ordinary closure of $f(\mathbf{C})$, and similarly for x.

I have tried to make Brody's argument work to prove this, but without success.

In analogy with Green's Theorem 2.6, I would make the:

Conjecture 2.7. *The complement of the exceptional set is hyperbolic.*

In the introduction, we defined the algebraic exceptional set $\mathrm{Exc}_{\mathrm{alg}}(X)$. To make the terminology functorial with respect to the ideas, we could define X to be **algebraically hyperbolic** if every rational map of $\mathbf{P}^1$ or an abelian variety into X is constant. (Actually, since $\mathbf{P}^1$ is a rational image of an elliptic curve, we could avoid mentioning $\mathbf{P}^1$ in this definition.) Then we say that X is **pseudo algebraically hyperbolic** if the algebraic exceptional set is a proper subset. Conjecturally, all these pseudo conditions are equivalent. Until the equivalences are proved, one can make various subsidiary conjectures, for instance that the complement of the algebraic exceptional set is hyperbolic.

§3. Volume forms; the Ricci (Chern) form and the Griffiths function

We shall be interested in one-dimensional manifolds to the extent that they shed light on the higher-dimensional case, in many ways. We give here basic definitions of the "curvature" to be used.

Let X be a complex manifold and let L be a line bundle. A **metric** ρ on L consists in giving a norm on each fiber, varying smoothly. If U is the open set of a (coordinate) chart, and s is a section represented by the function s_U on U, then the norm is represented by a function ρ_U such that

$$|s|^2 = |s_U|^2/\rho_U,$$

where $|s_U|$ is the ordinary absolute value.

As usual, we let

$$d = \partial + \bar{\partial} \quad \text{and} \quad d^c = \frac{1}{2\pi\sqrt{-1}}\,\frac{1}{2}(\partial - \bar{\partial}).$$

Then

$$dd^c = \frac{\sqrt{-1}}{2\pi}\partial\bar{\partial}.$$

We define the **Chern form** of the metric to be the real (1, 1)-form

$$c_1(\rho) = -dd^c \log|s|^2 = dd^c \log \rho_U \quad \text{on } U.$$

Let Ψ be a volume form on X. This is the same as a metric on the canonical line bundle K_X. In terms of complex coordinates $z_1, \dots, z_n$, such a form is one which can be written

$$\Psi(z) = \rho(z)\Phi(z) \quad \text{where } \Phi(z) = \prod_{j=1}^{n} \frac{\sqrt{-1}}{2\pi}\, dz_j \wedge d\bar{z}_j,$$

and ρ is real > 0. In practice one often has

$$\rho(z) = \sigma(z)|g(z)|^2,$$

where g is holomorphic invertible, and σ is real > 0. We define the **Ricci form** of Ψ to be the **Chern form of this metric** on K_X, so that

$$\mathrm{Ric}(\Psi) = c_1(\Psi) = dd^c \log \rho = dd^c \log \sigma.$$

Since a holomorphic change of charts changes the volume form precisely by a factor $g\bar{g}$ where g is holomorphic, and since

$$dd^c \log(g\bar{g}) = 0,$$

it follows that $\mathrm{Ric}(\Psi)$ is independent of the choice of complex coordinates, and defines a real (1, 1)-form. In important cases, this real (1, 1)-form will be positive, but we do not yet assume this. (By **positive**, we mean strictly positive throughout.)

Remark. *If C is a constant, then*

$$\mathrm{Ric}(C\Psi) = \mathrm{Ric}(\Psi).$$

If u is a positive function, then

$$\mathrm{Ric}(u\Psi) = \mathrm{Ric}(\Psi) + dd^c \log u.$$

Both assertions are trivial from the definition.

A 2-form commutes with all forms. By the nth power

$$\mathrm{Ric}(\Psi)^n$$

we mean the nth exterior power. Then $\mathrm{Ric}(\Psi)^n$ is an (n, n)-form, and in particular a top degree form on X. Since Ψ is a volume form, there is a unique function G on X such that

$$\frac{1}{n!}\mathrm{Ric}(\Psi)^n = G\Psi.$$

We may also write symbolically

$$G = \frac{1}{n!}\mathrm{Ric}(\Psi)^n/\Psi.$$

We call G the **Griffiths function**, associated with the original Ψ, so we may write

$$G = G_\Psi \quad \text{or} \quad G(\Psi), \quad \text{or} \quad Gr(\Psi).$$

The letter G also suggests Gauss, as we now see.

Special Case. *Let* $\dim X = 1$, *and let z be a complex coordinate. Then the Griffiths function is given by*

$$G_\Psi(z) = \frac{1}{\rho}\frac{\partial^2 \log \rho}{\partial z\, \partial \bar{z}}.$$

Proof. This follows immediately from the definition.

The function $-G$ is called the **Gauss curvature**. Note that when $\dim X = 1$, then

$$G = \mathrm{Ric}(\Psi)/\Psi.$$

Classically, up to Chern and Kobayashi, the Gauss curvature is used. Griffiths started using a notation in which the minus sign is systematically obliterated (see his footnote in **[Gri]**), and I strongly approve.

Example 3.1. *Let $X = \mathbf{D}_a$ be the disc of radius a with the volume form*

$$\Psi_a = \frac{2a^2}{\left(a^2 - |z|^2\right)^2}\frac{\sqrt{-1}}{2\pi}\, dz \wedge d\bar{z}.$$

Then

$$\mathrm{Ric}(\Psi_a) = \Psi_a \quad \text{and so} \quad G(\Psi_a) = 1.$$

Proof. Immediate from the definitions.

In classical terms, the Gauss curvature of the disc is -1. We put the factor $2a^2$ in the definition of Ψ_a so that we would come out with $G = 1$ and Gauss curvature -1 for the hyperbolic disc. We may call Ψ_a the **normalized hyperbolic form** on the disc $\mathbf{D}_a$.

Example 3.2. Similarly on the **polydisc**, let $a = (a_1, \ldots, a_n)$ be a sequence of positive numbers;

$$\mathbf{D}_a^n = \mathbf{D}^n(a) = \mathbf{D}_{a_1} \times \cdots \times \mathbf{D}_{a_n}.$$

We let $\Psi_a^{(n)}$ be the product of the normalized hyperbolic forms on each factor, so that

$$\Psi_a^{(n)}(z_1,\ldots,z_n) = \prod_{j=1}^{n} \Psi_a(z_j),$$

where the product is of course the alternating product. Again we call $\Psi_a^{(n)}$ the **normalized hyperbolic volume form** on the polydisc. For this volume form we have

$$\frac{1}{n!}\operatorname{Ric}\left(\Psi_a^{(n)}\right)^n = \Psi_a^{(n)} \quad \text{so} \quad G\left(\Psi_a^{(n)}\right) = 1.$$

Indeed, the variables separate, and

$$\operatorname{Ric}\left(\Psi_a^{(n)}\right)(z_1,\ldots,z_n) = \sum_j \operatorname{Ric}\Psi_{a_j}(z_j).$$

Taking the nth power and using Example 3.1 in one variable, we get the desired relationship.

Unless otherwise specified, all the numbers $a_1,\ldots,a_n$ will be assumed equal to the same number a, in which case we also write $\Psi_a^{(n)}$ for the volume form.

Example 3.3. In the case when $G(\Psi)$ is constant and the Ricci form is positive, the manifold is called **Einsteinian**. A positive constant multiple of the Ricci form is then taken as defining a hermitian metric, which is called an **Einstein-Kähler metric**.

Functoriality 3.4. *Let X, Y have the same dimension. Let $f\colon Y \to X$ be a holomorphic mapping. Let Ψ_X be a volume form on X. Then*

$$\operatorname{Ric}(f^*\Psi_X) = f^*\operatorname{Ric}(\Psi_X)$$

wherever $f^*\Psi_X$ is positive.

If Ψ_Y is a volume form on Y then there is a function $u \geqq 0$ such that $f^*\Psi_X = u\Psi_Y$, and we have

$$\operatorname{Ric}(f^*\Psi_X) = \operatorname{Ric}(\Psi_Y) + dd^c \log u \quad \text{wherever } u \neq 0.$$

Both assertions are immediate from the definitions.

Maximum Principle 3.5. *Let u be a real function > 0 on Y. Let $y_0 \in Y$ be a point such that $u(y_0)$ is a maximum. Then*

$$(dd^c \log u)(y_0) \leqq 0.$$

Proof. Given a complex tangent vector v in the complex tangent space at y_0 there exists an imbedding $f\colon U \to Y$ of an open disc U centered at 0 in $\mathbf{C}$ and a tangent vector (complex number) w such that $f(0) = y_0$ and such that $df(0)w = v$. By pull-back, it suffices to prove the negativity of the form $dd^c \log(u \circ f)$ at 0. Hence without loss of generality, we may assume $X = U$. With respect to a complex coordinate $z = x + iy$, $dd^c \log u$ is represented on U by

$$\left(\frac{\partial}{\partial x^2} + \frac{\partial}{\partial y^2}\right)(\log u)\frac{1}{2\pi}\, dx \wedge dy,$$

which is $\leqq 0$ at a maximum for u by elementary calculus.

Lemma 3.6. *Let* $\omega_1, \ldots, \omega_n$ *be positive* $(1,1)$*-forms on the complex manifold* X *of dimension* n. *Then*

$$\omega_1 \wedge \cdots \wedge \omega_n$$

is a volume form. Furthermore, if $\omega_i \leqq \eta_i$ *for* $i = 1, \ldots, n$ *then*

$$\omega_1 \wedge \cdots \wedge \omega_n \leqq \eta_1 \wedge \cdots \wedge \eta_n.$$

Proof. Left to the reader.

In particular, a positive $(1,1)$-form ω defines a hermitian metric, and $\omega^n/n!$ is a volume form, said to be **associated with the metric, or with** ω.

On a compact subset of X, two positive forms (volume or $(1,1)$) are of the same order of magnitude. That is, given, say, two volume forms Ψ_1 and Ψ_2 on X, and K a compact set, there exists a number $c > 0$ such that

$$\Psi_1 \leqq c\Psi_2 \quad \text{on } K.$$

This will be applied for instance when X itself is compact, and when Ψ_1, Ψ_2 are Ψ and $\mathrm{Ric}(\Psi)^n$ in case $\mathrm{Ric}(\Psi)$ is positive.

Remark. In this section, for simplicity of language, we limited ourselves to volume forms in the strict sense, that is "positive" meant "strictly positive". In the next section, we shall relax this condition and discuss pseudo volume forms. In important applications, it turns out that $\Psi \geqq 0$ but $\mathrm{Ric}(\Psi) > 0$. Thus it may be useful to consider the inverse of the Griffiths function, namely, when $\mathrm{Ric}(\Psi)$ is positive, to consider

$$\Psi \Big/ \frac{1}{n!} \mathrm{Ric}(\Psi)^n,$$

and give this an appropriate name.

§4. Distance and measure decreasing maps

We begin with a one-dimensional result in a simple context to see the basic pattern of proof.

Theorem 4.1 (Ahlfors' Lemma). *Let* $f\colon \mathbf{D}_a \to X$ *be a holomorphic map of the disc of radius* a *into a Riemann surface (one-dimensional complex manifold). Let* Ψ_X *be a volume form on* X *and* Ψ_a *the normalized hyperbolic form on* $\mathbf{D}_a$. *Assume that there exists a number* $B > 0$ *such that* $G(\Psi_X) \geqq B$. *Then*

$$f^*\Psi_X \leqq (1/B)\Psi_a.$$

Proof. Write $f^*\Psi_X = u\Psi_a$, with a function u. We take two steps.

First step. We reduce to the case when u has a maximum in $\mathbf{D}_a$. Let $0 < t < a$. Then $\Psi_t \to \Psi_a$ as $t \to a$. Let u_t be the function such that

$$f^*\Psi_X = u_t\Psi_t \text{ on } \mathbf{D}_t.$$

Then for each $z \in \mathbf{D}_a$, $u_t(z) \to u(z)$ as $t \to a$. Write

$$f^*\Psi_X = h\frac{\sqrt{-1}}{2\pi}\, dz \wedge d\bar{z} \quad \text{on } \mathbf{D}_a.$$

Then h is bounded on $\mathbf{D}_t^{\mathrm{cl}}$ and

$$u_t(z) = h(z)\big(t^2 - |z|^2\big)/2t^2,$$

so $u_t(z) \to 0$ as $|z| \to t$. Hence u_t has a maximum in $\mathbf{D}_t$, and it suffices to prove the inequality of the theorem for u_t, as desired.

Second step. Assume $u(z_0)$ is a maximum for u in $\mathbf{D}_a$. If $u(z_0) = 0$, we are done, because $u = 0$. Suppose $u(z_0) \neq 0$. Then f restricts to a local isomorphism $f\colon U \to X$ on a neighborhood U of z_0. Then by hypothesis,

$$\Psi_a + dd^c \log U = f^* \operatorname{Ric}(\Psi_X) \geqq Bf^*\Psi_X.$$

The function $\log u$ has a local maximum at z_0, and hence

$$(dd^c \log u)(z_0) \leqq 0$$

by the maximum principle 3.5. This proves that

$$\Psi_a(z_0) \geqq Bf^*\Psi_X(z_0),$$

and therefore $u(z_0) \leqq 1/B$. Since $u(z_0)$ is a maximum value, this also proves the theorem.

The first higher-dimensional version of the Schwarz-Ahlfors lemma is apparently due to Grauert-Reckziegel [**G-R**] and Dinghas [**Di**]. Extensions and clarifications of the differential geometric conditions under which the result is true were then given by Chern [**Ch**], Griffiths [**Gri**], Kobayashi [**Ko 1, Ko 2**], and Reckziegel [**Re**]. In [**G-R**], some basic properties are proved for a notion which extends hyperbolicity in one direction, involving "length functions" (called "differentialmetrik" in [**G-R**]), which will be briefly mentioned in §7. Such length functions have the advantage of applying also to singular analytic spaces as shown in Reckziegel's thesis [**Re**]. To keep the language simpler so that we can use the better-known formalism of differential forms, I shall stick to the case of manifolds. For instance, I reproduce the following two theorems as in Kobayashi. However, the analogues occur in [**G-R**] and [**Re**], who work with a differential geometric definition of hyperbolicity, in the more general context of length functions.

Theorem 4.2. *Let X be a complex manifold with hermitian structure defined by a positive $(1,1)$-form ω_X. Assume that there is a constant c such that for every holomorphic map $f\colon \mathbf{D} \to X$ we have*

$$f^*\omega_X \leqq c\Psi_1.$$

Then X is hyperbolic.

Proof. After multiplying ω_X by a suitable constant, we may assume that $c = 1$. Then f is distance decreasing, and in particular the Kobayashi distance is bounded from below by the hermitian distance, which is non-trivial, so the manifold is hyperbolic.

The above theorem is then applied to the following situation.

Theorem 4.3. *Let X be a complex hermitian manifold, with hermitian metric associated with the positive* $(1,1)$*-form* ω_X. *For each one-dimensional immersed submanifold Y of X, let $\omega_Y = \omega_X | Y$, so ω_Y is a volume form on Y. Assume that there is a number $B > 0$ such that $G(\omega_Y) \geqq B$ for all such Y. Let*

$$f \colon \mathbf{D}_a \to X$$

be a holomorphic map. Then

$$f^*\omega_X \leqq (1/B)\Psi_a,$$

and X is hyperbolic.

Proof. Again we reduce the statement to the case when $f^*\omega_X = u\Psi_a$ and u has a maximum $u(z_0)$ in $\mathbf{D}_a$. If $u(z_0) = 0$, we are done. If $u(z_0) \neq 0$ then f is a local isomorphism at z_0, and $f\colon U \to f(U) = Y$ thus gives an isomorphism of a neighborhood of z_0 with a one-dimensional complex submanifold of X. We can now apply the hypothesis and Theorem 4.1 to complete the proof of the first statement. That X is hyperbolic follows by applying the criterion of Theorem 4.2.

In [**Ko 2**], Chapter IX, Problem 5, Kobayashi raises the problem whether the converse is true. We phrase this here as

Problem 4.4. *Let X be a non-singular hyperbolic variety. Does there exist a positive* $(1,1)$*-form ω and a number $B > 0$ such that for every one-dimensional immersed submanifold Y in X, we have*

$$G_{\omega|Y} \geqq B.$$

This problem involving a $(1,1)$-form (condition 4 in the introduction) is the stronger version of an alternative converse to Theorem 4.3, in which this $(1,1)$-form is replaced by a length function, originally introduced in different contexts of hyperbolicity by Grauert-Reckziegel [**G-R**] and Kobayashi [**Ko 4**]. We shall discuss this notion and an application in §7. For a one-dimensional complex manifold the notion of length function coincides with that of a hermitian metric, and the point is that Y in Problem 4.4 is one-dimensional. Thus an intermediate problem would be to show that if X is hyperbolic, then there exists a length function whose Griffiths function satisfies the desired inequality. Recall here that by our conventions, a variety is projective, so compact.

Next we have a result in the equidimensional case which evolved from Dinghas [**Di**] and Chern [**Ch**] to Kobayashi [**Ko 1**], see also Griffiths [**Gri**] and Kobayashi-Ochiai [**K-O**]. The proof will repeat the pattern of the proof of Ahlfors' Lemma.

Theorem 4.5. *Let X be a compact complex manifold. Let Ψ_X be a volume form such that* $\mathrm{Ric}(\Psi_X)$ *is positive. Let* $\dim X = n$. *Let $\Psi_a^{(n)}$ be the normalized hyperbolic volume form on $\mathbf{D}_a^n$. Then there exists a constant $c > 0$ such that for all holomorphic maps*

$$f\colon \mathbf{D}_a^n \to X$$

we have

$$f^*\Psi_X \leqq c\Psi_a^{(n)}.$$

Proof. Let $0 < t < a$. By reproducing exactly the argument in Step 1 of Theorem 4.1 for the volume forms, we reduce the proof to the case when

$$f^*\Psi_X = u\Psi_a^{(n)}$$

and $u \geqq 0$ has a maximum at z_0 in $\mathbf{D}_a^n$. Again if this maximum is 0, we are done, so we may assume the maximum $\neq 0$. Then

$$dd^c \log u = f^* \operatorname{Ric}(\Psi_X) - \operatorname{Ric}(\Psi_a^{(n)})$$

wherever $u \neq 0$, and at the maximum point we know by the maximum principle 3.5 that the left side is $\leqq 0$. Hence we find

$$f^* \operatorname{Ric}(\Psi_X)(z_0) \leqq \operatorname{Ric}(\Psi_a^{(n)})(z_0).$$

We take the nth power of each side. The divided nth power $(1/n!)\operatorname{Ric}(\Psi_a^{(n)})^n$ is just $\Psi_a^{(n)}$, and since we assumed $\operatorname{Ric}(\Psi_X)$ positive, we have trivially

$$\Psi_X \leqq c\operatorname{Ric}(\Psi_X)^n \quad \text{for some constant } c > 0.$$

Hence

$$f^*\Psi_X(z_0) \ll f^* \operatorname{Ric}(\Psi_X)^n(z_0) \ll \Psi_a^{(n)}(z_0),$$

where the sign $\ll$ means "less than a constant times". The constant depends only on the given form Ψ_X and X, but is independent of f. This shows that $u(z_0)$ is bounded by such a constant, and since $u(z_0)$ is a maximum for u, this proves the theorem.

Remarks. As pointed out by Kobayashi in his book **[Ko 2]**, the statements generalize when the polydiscs are replaced by various bounded domains. What we needed was a starlike property, to use the trick with $0 < t < a$, and the fact that the hyperbolic metric goes to infinity toward the boundary. Then the same arguments go through. This is important for applications to bounded (symmetric) domains.

The statement will be extended to pseudo canonical varieties in the next section.

Also, the same proof and conclusion hold in the non-compact case, provided there exists an inequality $\Psi_X \ll \operatorname{Ric}(\Psi_X)^n$. For applications in the next section, we give the appropriate formulation of the result under more general conditions.

Let X be a complex manifold of dimension n. By a **pseudo volume form** Ψ or Ψ_X we shall mean a continuous (n, n)-form which is C^∞ outside a proper analytic subset, and which locally in terms of complex coordinates can be expressed as

$$\Psi(z) = |g(z)|^{2q} h(z)\Phi(z),$$

where

q is some fixed rational number > 0;

g is holomorphic not identically zero;

h is C^∞ and > 0;

and

$$\Phi(z) = \prod_{i=1}^{n} \frac{\sqrt{-1}}{2\pi} dz_i \wedge d\bar{z}_i.$$

The definition is a variation of other possible definitions which weaken the conditions of a volume form, and is adjusted for the applications we have in mind here. Zeros are allowed on a proper analytic set.

We can define $\operatorname{Ric}(\Psi)$ for a pseudo volume form just as we did for a volume form, by the formula

$$\operatorname{Ric}(\Psi) = dd^c \log h.$$

Since g is assumed holomorphic, $dd^c \log|g|^{2q} = 0$ wherever $g \neq 0$.

Theorem 4.5′. *Let X be a complex manifold. Let Ψ_X be a pseudo volume form such that* $\operatorname{Ric}(\Psi_X)$ *is positive and such that there exists a constant $c' > 0$ for which*

$$\Psi_X \leqq c' \operatorname{Ric}(\Psi_X)^n.$$

(Such a c' exists if X is compact.) Then there exists a constant $c > 0$ such that for all holomorphic maps

$$f\colon \mathbf{D}_a^n \to X$$

we have

$$f^*\Psi_X \leqq c\Psi_a^{(n)}.$$

The proof is identical with the previous proof. The arguments are valid under the weaker assumptions.

We now pass to measure theoretic considerations.

Let X be a complex manifold, with volume or pseudo volume form Ψ. Then Ψ defines a positive functional on $C_c(X)$ (continuous functions with compact support) by

$$\varphi \mapsto \int_X \varphi\Psi.$$

Hence there is a unique regular positive measure μ_Ψ such that for all $\varphi \in C_c(X)$ we have

$$\int_X \varphi\Psi = \int_X \varphi \, d\mu_\Psi.$$

Let Z, X be complex analytic spaces. We take for granted that if $f\colon Z \to X$ is analytic, and U open in Z, then $f(U)$ is Borel measurable in X, in fact equal to a countable union of analytic subspaces of X. We shall assume that there is a countable sequence of analytic maps $\{f_i\}$ such that $\bigcup f_i(Z)$ covers X. We let

μ_Z be a positive regular Borel measure on Z. In the application, we shall take

$$Z = \mathbf{D}_1^n = n\text{-fold product of the unit disc;}$$
$$\mu_Z = \text{hyperbolic metric } \Psi_1^{(n)}.$$

Let B be measurable in X, and consider sequences f_i: $Z \to X$ of analytic maps, and open sets U_i in Z such that

$$B \subset \bigcup f_i(U_i).$$

We define the **Kobayashi measure** (with respect to μ_Z)

$$\mu_{\mathrm{Kob},X}(B) = \inf \sum_{i=1}^{\infty} \mu_Z(U_i),$$

where the inf is taken over all sequences $\{f_i\}$ and sets $\{U_i\}$ prescribed above. It is an exercise in basic techniques of measure theory to show that $\mu_{\mathrm{Kob},X}$ is a measure.

Since the open sets generate the σ-algebra of Borel measurable sets, it follows that if A is measurable in Z and f is holomorphic, then $f(A)$ is measurable. Furthermore, a regular measure satisfies the property that the measure of a set is the inf of the measures of the open sets containing it. Hence in the definition of the Kobayashi measure, instead of taking open sets U_i we could take measurable sets A_i in Z.

Let Y, X be complex analytic spaces, and let f: $Y \to X$ be an analytic map. Let μ_X, μ_Y be regular Borel measures on X and Y. We say that f is **measure decreasing** if for every open set U in Y we have

$$\mu_X(f(U)) \leqq \mu_Y(U)$$

or equivalently, for every measureable set A in Y we have

$$\mu_X(f(A)) \leqq \mu_Y(A).$$

Example. Let X, Y be complex manifolds with $\dim X = \dim Y$. Let f: $Y \to X$ be holomorphic. Let Ψ_X, Ψ_Y be volume or pseudo volume forms on X and Y respectively, defining measures μ_X, μ_Y. Assume that

$$f^*\Psi_X \leqq \Psi_Y.$$

Then f is measure decreasing. Indeed, the set of points $y \in Y$ such that $df(y)$ is singular is an analytic subset S, and $f(S)$ has measure 0. On the open complement of S, one sees at once that f is measure decreasing, so f is measure decreasing. This can be applied to the cases of Theorems 4.5 and 4.5′, after multiplying Ψ_X with a sufficiently small positive constant.

From now on, we let $Z = \mathbf{D}_1^n$ with the normalized hyperbolic measure $\mu_1^{(n)}$ corresponding to the volume form $\Psi_1^{(n)}$. For every complex manifold X we then obtain the Kobayashi measure $\mu_{\mathrm{Kob},X}$.

The following properties are immediate, concerning measure decreasing maps.

MD 1. Let $f\colon Y \to X$ be holomorphic, and $\dim Y = \dim X = n$. Then f is measure decreasing for the Kobayashi measures.

MD 2. Let ν be a measure on X such that every holomorphic map $f\colon \mathbf{D}_1^n \to X$ is measure decreasing from $\mu_1^{(n)}$ to ν. Then $\nu \leqq \mu_{\text{Kob},X}$.

MD 3. If all holomorphic maps $f\colon \mathbf{D}_1^n \to X$ are measure decreasing, and if $\mu_X(V) > 0$ for all open sets V in X, then $\mu_{\text{Kob},X}(V) > 0$ for all open V in X.

We define X to be **measure hyperbolic** if $\mu_{\text{Kob},X}(V) > 0$ for all non-empty open sets V in X.

Corollary 4.6. *If X is a compact manifold and* $\text{Ric}(\Psi_X)$ *is positive then X is measure hyperbolic.*

Proof. After multiplying Ψ_X by a sufficiently small constant, we see from Theorem 4.5 that holomorphic maps as above are measure decreasing, and hence that X is measure hyperbolic.

In order to avoid mentioning constants, one could define a map to be **essentially volume decreasing** by allowing a positive constant in comparing the volumes.

We also formulate the corollary for pseudo volume forms.

Corollary 4.6′. *Let X be a complex manifold, and let Ψ_X be a pseudo volume form. Assume that* $\text{Ric}(\Psi_X)$ *is positive, and that there exists $c' > 0$ such that*

$$\Psi_X \leqq c' \operatorname{Ric}(\Psi_X)^n.$$

(If X is compact, then c' always exists.) Then X is measure hyperbolic.

For our purposes here, I quote without proof Theorem 1.10 of Kobayashi's Chapter IX:

Theorem 4.7. *Let X be a complex manifold which is hyperbolic. Then X is measure hyperbolic.*

Then I make the conjecture

Conjecture 4.8. *Let X be a variety. Then X is hyperbolic if and only if all subvarieties of X (including X itself) are measure hyperbolic.*

The implication in one direction comes from Kobayashi's theorem (which is presumably valid also for varieties, even singular, with the same proof he gives in his book), and from Brody's characterization of hyperbolic varieties by the condition that they admit no complex lines (non-constant). This condition also applies to subvarieties. Thus the content of the conjecture lies in the converse: if all subvarieties are measure hyperbolic, then X is hyperbolic.

In the next section, we look at the algebraic geometric context for this conjecture, and its relation with another conjecture of Kobayashi.

In connection with algebraic geometry, it is unknown if the property of being measure hyperbolic is birationally invariant. To deal with this problem, Yau [**Ya**] uses meromorphic maps instead of holomorphic maps to define a variant of the Kobayashi measure. He then has to prove the functorial properties concerning pull-backs of forms, but once this is done, his definition makes the birational invariance obvious.

§5. Pseudo canonical varieties (general type)

With the simultaneous appearance of the papers by Griffiths [**Gri**] and Kobayashi-Ochiai [**K-O**], it was realized that arguments which applied to compact manifolds with positive Ricci form, combined with results of Kodaira, also apply to varieties where this condition is weakened, and which classically are called "of general type". Roughly speaking, given a certain property, the weakening of this property obtained by requiring that it holds only outside a proper algebraic subset may be called its **pseudofication**. The purpose of this section is to carry out the arguments concerning this more general case. The above papers are very similar. I follow especially [**K-O 1**].

To begin, recall a theorem of Kodaira which says that if a compact complex manifold X has a metrized line bundle with positive Chern form then X admits a projective imbedding, cf. Griffiths-Harris, Chapter II, §4. Let L be a line bundle on X. We abbreviate the tensor product $L^{\otimes m}$ by L^m. We say that L is **ample** if there exists some m such that a basis of sections $(s_0, \ldots, s_N)$ of $H^0(X, L^m)$ generates L^m at every point and give a projective imbedding

$$\varphi_m = (s_0, \ldots, s_N): X \to \mathbf{P}^N.$$

We say that L is **very ample** if we can take $m_0 = 1$ in the above condition, so already sections of $H^0(X, L)$ give the imbedding.

Even if we do not get a projective imbedding by means of a basis for the sections, we still get a rational map into $\mathbf{P}^N$. We let

$$m \underset{\text{div}}{\to} \infty$$

denote the property that m tends to infinity, ordered by divisibility. We say that L is **pseudo ample** if φ_m is birational, that is φ_m gives a projective imbedding of a non-empty Zariski open subset, for m large ordered by divisibility.

Suppose X non-singular, and let K_X be the canonical bundle. Classically, X has been called canonical if K_X is very ample [**Gri**]. But for the same reason that Grothendieck changed the meaning of "ample" to what it is now, it seems more fruitful to say that X is **canonical** if K_X is ample, and **very canonical** if K_X is very ample. On the other hand, if K_X is pseudo ample, then X is usually said to be of general type, but *with the support of Griffiths*, I shall say that X is

pseudo canonical to make the terminology functorial with respect to the ideas. Finally, if X is singular, we say that X is of **general type** or **pseudo canonical** if some desingularization has this property. We recall a basic elementary fact.

If X, Y are non-singular, then a birational map $X \to Y$ induces an isomorphism

$$H^0(X, K_X^m) \xrightarrow{\approx} H^0(Y, K_Y^m)$$

for every positive integer m.

Proof. See Hartshorne, *Algebraic Geometry*, Chapter II, Theorem 8.19. The proof given there when $m = 1$ works in general. The idea is that a section of K_X^m can be represented globally birationally—for instance, as

$$g(dg_1 \wedge \cdots \wedge dg_n)^m,$$

where $g, g_1, \ldots, g_n$ are rational functions. If $g_1, \ldots, g_n$ are local parameters at a point, then g is in the local ring at that point. One now uses the fact that if a rational function is not in the local ring of a point, then it has a divisorial pole passing through that point. Such a pole induces a point on any complete model X or Y. This shows that if a rational form as above gives rise to a section of K_X^m for one model, then it must give rise to a section of K_Y^m for any other model.

We shall be interested in the dimension

$$h^0(X, L^m) = \dim H^0(X, L^m)$$

for various line bundles L, starting with the canonical bundle, but involving other bundles as in §7. In speaking of estimates, we use the standard notation of number theorists (Vinogradov)

$$A(m) \ll B(m) \quad \text{for } m \to \infty$$

to mean that there is a constant c such that $A(m) \leqq cB(m)$ for all m sufficiently large. If the going to infinity is by divisibility ordering, then sufficiently large is according to this ordering. Following **[K-O 1]** and its addendum using Kodaira's technique, we shall now construct a pseudo volume form on a non-singular pseudo canonical variety, with positive Ricci form. We recall two lemmas from algebraic geometry.

Lemma 5.1. *Let X be a variety of dimension n. Let D be a divisor on X. Then*

$$h^0(mD) = \dim H^0(X, mD) \ll m^n \quad \textit{for } m \to \infty.$$

Proof. Let E be a divisor which is ample and such that $D + E$ is ample. Then we have an inclusion

$$H^0(mD) \subset H^0(mD + mE), \text{ so } h^0(mD) \leqq h^0(mD + mE).$$

Furthermore, if $E' = D + E$ is ample, then

$$h^0(mE') = \chi(mE') \quad \text{for } m \text{ large}$$

because the higher cohomology groups vanish for m large, and the Euler characteristic is a polynomial in m of degree $\leqq n$, thus proving the lemma.

Lemma 5.2. *Let X be a non-singular variety of dimension n. Let E be very ample on X, and let D be a divisor on X such that*

$$h^0(mD) \gg m^n \quad \text{for } m \underset{\text{div}}{\to} \infty.$$

Then

$$h^0(mD - E) \gg m^n \quad \text{for } m \underset{\text{div}}{\to} \infty,$$

in particular, $mD - E$ is linearly equivalent to an effective divisor.

Proof. Without loss of generality, we may replace E by any divisor in its class, and thus we may assume that E is an irreducible non-singular subvariety of X (a hyperplane section, in fact). We have the exact sequence

$$0 \to \mathcal{O}(mD - E) \to \mathcal{O}(mD) \to \mathcal{O}(mD)|E \to 0,$$

whence the exact cohomology sequence

$$0 \to H^0(X, mD - E) \to H^0(X, mD) \to H^0(E, \mathcal{O}(D)|E)^{\otimes m})$$

noting that $\mathcal{O}(mD)|E = (\mathcal{O}(D)|E)^{\otimes m}$. Applying the first lemma to this invertible sheaf on E we conclude that the dimension of the term on the right is $\ll m^{n-1}$, so $h^0(X, mD - E) \gg m^n$ for m large, and in particular is positive for m large, whence the lemma follows.

Suppose that X is pseudo canonical. Then $h^0(X, K_X^m) \gg m^n$ for m large, so we can apply the lemma. Let L be a very ample line bundle on X. We shall obtain a projective imbedding of X by means of *some* of the sections in $H^0(X, K_X^m)$. By Lemma 5.2, for m large there exists a non-trivial holomorphic section α of $K_X^m \otimes L^{-1}$. Let $\{s_0, \dots, s_N\}$ be a basis of $H^0(X, L)$. Then

$$\alpha \otimes s_0, \dots, \alpha \otimes s_N$$

are linearly independent sections of $H^0(X, K_X^m)$. Since $(s_0, \dots, s_N)$ gives a projective imbedding of X into $\mathbf{P}^N$ because L is assumed very ample, it follows that $\alpha \otimes s_0, \dots, \alpha \otimes s_N$ vanish simultaneously only at the zeros of α, but *nevertheless give the same projective imbedding*, which is determined only by their ratios. Then

$$\alpha\bar{\alpha} \otimes \sum s_j \otimes \bar{s}_j$$

may be considered as a section of

$$(K_X^m L^{-1}) L \otimes (\bar{K}_X^m \bar{L}^{-1}) \bar{L} = K_X^m \otimes \bar{K}_X^m,$$

and can be locally expressed in terms of complex coordinates in the form

$$|g(z)|^2 \sum_{j=1}^{n} |g_j(z)|^2 \Phi(z)^{\otimes m},$$

where as usual $\Phi(z)$ is the standard euclidean volume form on $\mathbf{C}^n$, while $g(z)$, $g_0(z), \dots, g_N(z)$ are local holomorphic functions representing α, $s_0, \dots, s_N$ respectively.

Let

$$h(z) = \left(\sum_{j=1}^{n} |g_j(z)|^2 \right)^{1/m}.$$

Then there is a unique pseudo volume form Ψ *on* X *which has the local expression*

$$\Psi(z) = |g(z)|^{2/m} h(z) \Phi(z).$$

Furthermore $\mathrm{Ric}(\Psi)$ *is positive, because* $\mathrm{Ric}(\Psi)$ *is the pull-back of the Fubini-Study form on* $\mathbf{P}^N$ *by the projective imbedding.*

In particular, we have proved:

Theorem 5.3 (Kodaira, Kobayashi-Ochiai). *Let* X *be a non-singular pseudo canonical variety. Then* X *admits a pseudo volume form* Ψ_X *with* $\mathrm{Ric}(\Psi_X)$ *positive, and* X *is measure hyperbolic.*

As to the converse, we have on the one hand:

Conjecture 5.4 (Kobayashi, see [**Ko 2**, Chapter IX]). *If* X *is measure hyperbolic, then* X *is pseudo canonical.*

This would give the neat statement that

A non-singular variety is pseudo canonical if and only if it is measure hyperbolic.

Kobayashi's conjecture is known for surfaces, through the paper of Green-Griffiths [**Gr-Gr**], completed in one remaining case (arising from the classification of surfaces) by Bogomolov and Mumford. Cf. the appendix of Mori-Mukai [**Mo-Mu**].

On the other hand, the converse in the differential geometric context is also a problem, raised by Kobayashi [**Ko 3**, Theorem 7.1 and p. 377], namely:

If X *is non-singular, and there exists a pseudo volume form with positive Ricci form, is* X *pseudo canonical?*

This would be a pseudofication of Kodaira's imbedding theorem.

Conjecture 5.5 (Green-Griffiths). *Let* X *be a pseudo canonical variety. Let* $f\colon \mathbf{C} \to X$ *be holomorphic non-constant. Then the image of* f *is contained in a proper subvariety.*

For more precise information on this, see §8.

I would conjecture:

Conjecture 5.6. *A variety* X *is hyperbolic if and only if every subvariety (including* X *itself) is pseudo canonical.*

This would give an *algebraic* condition characterizing hyperbolicity on algebraic varieties. If X is hyperbolic then X is measure hyperbolic by Kobayashi's Theorem 4.7, and if X is a surface then X is pseudo canonical as mentioned above (in general it is Conjecture 5.4). Conversely, if X is pseudo canonical, then the Green-Griffiths conjecture applies, and if all subvarieties are assumed pseudo canonical, then by that conjecture X is hyperbolic; but the question remains even on surfaces.

Conjecture 5.6 is consistent with known results about the following special example. Let X be a subvariety of an abelian variety, which does not contain the translate of a subtorus of dimension > 0. Then by Theorems 1 and 1′ of [**Ya**] (using fundamental results of Iitaka), it follows that X is pseudo canonical. In particular, every subvariety of X is also pseudo canonical, and we already know by Green's Theorem 2.4 that X is hyperbolic.

We now come to the diophantine connection.

Conjecture 5.7. *Let X be a pseudo canonical variety. Then the rational points in every finitely generated field F are not Zariski dense.*

The history of this conjecture will be recalled in an appendix. Here it is relevant to note that Noguchi proved the analogue of this conjecture for algebraic families, under the hypothesis that the cotangent bundle is ample. A vector bundle E is defined to be **ample** if the tautological line bundle on the associated projective bundle $\mathbf{P}E$ is ample, or equivalently, if $\mathcal{E}$ is the sheaf of sections, then for any coherent sheaf $\mathcal{F}$, $\mathcal{F} \otimes S^m\mathcal{E}$ is generated by its global sections for m sufficiently large. For an excellent foundational discussion of ample vector bundles, see Hartshorne [**Ha**]. Ampleness implies that for m large, a basis for the global sections of S^mE give a projective imbedding in the Grassmanian, but in dimension > 1 the converse is not true. We postpone to the next section a discussion of the condition of ampleness and other conditions, to have an appropriate structural setting for them.

We note that Conjectures 5.6 and 5.7 imply Conjecture 2.3 that a hyperbolic variety is mordellic. Thus Conjecture 5.6 replaces the complex analytic condition by an algebraic one. Vojta, in connection with his translation of Nevanlinna theory into number theory [**Vo**] conjectures that for a pseudo canonical variety over a number field, the proper algebraic subset which may contain infinitely many rational points can be taken independently of all number fields, which is a refinement of Conjecture 5.7 over number fields. But I conjecture the more precise statement expressed geometrically as follows. Recall that the **exceptional set** $\mathrm{Exc}(X)$ is the Zariski closure of all images of non-constant holomorphic maps of $\mathbf{C}$ into X.

Conjecture 5.8. *Let X be pseudo canonical. Then* $\mathrm{Exc}(X) \neq X$ *and the complement of* $\mathrm{Exc}(X)$ *is mordellic.*

We shall describe one possible approach to the exceptional set in §8.

Another approach may come from Theorem 2 of [**Ax**] already mentioned in the proof of Theorem 2.6. Transposed to the present case, Ax's technique may show how holomorphic curves yield integrable submanifolds, giving insight into the Zariski closure as in Ax's theorem.

We observe that there has been a systematic pattern of conditions which admit a weakening, like volume forms or ampleness which weaken to pseudo volume forms and pseudo ampleness. Recall that we define a variety X to be **pseudo hyperbolic** if the exceptional set is a proper subset. Similarly, we define X to be **pseudo mordellic** if there exists a proper algebraic subset Y such that $X - Y$ is mordellic. Then by Conjecture 5.8, pseudo canonical implies pseudo hyperbolic, and also implies pseudo mordellic. Furthermore, we can take $Y = \text{Exc}(X)$. The question arises under what conditions does pseudo mordellic imply pseudo canonical, and which varieties are pseudo mordellic without being pseudo canonical, or which are mordellic without being hyperbolic. Are there any? My guess is no.

Conjecture 5.9. *If X is pseudo mordellic then X is pseudo canonical.*

The modular varieties are pseudo canonical in high dimension. I am told by experts that the structure of the exceptional set and of the boundary components after desingularization is mostly a mystery. The literature on moduli varieties is open ended, so I just mention two papers by Harris-Mumford [**Ha-M**] and Tai [**Ta**] proving the pseudo ampleness. It should be emphasized that although the open part of the moduli space which is a quotient of a bounded domain is hyperbolic (both in the Brody and the Kobayashi sense), nevertheless in general, its compactification is not. There is a very strong tendency for this compactification to contain rational curves, or non-constant holomorphic images of **C**, which must therefore intersect the boundary at infinity.

Finally we observe that the property in Conjecture 5.7 is definitely weaker than the property in Conjecture 5.8. For example, let $X = \mathbf{P}^1 \times C$, where C is a curve of genus 2, defined over a number field. Then for every number field F, $X(F)$ is not Zariski dense in X by Falting's theorem, but X is not pseudo canonical. This phenomenon has to do with the canonical class being somewhat less than pseudo ample, e.g. having less than maximal Kodaira dimension.

At this point, we have stated and discussed one by one the problems and theorems summarized in the introduction. If (as I hope) all the conjectures made there are true, then they give a very clear picture of the relations between the diophantine property of having finitely or infinitely many rational points, and the other properties coming from algebraic geometry, differential geometry, and measure theory, because of the necessary and sufficient conditions in all possible directions, except possibly for $\mathbf{3} \Rightarrow \mathbf{4}$.

Warning. We have applied the property of being mordellic to an open algebraic subset of a variety, when discussing pseudo canonical and pseudo mordellic varieties. We remind the reader that for open subsets of varieties, hyperbolic is not equivalent to mordellic, as one sees from Theorem 2.6. Green has also constructed an algebraic example of a Zariski open set which is Brody hyperbolic but not Kobayashi hyperbolic. The "hyperbolicity" of affine varieties has to do with the finiteness of "integral points".

§6. Minimal models

In this section, we mention still another connection. The reader interested in pursuing the previous ideas can skip immediately to the next section.

The property of being pseudo canonical allows for exceptional subvarieties on a variety, which may have infinitely many rational points. One tries to get rid of these by blowing them down, and thus we can also consider hyperbolicity in the following context. We say that a variety is **absolutely minimal** if every birational map $f\colon Y \to X$ with Y non-singular is a morphism. I reproduce here part of Theorem 2.1 of [**Ko 2**, Chapter VIII].

Theorem 6.1. *Every hyperbolic variety is absolutely minimal.*

Under what conditions is the converse true for pseudo canonical varieties? I once hoped that a minimal model would be hyperbolic in order to show that Conjecture 2.3 implies 5.8, but that it is not true in general. Artin pointed out to me that one can construct examples (more or less generically) as follows. We consider a hypersurface of degree 5 in $\mathbf{P}^3$, defined by the general polynomial equation

$$P(x_0, \ldots, x_3) = 0.$$

We impose the condition that it contains a rational curve of degree 5 with simple nodes, say six nodes (to make it have genus 0), and adjust coordinates so that the curve is defined by the equation

$$x_0 = 0.$$

The general solution to this gives a non-singular surface which is absolutely minimal. There cannot be a hyperbolic model of its function field since by Kobayashi's Theorem 5.10 the map from the quintic surface to such a model would be a morphism, and on the other hand, the rational curve cannot be blown down since it is numerically positive.

Concerning the existence of minimal models, it may be useful for the reader to recall here two known results. In the case of surfaces, one has the following theorem of Mumford [**Mu**].

Let X be a non-singular pseudo canonical surface. Let

$$R = \bigoplus_{m=0}^{\infty} H^0(X, K_X^m)$$

and let $X' = \mathrm{Proj}(R)$. Then X' is a normal surface with a finite number of rational double points. A minimal desingularization of X' exists, and it is an absolutely minmal model of X.

For other theorems concerning the canonical ring of surfaces, see Bombieri [**Bom**]. In arbitrary dimensions, we have the following theorem of Kobayashi [**Ko 2**], Chapter VIII, Theorem 3.6.

Let X be a non-singular variety such that K_X^m has no base point for some m, that is, given $x \in X$ there is a section s of K_X^m with $s(x) \neq 0$. Then X is relatively minimal.

The recent literature on minimal models and canonical singularities is very extensive, and there is no question of attempting to survey it here. It may help the reader just to refer to Kawamata [**Ka 1**, **Ka 2**], for instance, and follow up the bibliographies at the end of these papers.

§7. Length functions, ampleness and hyperbolicity

We have now met several conditions reflecting a certain type of behavior having to do with "positivity":

Ample cotangent bundle;
Hyperbolic;
Ample canonical class;
Pseudo ample canonical class (general type);
Measure hyperbolic.

Previously stated theorems and conjectures already gave some relations between them, notably the last three. Of course we trivially have

$$\begin{aligned}\text{Ample cotangent bundle} &\Rightarrow \text{Ample canonical class}\\ &\Rightarrow \text{Pseudo ample canonical class.}\end{aligned}$$

None of the reverse implications are true in general. We shall now discuss how to fit hyperbolicity in this chain of implications.

Theorem 7.1 (Kobayashi [Ko 4] Corollary 6.3). *Let X be a non-singular variety whose cotangent bundle is ample. Then X is hyperbolic.*

Kobayashi's proof is to show that the formalism of Ricci forms and the Schwarz lemma works for something more general than volume forms, pseudo volume forms and hermitian metrics, namely Finsler structures, which are called "Differentialmetrik" in Grauert-Reckziegel [**G-R**]. I have found that the name "Finsler structure" acts as a psychological deterrent for a very simple notion which generalizes that of a norm, and which has proved useful in practice. In agreement with Griffiths, let us define a **length function** on a complex vector bundle E over X to be a function into the reals $\geqq 0$,

$$l\colon E \to \mathbf{R}_{\geqq 0}$$

such that for all complex numbers c and $v \in E$ we have:

LF 1. l is continuous, and C^∞ outside the zero section of E.
LF 2. $l(v) = 0$ if and only if $v = 0$.
LF 3. l is absolutely homogeneous of degree 1, which means

$$l(cv) = |c|l(v).$$

A "Finsler structure" is the square of a length function. It does away with the triangle inequality, and weakens the smoothness conditions of a hermitian metric. It is a hermitian metric if the rank of E is 1. Instead of writing $l(v)$, we shall also write more suggestively

$$l(v) = |v|,$$

with the usual absolute value sign.

Let $\pi\colon \mathbf{P}E \to X$ be the projective bundle of lines in E. Then $\mathbf{P}E$ has the tautological line bundle L on π^*E. As Kobayashi remarks, the length function on E then defines a length function on L, which is a smooth hermitian metric because L is one-dimensional. Then Kobayashi defines the curvature and proves for this situation the standard theorems for hermitian metrics.

For the convenience of the reader, I reproduce an argument of Griffiths giving Kobayashi's Theorem 7.1 directly. So assume that the cotangent bundle $T^\vee$ is ample. Then there is an appropriate symmetric power $S^m T^\vee$ of the cotangent bundle such that a basis of its sections $s_0, \ldots, s_N$ gives an imbedding of X in the Grassmanian. In fact, if $\mathbf{P}T$ is the projective bundle of lines in the tangent bundle, then we get a projective imbedding

$$\varphi_m \colon \mathbf{P}T \to \mathbf{P}^N$$

by

$$v \mapsto \left(s_0(v^m), \ldots, s_N(v^m)\right).$$

On the hyperplane bundle of $\mathbf{P}^N$ we have the **tautological** (or **projective**) **metric** defined in terms of the projective coordinates $w_0, \ldots, w_N$ by the function

$$\rho_i(w_0, \ldots, w_N) = \sum_{j=0}^{N} |w_j/w_i|^2 \text{ on the open set } w_i \neq 0.$$

Its Chern form is positive and defines the Fubini-Study metric. The inverse image φ_m^* of this tautological metric is a metric ρ on L which defines a length function l on T by

$$|v|^{2m} = \sum_{j=0}^{N} |s_j(v^m)|^2.$$

We let $c_1(\rho)$ be the Chern form of the above metric on L, so that $c_1(\rho)$ is a positive $(1,1)$-form on $\mathbf{P}T$. Then by compactness, there is a constant B such that

$$c_1(\rho) \geqq B\pi^*(l^2).$$

A non-constant holomorphic map $f \colon \mathbf{D} \to X$ can be lifted to a map $f_{\mathbf{P}} \colon \mathbf{D} \to \mathbf{P}T$, making the following diagram commutative.

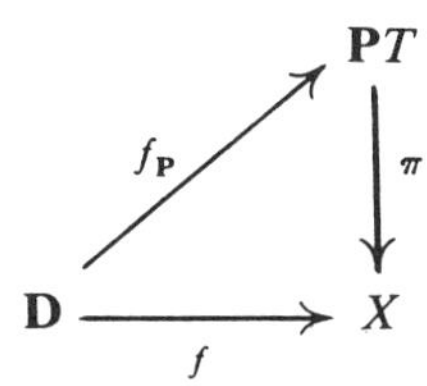

At a point, the lifting is given by $z \mapsto (f(z), df(z)v)$ wherever $df(z) \neq 0$, and v is any non-zero complex tangent vector. The extension to the set of isolated singular points is due to the fact that a meromorphic map of an open subset of $\mathbf{C}$ into a projective variety is in fact holomorphic, because at any point, one can factor out a common factor where all projective coordinates may have a common zero or pole. Then $f^*(l^2) = f_{\mathbf{P}}^*\pi^*(l^2)$, so

$$f_{\mathbf{P}}^* c_1(\rho) \geqq Bf^*(l^2).$$

Thus the length function behaves just like a hermitian metric, and its pull-back to $\mathbf{D}$ is a hermitian pseudo metric on $\mathbf{D}$ whose Chern form is positive, and whose Griffiths function is $\geqq B$. The analogues of Theorems 4.2 and 4.3 then work in the present case, and we conclude that X is hyperbolic.

In general, Kobayashi defines the Ricci form of a length function, and discusses systematically how it relates to the Ricci form on the tautological line bundle of an arbitrary holomorphic bundle. See his Theorems 4.1 and 5.1 of [**Ko 4**]. These theorems lead into questions of the existence of (1, 1)-forms or length functions in the direction of a converse for Theorem 4.3. One problem lies in the smoothness of the length function. In Royden [**Ro**], the length function associated with the Kobayashi distance on a hyperbolic variety is only upper semicontinuous.

Concerning the canonical class itself, as far as I know, the following is an open question.

Conjecture 7.2. *If X is non-singular and hyperbolic then the canonical class is ample.*

Sommese tells me that hyperbolic *together with pseudo canonical* does imply ample canonical class in dimenion $\leqq 3$. Conjecture 5.6 would show that the hypothesis of pseudo canonical is redundant. Conversely:

A non-singular hypersurface X of degree $d \geqq n + 2$ in $\mathbf{P}^n$ has ample canonical class.

Proof. The adjunction formula immediately implies that the canonical sheaf on X is $\mathcal{O}_X(-n - 1 + d)$, and $d \geqq n + 2$ is precisely the condition which makes the canonical sheaf ample.

Since the Fermat hypersurfaces

$$x_0^d + \cdots + x_n^d = 0$$

contain lines, we see that in general the condition that X has ample canonical class does not imply X hyperbolic. When the canonical sheaf is not ample, then the hypersurface has a tendency to have lines. Changes of behavior presumably occur for $d = n + 1, n, n - 1$.

Let X be a hypersurface of degree d in $\mathbf{P}^n$. If $d \leqq n - 1$ then X contains a line through every point.

This result is classical and easy. Given a point P we can represent it as the origin $P = (0, \ldots, 0)$ on affine n-space $\mathbf{A}^n$. Let $f = 0$ be the affine equation for X on this space. Decompose

$$f = f_1 + \cdots + f_d$$

into its homogeneous components of degrees $1, \ldots, d$. Then we get a line through P in $\mathbf{P}^n$ if we can solve simultaneously the homogeneous equations

$$f_1 = \cdots = f_d = 0 \quad \text{in } \mathbf{P}^{n-1}.$$

The condition $d \leqq n - 1$ is precisely the inequality which guarantees that this can always be done, for trivial dimensional reasons.

The existence of rational curves in a variety with canonical class which is not numerically effective has long been a subject of interest to algebraic geometers, and has received especially significant impetus in the last few years through the work of Mori, see for instance [**Mo**]. Such algebraic geometers work over an algebraically closed field, and there is of course the diophantine question of whether rational curves are defined over a given field of definition for the variety or hypersurface. One of Mori's theorems states:

If X is a non-singular projective variety, and the tangent bundle is ample, then X is birationally equivalent to projective space (previously Hartshorne's conjecture).

This is valid over the algebraic closure of a field of definition, and applies to hypersurfaces when $d \leqq n$. Over a given field of definition, the variety may not have any rational points, or may be only unirational, that is, rational image of projective space.

Green [**Gr 1**] showed that the only non-constant holomorphic maps of $\mathbf{C}$ into a Fermat hypersurface of degree d in $\mathbf{P}^n$ are the "Fermat" ones if $d \geqq n^2$. For instance, for a given subset of indices, say $0, \ldots, r$ ($r \leqq n-2$), let $c_1, \ldots, c_r$ be numbers such that

$$1 + c_1^d + \cdots + c_r^d = 0.$$

Then $(t, c_1 t, \ldots, c_r t)$ gives a Fermat line, and all Fermat lines are constructed in this way for each partition of the set of indices. Note that the existence of Fermat lines over a given field, say the rational numbers, amounts to finding rational points on the Fermat hypersurface itself, possibly with fewer variables.

Euler was already concerned with the problem of finding rational curves, that is solving the Fermat equation with polynomials. Swinnerton-Dyer [**Sw-D**] gives explicit examples of rational curves over the rationals, on

$$x_0^5 + \cdots + x_5^5 = 0.$$

Here X has degree $d = n = 5$, and thus the canonical class is very non-ample. Swinnerton-Dyer says: "It is very likely that there is a solution in four parameters, or at least that there are an infinity of solutions in three parameters, but I see no prospect of making further progress by the methods of this paper." In general, I conjecture that if $d = n$ then the rational curves are Zariski dense, and even that X is unirational over $\mathbf{Q}$. Actually, when $d = n = 3$, the Fermat hypersurface has Ramanujan's taxicab rational point (1729 is the sum of two cubes in two different ways: 9, 10 and 12, 1). Joe Buhler has verified on the computer that this point satisfies the hypothesis of Theorem 12.11 in Manin's book *Cubic Forms*, Chapter II, and by that theorem, it follows that in this case, the Fermat hypersurface is indeed unirational over $\mathbf{Q}$. It would be worth while to treat systematically the Fermat hypersurfaces from the present point of view of algebraic geometry, for the existence of rational curves both geometrically and over $\mathbf{Q}$, and for the possibility of their being rational images of projective space for low degrees compared to n.

Kobayashi raised the following possibility [**Ko 2**], p. 71, and Chapter IX, Problem 4:

Let X be a generic complete intersection of hypersurfaces of degrees $d_1, \ldots, d_r$ in $\mathbf{P}^n$. Let $d = d_1 + \cdots + d_r$. If $d \geqq n+2$, then X is hyperbolic.

If X is a non-singular complete intersection of hypersurfaces of degrees $d_1, \ldots, d_r$ respectively, then the canonical sheaf on X is

$$\mathcal{O}_X(-n - 1 + d),$$

where $d = d_1 + \cdots + d_r$ is the sum of the degrees. Hence again the condition for the canonical class to be ample is $d \geqq n + 2$. When d is in a range where the Brody-Green example applies, then the hyperbolicity of the generic hypersurface of that degree would follow from Brody's theorem that hyperbolicity is an open condition, if one had an algebraic characterization of hyperbolicity.

§8. Jet differentials

Interest in classical questions raised by Bloch was revived by Ochiai [**Och**]. Both introduced jets into the picture. Bogomolov [**Bo**] then used symmetric differentials. This was followed by the work of Green-Griffiths [**Gr-Gr**] from the point of view of jet differentials and curvature. The reader will find an extensive bibliography at the end of [**Gr-Gr**]. I shall now describe jet differentials, and state a conjecture and results of Green-Griffiths which give greater insight into the exceptional locus of pseudo canonical varieties.

Let X be a non-singular variety, and let $x \in X$. Let $\mathbf{D}_r$ be a disc as usual, and consider germs of holomorphic mappings $f\colon \mathbf{D}_r \to X$ such that $f(0) = x$. In local holomorphic coordinates, such an f is given by a convergent power series

$$f(z) = a_0 + a_1 z + a_2 \frac{z^2}{2!} + \cdots$$

with $a_k = a_k(f) = f^{(k)}(0) \in \mathbf{C}^n$. Given two such maps f, g we say that they **osculate to order** k if

$$f^{(i)}(0) = g^{(i)}(0) \quad \text{for } i = 0, \ldots, k.$$

Equivalence classes of such germs will be called jets of order k at x, and the set of such classes will be denoted by $J_k(X)_x$. We let

$$J_k(X) = \bigcup_{x \in X} J_k(X)_x.$$

Taking obvious charts makes $J_k(X)$ into a complex manifold of dimension $n + kn$, and if U is an open set in X with holomorphic coordinates, then this choice of coordinates gives a holomorphic isomorphism

$$J_k(U) \approx U \times \mathbf{C}^{kn}.$$

Note that even though the fiber of $J_k(X)$ at x is complex analytically isomorphic to $\mathbf{C}^{kn}$, nevertheless $J_k(X)$ is not a vector bundle. There is a natural projection

$$J_{k+1}(X) \to J_k(X)$$

whose fibers are affine bundles, and whose associated vector bundle is the tangent bundle $T(X)$.

There is a natural action of $\mathbf{C}^*$ on $J_k(X)$. For $t \in \mathbf{C}^*$, in terms of coordinates as above, the action of t is represented by

$$t(a_0, a_1, \ldots, a_k) = (a_0, ta_1, t^2 a_2, \ldots, t^k a_k).$$

We may call this a **weighted action** (cf. Dolgachev [**Do**]). We let

$$J_k^*(X) = \text{non-constant jets, that is those with some } a_j \neq 0 \text{ for } 1 \leqq j \leqq k.$$

$$P_k(X) = J_k^*(X)/\mathbf{C}^* \quad \text{with projection } \pi\colon P_k(X) \to X.$$

We call $P_k(X)$ the **weighted projective bundle** over X. Its fibers may be singular because the action of $\mathbf{C}^*$ has fixed points.

We record here that the projection $\pi\colon P_k(X) \to X$ *is proper.*

Of course, one can define the weighted projective bundle purely algebraically, but we have put ourselves immediately in the complex analytic context because of the specific applications which follow.

Let $f\colon \mathbf{D}_r \to X$ be holomorphic. Let $f(z) = x$. Then by taking the Taylor expansion of f up to order k at the given point, in terms of a chart at x, we can define a k-jet

$$j_k(f)_{z,x}$$

which is an element of $J_k(X)_x$. In this way we obtain a map

$$\mathbf{D}_r \to J_k(X).$$

One can put the analogue of length functions on the bundle $P_k(X)$, taking the weighted action of $\mathbf{C}^*$ into account. Green-Griffiths then consider the analogue of the Chern form

$$dd^c \log|j_k(f)|^2$$

for pseudo length functions on the jet set, and maps f as above. Comparing this form with $|j_k(f)|^2$ they extend the formalism of the Ricci form to this jet set. In such applications, they substitute negative curvature arguments (positivity of the analogue of the Griffiths function) for Nevanlinna theory as in Ochiai [**Och**]. See their comments, e.g. on p. 64.

In this context, Noguchi's theorem (mentioned in §5), and Kobayashi's Theorem 6.1 are jet setting to order 1.

Just as with ordinary projective space, there is a tautological sheaf $\mathscr{L}_k$ over $P_k(X)$ which may not be invertible. However, for all m divisible by $k!$ the tensor power $\mathscr{L}_k^m$ is invertible, and so can be identified with a line bundle. See [**Gr-Gr**], p. 46. Because of this, we can define **pseudo ampleness** for $\mathscr{L}_k$ just as we did previously, since the definition involves only large values of m ordered by divisibility.

Conjecture 8.1 (Green-Griffiths). *Let X be pseudo canonical. Then for k large, the tautological sheaf $\mathscr{L}_k$ is pseudo ample.*

We recall that this means that high tensor powers L_k^m have maximal Kodaira dimension, or equivalently, the rational map into projective space

$$\varphi_m = (s_0, \ldots, s_N)\colon P_k(X) \to \mathbf{P}^N$$

induced by a basis of $H^0(P_k(X), \mathscr{L}_k^m)$ is birational for m sufficiently large ordered by divisibility. The conjecture is proved in [**Gr-Gr**], Proposition 1.11, in the case of surfaces. Also Proposition 1.10 loc. cit. shows that the Euler characteristic of the higher-order jets comes out to be the "right" number in any dimension, thus giving evidence for the higher-dimensional conjecture.

We now suppose that k is sufficiently large. The birational map as above for m sufficiently large has a Bad locus, or base locus, consisting of points of indeterminacy, and the union of positive dimensional fibers of φ_m. Let $B_{k,m}$ be this base locus. Then $B_{k,m}$ stabilizes for m large (ordered by divisibility), so we write it B_k. Finally we let

$$S = \bigcap \pi(B_k)$$

be the **Green-Griffiths set.**

Theorem 8.2. *A non-constant holomorphic map* $\mathbf{C} \to X$ *is contained in the Green-Griffiths set.*

This theorem holds whether the Green-Griffiths set is a proper subset or not, and is proved in **[Gr-Gr]**, Corollary 2.8. (There is a misprint, the union sign should be an intersection.)

In particular, the exceptional set of Conjecture 5.8 is contained in the Green-Griffiths set S. I asked Green-Griffiths whether they might be equal. Green told me that the two sets are not equal in general. Certain Hilbert modular surfaces constructed by Shavel **[Sh]**, compact quotients of the product of the upper half-plane with itself, provide a counterexample which is hyperbolic, but such that the Green-Griffiths set is the whole variety. Thus the jet construction appears insufficient so far to characterize the exceptional set of Conjecture 5.8 completely algebraically. It becomes a problem to find conditions under which they are equal.

I also asked whether the complement $X - S$ is hyperbolic, and unlike the situation in Conjecture 5.8, Green told me he could prove it in this case.

Historical appendix: algebraic families

The various conditions which we have met, influencing the diophantine behavior of a variety, arose in the context of algebraic families. These played no role in the main part of this article, but played an important historical role, as follows.

Instead of a single variety defined over a field F_0, we consider an algebraic family of varieties $\{X_w\}$, depending on parameters w lying in some parameter variety W. Instead of rational points, we consider rational sections of W in the family. This point of view was taken systematically in **[La 0]**. One can then translate systematically the conjectures about the finiteness of rational points into conjectures about the finiteness of the number of sections. For instance, if the generic fiber X_w is hyperbolic and there exist infinitely many sections, then their Zariski closure must split birationally into a product $X_0 \times W$. In that case, a section comes from a rational map

$$W \to X_0.$$

If, for instance, $X_0 = C \times C$, where C is a curve of genus $\geqq 2$, and $W = C$, then there is a family of rational maps $W \to X_0$ such that $C \to C \times \{a\}$, with $a \in C$. But in any case, conjecturally, there is only a finite number of surjective rational maps of W onto X_0, and I told this conjecture to Kobayashi. When X_0 has dimension 1, and genus $\geqq 2$, this is a classical theorem of de Franchis. Kobayashi and Ochiai proved the generalization to higher dimension, for the split case of algebraic families, not under the original hypothesis that the generic fiber is hyperbolic, but under the condition that X_0 is "of general type" [**K-O 2**]. They conjectured that hyperbolic implies general type. This immediately made me conjecture that for a variety of general type, the set of rational points is not Zariski dense (unpublished). Noguchi [**No 1**] proved the general conjecture for non-split families under still a third possible condition, that the cotangent bundle is ample. Noguchi also reports that Bombieri made the above conjecture in lectures at Chicago in 1980. For more in this direction, see also [**No 2**].

In the case of an algebraic family $\{X_w\}$ with generic fiber of dimension 1—that is, a family of curves—all these conditions coincide, and are equivalent to the condition that the curve has genus $\geqq 2$. The finiteness of the number of sections if the family does not split was originally due to Manin [**Ma**]. Grauert [**Gra**] subsequently gave another proof for Manin's theorem, and Noguchi's method extends Grauert's. In this way, the deformation theory of algebraic varieties over the complex numbers influenced the diophantine theory of varieties over finitely generated fields over the rational numbers.

Riebesehl [**Ri**] took off from Grauert-Reckziegel [**G-R**], and gave a version of the diophantine result for algebraic families in the higher-dimensional case under a hypothesis of negative curvature (which he takes as his definition of hyperbolicity). He actually uses length functions, and handles the singular case, but he assumes that *all* the fibers have negative curvature, which is similar to "good reduction" everywhere. This is quite a restrictive assumption.

Despite this restriction, the method of proof is interesting for its own sake, and we sketch it here. Let $\pi\colon X \to W$ be the family. In various degrees of generality, Grauert-Reckziegel and Riebesehl prove the following fundamental fact. Let w be a point of the base. Then there exists a neighborhood V of w such that $\pi^{-1}(V)$ has a hyperbolic length function. It then follows from the Ahlfors-Schwarz lemma (the length decreasing property) that the family of sections is a normal family. From this one deduces at once that when W, X are imbedded in projective space, the projective degrees of the sections are bounded. This is the essential part of the diophantine result: sections are of bounded height.

There remains the question of splitting the family. One can then use the techniques from early days [**La 0**]. The boundedness of the degrees implies that the sections lie in only a finite number of Chow families. If one member of an irreducible component is a section, so is the generic member. Assuming that the family of sections is Zariski dense, it follows that there exists a variety T and a generically surjective rational map

$$f\colon T \times W \to X.$$

From this one would like to split X birationally, but Riebesehl splits only a curve, using results from **[La 0]**. More generally, it remains a problem to prove the following, which we state in a self-contained way, under weaker assumptions.

> *Let π: $X \to W$ be a generically surjective rational map, whose generic fiber is geometrically irreducible and pseudo canonical. Assume that there exists a variety T and a generically surjective rational map*
>
> $$f: T \times W \to X.$$
>
> *Then X is birationally equivalent to a product $X_0 \times W$.*

The study of algebraic families from the diophantine point of view thus put forth several conditions like hyperbolic, general type, ample cotangent bundle, negative curvature.

Manin's original proof introduced a different aspect of differential geometry into the question, what became known as the Gauss-Manin connection: differentiating with respect to the parameters of the family.

For convenience of language, it was useful to start with a variety defined over the complex numbers. In number theory, one usually starts with a variety defined over a field F_0 which is finite over the rationals, that is a number field. There was no need to do so here. However, a few remarks on the relations between number fields and arbitrary finitely generated fields may be useful to the reader. A priori, the statement:

(a) A variety of a certain type has only a finite number of rational points in every finitely generated field over **Q**.

is stronger than

(b) A variety of a certain type, defined by equations over a number field, has only a finite number of rational points in every number field.

However, for varieties of dimension 1 (curves), the two statements are equivalent in the case of curves of genus $\geqq 2$. This can be proved easily by a specialization argument based on a theorem of Néron, using Hilbert's Irreducibility Theorem, see **[La 2]**, Chapter IX, Theorem 6.2, and Chapter XII, Theorem 2.3, including an extension by Silverman. It would be interesting to study the higher-dimensional case from this point of view, which lies somewhere between algebraic families over the complex numbers, and single varieties defined over a number field.

Bibliography

[Ax] J. AX, *Some topics in differential algebraic geometry* II: *On the zeros of theta functions*, Amer. J. Math. **94** (1972), 1205–1213.

[Bl] A. BLOCH, *Sur les systèmes de fonctions uniformes satisfaisant à l'équation d'une variété algébrique dont l'irrégularité dépasse la dimension*, J. de Math. **V** (1926), 19–66.

[Bog] F. BOGOMOLOV, *Families of curves on a surface of general type*, Soviet Math. Dokl. **18** (1977), 1294–1297.

[Bom] E. BOMBIERI, *Canonical models of surfaces of general type*, Inst. Hautes Études Sci. Publ. Math **42** (1973), 171–219.

[Br 1] R. BRODY, *Intrinsic metrics and measures on compact complex manifolds*, thesis, Harvard, 1975.

[Br 2] R. BRODY, *Compact manifolds and hyperbolicity*, Trans. Amer. Math. Soc. **235** (1978), 213–219.

[Br-Gr] R. BRODY and M. GREEN, *A family of smooth hyperbolic hypersurfaces in P_3*, Duke Math. J. **44** (1977), 873–874.

[Ch] S. S. CHERN, *On holomorphic mappings of Hermitian manifolds of the same dimension*, Proc. Sympos. Pure Math., vol. 11, Amer. Math. Soc., Providence, R. I., 1968, pp. 157–170.

[Di 1] A. DINGHAS, *Ein n-dimensionales Analogon des Schwarz-Pickschen Flächensatzes fur holomorphe Abbildungen der komplexen Einheitskugel in eine Kähler-Manningfaltigkeit*, Festschrift zur Gedachtnisfeier für Karl Weierstrass, Westdeutscher Verlag, Cologne (1966), 477–494.

[Di 2] A. DINGHAS, *Über das Schwarzsche Lemma und verwandte Sätze*, Israel J. Math. **5** (1967), 157–169.

[Do] I. DOLGACHEV, *Weighted projective varieties*, Group Actions and Vector Fields (Proc. Polish–North Amer. Sem., Vancouver, 1981; J. B. Carrels, editor), Lecture Notes in Math., vol. 956, Springer-Verlag, Berlin and New York, 1982, pp. 34–71.

[E] D. A. EISENMAN, *Intrinsic measures on complex manifolds and holomorphic mappings*, Mem. Amer. Math. Soc. No. 96 (1970).

[Gr-Wu] I. GRAHAM and H. WU, *Some remarks on the intrinsic measures of Eisenman*, Trans. Amer. Math. Soc. **288** (1985), 625–660.

[Gra] H. GRAUERT, *Mordell's Vermutung über rationale Punkte auf algebraische Kurven und Funktionenkörper*, Inst. Hautes Études Sci. Publ. Math. **25** (1965), 131–149.

[G-P] H. GRAUERT and U. PETERNELL, *Hyperbolicity of the complement of plane curves*, Manuscripta Math. **50** (1985), 429–441.

[G-R] H. GRAUERT and H. RECKZIEGEL, *Hermitesche Metriken und normale Familien holomorpher Abbildungen*, Math. Z. **89** (1965), 108–112.

[Gr 1] M. GREEN, *Some Picard theorems for holomorphic maps to algebraic varieties*, Amer. J. Math. **97** (1975), 43–75.

[Gr 2] M. GREEN, *Holomorphic maps to complex tori*, Amer. J. Math. **100** (1978), 615–620.

[Gr-Gr] M. GREEN and P. GRIFFITHS, *Two applications of algebraic geometry to entire holomorphic mappings*, The Chern Symposium 1979, (Proc. Internat. Sympos., Berkeley, Calif., 1979), Springer-Verlag, New York, 1980, pp. 41–74.

[Gri] P. GRIFFITHS, *Holomorphic mapping into canonical algebraic varieties*, Ann. of Math. (2) **98** (1971), 439–458.

[Ha-M] J. HARRIS and D. MUMFORD, *On the Kodaira dimension of the moduli space of curves*, Invent. Math. **67** (1982), 23–86.

[Ha] R. HARTSHORNE, *Ample vector bundles*, Inst. Hautes Études Sci. Publ. Math., **29** (1966), 319–350.

[K-K] P. KIERNAN and S. KOBAYASHI, *Holomorphic mappings into projective space with lacunary hyperplanes*, Nagoya Math J. **50** (1973), 199–216.

[Ka 1] Y. KAWAMATA, *The cone of curves of algebraic varieties*, Ann. of Math. (2) **119** (1984), 603–633.

[Ka 2] Y. KAWAMATA, *Pluricanonical systems on minimal algebraic varieties*, Invent. Math. **79** (1985), 567–588.

[Ko 1] S. KOBAYASHI, *Volume elements, holomorphic mappings and Schwarz lemma*, Proc. Sympos. Pure Math., vol. 11, Amer. Math. Soc., Providence, R. I., 1968, pp. 253–260.

[Ko 2] S. KOBAYASHI, *Hyperbolic manifolds and holomorphic mappings*, Marcel Dekker, New York, 1970.

[Ko 3] S. KOBAYASHI, *Intrinsic distances, measures and geometric function theory*, Bull. Amer. Math. Soc. **82** (1976), 357–416.

[Ko 4] S. KOBAYASHI, *Negative vector bundles and complex Finsler structures*, Nagoya Math. J. **57** (1975), 153–166.

[K-O 1] S. KOBAYASHI and T. OCHIAI, *Mappings into compact complex manifolds with negative first Chern class*, J. Math. Soc. Japan **23** (1971), 137–148.

[K-O 2] S. KOBAYASHI and T. OCHIAI, *Meromorphic mappings into compact complex spaces of general type*, Invent. Math. **31** (1975), 7–16.

[La 0] S. LANG, *Integral points on curves*, Inst. Hautes Études Sci. Publ. Math. **6** (1960), 27–43.

[La 1] S. LANG, *Higher dimensional diophantine problems*, Bull. Amer. Math. Soc. **80** (1974), 779–787.

[La 2] S. LANG, *Fundamentals of diophantine geometry*, Springer-Verlag, 1983.

[La 3] S. LANG, *Introduction to transcendental numbers*, Addison-Wesley, 1966.

[Ma] J. MANIN, *Rational points of algebraic curves over function fields*, Izv. Akad. Nauk **27** (1963), 1395–1440.

[Mo] S. MORI, *Threefolds whose canonical bundles are not numerically effective*, Ann. of Math. (2) **116** (1982), 133–176.

[Mo-Mu] S. MORI and S. MUKAI, *The uniruledness of the moduli space of curves of genus* 11, Algebraic Geometry Conference, Tokyo-Kyoto, 1982, Lecture Notes in Math., vol. 1016, pp. 334–353.

[Mu] D. MUMFORD, *The canonical ring of an algebraic surface*, Ann. of Math. (2) **76** (1962), 612–615.

[No 1] J. NOGUCHI, *A higher-dimensional analogue of Mordell's conjecture over function fields*, Math. Ann. **258** (1981), 207–212.

[No 2] J. NOGUCHI, *Logarithmic jet spaces and extensions of de Franchis' theorem* (to appear).

[Och] T. OCHIAI, *On holomorphic curves in algebraic varieties with ample irregularity*, Invent. Math. **43** (1977), 83–96.

[Re] H. RECKZIEGEL, *Hyperbolische Raüme und normale Familien holomorpher Abbildungen*, Dissertation, Göttingen, 1967.

[Ri] D. RIEBESEHL, *Hyperbolische Komplex Raüme und die Vermutung von Mordell*, Math. Ann. **257** (1981), 99–110.

[Ro] H. ROYDEN, *Remarks on the Kobayashi metric*, Several Complex Variables. II (Proc. Internat. Conf. Univ. Maryland, College Park, Md., 1970), Lecture Notes in Math., vol. 185, Springer-Verlag, 1971, pp. 125–137.

[Sh] I. SHAVEL, *A class of algebraic surfaces of general type constructed from quaternion algebras*, Pacific J. Math. **76** (1978), 221–245.

[Si] C. L. SIEGEL, *Über einige Anwendungen diophantischer Approximationen*, Abh. Preuss. Akad. Wiss. Phys. Math. Kl. (1929), 41–69.

[Sw-D] P. SWINNERTON-DYER, *A solution of* $A^5 + B^5 + C^5 = D^5 + E^5 + F^5$, Proc. Cambridge Philos. Soc. 48 (1952), 516–518. (*Note*: Under the name of P. S. Dyer, see also the case of degree 4 in J. London Math. Soc. 1943, extending the work of Euler.)

[Ta] Y.-S. TAI, *On the Kodaira dimension of the moduli space of abelian varieties*, Invent. Math. **68** (1982), 425–439.

[Vo] P. VOJTA, Springer Lecture Notes, to appear; see also his thesis, Harvard, 1984.

[Wo] J. WOLFART, *Taylorentwicklungen automorpher Formen und ein Transzendenzproblem aus der Uniformisierungs theorie*, Abh. Math. Sem. Univ. Hamburg **54** (1984), 25–33.

[W-W] J. WOLFART and G. WUSTHOLZ, *Der Überlagerungsradius algebraischer Kurven und die Werte der Betafunktion an rationalen Stellen* (to appear).

[Ya] S. T. YAU, *Intrinsic measures on compact complex manifolds*, Math. Ann. **212** (1975), 317–329.

DEPARTMENT OF MATHEMATICS, YALE UNIVERSITY, NEW HAVEN, CONNECTICUT 06520

Contemporary Mathematics
Volume 67, 1987

DIOPHANTINE PROBLEMS IN COMPLEX HYPERBOLIC ANALYSIS

Serge Lang

ABSTRACT. We discuss two results concerning holomorphic maps into hyperbolic complex manifolds, or hyperbolically imbedded manifolds. One is an extension of a theorem of Noguchi, dealing with the uniform convergence of certain sequences of mappings; the other is a theorem of Cartan, the Second Main Theorem of Nevanlinna theory in the context of maps into projective space and the complement of certain hyperplanes. We improve Cartan's theorem by giving a better form to the ramification term.

0. INTRODUCTION. After Manin's proof of the Mordell conjecture in the function field case [Ma], Grauert not only gave another proof, but also gave an idea for still another over the complex numbers, which developed from Grauert-Reckziegel [G-R] to Riebesehl's recent paper [Ri].

On the other hand, in [L 2], partly motivated by similar considerations (like [Ko], Theorem 3.2 of Chapter V), I conjectured that a hyperbolic variety is mordellic, i.e. has only a finite number of rational points in every finitely generated field over the rationals. "Hyperbolic" here means Kobayashi hyperbolic [Ko]. As usual, the conjecture has a function field analogue. Noguchi has written recently a series of papers dealing with the function field case, under various hypotheses, and especially in the last one he pushes Grauert's idea much further, in the context of Kobayashi hyperbolicity.

I first review some definitions concerning hyperbolic spaces, and then prove an extension of Noguchi's theorem. I discuss afterward more precisely (and technically) in what ways it is an extension, and I also indicate how it is related to the Mordell conjecture. This is of interest to arithmeticians, but the extension also takes place in the analytic context of a theorem of Kwack, which we shall also recall.

In the second part of the paper I state an extension of a theorem of

1980 Mathematics Subject Classification (1985 Revision), 11D99, 11J99 14D10, 30D35, 32A20, 53C99

Cartan [Ca] having to do with diophantine questions in the context of Nevanlinna theory. Vojta [Vo] had the great insight to draw an analogy between the Second Main Theorem of this theory and the theory of heights, Roth's theorem, Schmidt's theorem, and the Mordell conjecture as well as other diophantine questions of algebraic number theory. I shall indicate briefly how the extension of Cartan's theorem fits in Vojta's program.

1. HYPERBOLICITY AND NOGUCHI'S THEOREM. Let X be a complex space and let $\mathbb{D}$ be the unit disc with its hyperbolic metric. The Kobayashi semi distance d_X is defined as follows. Given $x, y \in X$ we consider finite sequences of holomorphic maps

$$f_i : \mathbb{D} \longrightarrow X, \quad (i = 1,\ldots,m),$$

points p_i, q_i in $\mathbb{D}$ such that $f_i(q_i) = f_{i+1}(p_i)$, and $f_1(p_1) = x$, $f_m(q_m) = y$. In other words, we join x to y by a chain of holomorphic discs. We let $d_{\mathbb{D}}$ be the hyperbolic distance on $\mathbb{D}$. We define the Kobayashi semi distance to be

$$d_X(x,y) = \inf \sum_{i=1}^{m} d_{\mathbb{D}}(p_i, q_i)$$

where the inf is taken for all choices of such chains of discs joining x to y. If for instance $X = \mathbb{C}$ then $d_X(x,y) = 0$ for all x, y. It is immediate that holomorphic maps are Kobayashi semi distance decreasing. Furthermore, if $X = \mathbb{D}$ then the Kobayashi semi distance is the usual hyperbolic (Poincaré-Lobatchevsky) distance.

We say that X is Kobayashi hyperbolic if d_X is a distance, that is $d_X(x,y) > 0$ for $x \neq y$.

Let X be a complex subspace of Y. We say that X is hyperbolically imbedded in Y if X is hyperbolic, and given two sequences $\{x_n\}$, $\{y_n\}$ in X converging to points x, y in $\bar{X}$ respectively, if $d_X(x_n, y_n) \longrightarrow 0$ then $x = y$.

Let Z be a complex manifold. By a length function H on Z (or on its tangent bundle TZ) we mean a function

$$H : TZ \longrightarrow \mathbb{R}_{\geqq 0}$$

such that $H(v) > 0$ for all $v \in TZ$, $v \neq 0$, and

$$H(cv) = |c|\, H(v) \quad \text{for } c \in \mathbb{C}.$$

Finally we require that H be continuous. The norm associated with a hermitian metric is a length function.

If H is a length function, then we can define the length of curves in the usual way, and we can then define a semi distance d_H, whereby $d_H(x,y)$ is the inf of the lengths of curves joining x to y.

If Y is a complex space rather than a complex manifold, then one can define the similar notion for Y, either by accepting to imbed Y globally in a complex manifold (which is OK in all the applications I know of), or by localizing the construction of a length function, using only such local imbeddings. I don't want to go into such technical details at this point.

It is easy to show that if X is hyperbolically imbedded in Y, then there exists a length function H on Y such that $d_H \leqq d_X$ on X. If X is relatively compact in Y, then for every length function H, we have $d_H \ll d_X$, the sign $\ll$ meaning that there exists a constant $C > 0$ such that $d_H \leqq Cd_X$. In other words, the Kobayashi distance on X is bounded from below by a hermitian distance on an ambient global complex manifold.

Let d be a distance function on a set, which is thus a metric space. Then one associates with d the Hausdorff measure μ_d^m for every positive integer m. Suppose now that M is a complex hermitian manifold of dimension m, with its positive (1,1)-form ω, and volume form $\Omega = \omega^m/m!$. Then we also have the measure μ_Ω associated with Ω, and it is a fact that

$$\mu_\Omega = \mu_d^{2m},$$

where d is the Riemannian distance associated with the hermitian metric. This is basically standard for euclidean space, and the proof for manifolds is essentially the same.

We shall also need a foundational fact due to Lelong [Le 1] and [Le 2]. For an elegant exposition, see also Griffiths [Gr], p. 13.

LELONG'S THEOREM. *Let μ_{2m} be the euclidean measure induced on an m-dimensional complex subspace X of $\mathbb{C}^N$. Let $\bar{X}$ be the closure, and assume that $\bar{X}-X$ lies outside the ball $B(R_0)$, say centered at the origin. Assume that $0 \in X$. Then for $r < R_0$ we have*

$$\mu_{2m}(X_r) \geqq v_{2m} r^{2m},$$

where v_{2m} is the Lebesgue volume of the unit ball in $\mathbb{R}^{2m}$, and $X_r = X \cap B(r)$ is the part of X in the ball of radius r.

Next we recall Kwack's theorem as extended by Kobayashi and Kiernan.

THE K^3-THEOREM, KIERNAN-KOBAYASHI-KWACK. *Let* A *be a divisor with normal crossings on a complex manifold* M. *Let* $X \subset Y$ *be a relatively compact hyperbolically imbedded complex subspace. Then every holomorphic map*

$$f: M-A \longrightarrow X$$

extends to a holomorphic map of M *into* Y.

This theorem was proved by Kwack [Kw] when $X = Y$ is compact. In this case, A can be arbitrary, no assumption on the nature of the singularities is needed since one can extend the map to M-Sing(A), then to M-Sing(Sing(A)), and so forth. Kobayashi introduced the concept of "hyperbolically imbedded", and generalized Kwack's result to the case when X is hyperbolically imbedded in Y, and A is non-singular. Kiernan treated the case when A has simple normal crossings (normal crossings for short), cf. [Ko 3] p. 93 as compared to [Ki 3]. The method of proof is the same as Kwack's, and part of the method goes back to Grauert-Reckziegel.

THEOREM 1. *Let:*

$X \subset Y$ *be a relatively compact, hyperbolically imbedded complex subspace;*

M *be a complex manifold of dimension* m;

A *be a divisor with normal crossings in* M.

Let

$$f_n: M-A \longrightarrow X$$

be a sequence of holomorphic maps, which converges uniformly on compact subsets of M-A *to a holomorphic map* $f: M-A \longrightarrow X$. *Let* $\bar{f}_n$, $\bar{f}$ *be their holomorphic extensions from* M *into* Y. *Then the sequence* $\{\bar{f}_n\}$ *converges uniformly to* $\bar{f}$ *on every compact subset of* M *itself*.

Before we give the proof, we point to the example when $M = \mathbf{D}$ is the unit disc, A is the origin, so $M-A = \mathbf{D}^*$ is the punctured unit disc. Then $m = 1$, the Hausdorff measure associated with the Kobayashi distance on $\mathbf{D}^*$ is 2-dimensional, and is the hyperbolic measure. Observe that the Hausdorff measure of punctured discs around the origin tends to 0 as the radius tends to 0. This is easily proved by representing the disc as the quotient of the upper half plane, and using the hyperbolic measure on the upper half plane.

We shall now prove the theorem. The question of convergence arises in the neighborhood of a point $a \in A$. Without loss of generality, we may assume that complex coordinates $z_1,\ldots,z_m$ are chosen such that

$$M = \underline{\underline{D}}^m, \qquad a = 0, \qquad \text{and } A \text{ is defined by } z_1 \cdots z_p = 0.$$

Assume first that A is defined by $z_1 = 0$, so

$$A = \underline{\underline{D}}^* \times \underline{\underline{D}}^{m-1}.$$

Neighborhoods of 0 are given by

$$U_{k,r} = \underline{\underline{D}}_{1/k} \times \underline{\underline{D}}_r^{m-1} \qquad \text{with } 0 < r < 1.$$

We let $S_{1/k}$ denote the circle of radius $1/k$. Let $f(0) = y_0$. We let W denote a small open neighborhood of y_0 in Y. We can identify $W \subset B(y_0,1)$ with a complex subspace of a ball of radius 1 in some $\underline{\underline{C}}^N$. We write coordinates of $z \in \underline{\underline{D}}^m$ as

$$z = (z_1, z') \qquad \text{with } z_1 \in \underline{\underline{D}} \text{ and } z' \in \underline{\underline{D}}^{m-1}.$$

If the sequence $\{\bar{f}_n\}$ does not converge uniformly on some neighborhood of 0, then we can pick W as above such that, given k, r there are infinitely many n for which

$$\bar{f}_n(U_{k,r}) \not\subset W.$$

For each $w \in B(y_0,1)$ and $t > 0$ we let $B(w,t)$ be the open ball of radius t and center w, and we let $S(w,t)$ be the sphere of center w and radius t. There exists k_0 and r_0 such that

$$\bar{f}(\bar{\underline{\underline{D}}}_{1/k_0} \times \underline{\underline{D}}_{r_0}^{m-1}) \subset B(y_0, 1/8)$$

simply by the continuity of $\bar{f}$. Since $\{\bar{f}_n\}$ converges uniformly to $\bar{f}$ on $S_{1/k} \times \bar{\underline{\underline{D}}}_{r_0}^{m-1}$, there exists a subsequence $\{f_{n_k}\}$ and a sequence

$\{z'_k\}$ of points $z'_k \in \underline{\underline{D}}^{m-1}_{r_0}$ such that

$$\lim z'_k = 0$$

and

(1) $f_{n_k}(S_{1/k}, z'_k) \subset B(y_0, 1/4)$

(2) $\bar{f}_{n_k}(\underline{\underline{D}}_{1/k}, z'_k) \not\subset B(y_0, 1)$.

Hence for each $k \geqq k_0$ there is a point $z_{1k} \in \underline{\underline{D}}_{1/k}$ such that

(3) the point $x_k = \bar{f}_{n_k}(z_{1k}, z'_k)$ lies on $S(y_0, 1/2)$.

Let

$$E_k = \bar{f}_{n_k}(\underline{\underline{D}}_{1/k}, z'_k) \cap B(x_k, 1/8).$$

Then E_k is a one-dimensional complex subspace of $B(x_k, 1/8)$. We note that $f_{n_k}(S_{1/k}, z'_k)$ lies outside $B(x_k, 1/8)$. By Lelong's theorem, it follows that the euclidean measure of E_k satisfies

$$c_0 \leqq \mu^2_{\text{euc}}(E_k) \quad \text{for all } k = 1, 2, \ldots .$$

On the compact set $\bar{B}(y_0, 1) \cap Y$ all length functions are equivalent, i.e. each is less than a constant times the other, so the corresponding distance functions are equivalent, and so are the corresponding Hausdorff measures. If H is a length function on Y, with distance d_H, we let:

μ^2_H = 2-dimensional Hausdorff measure defined by d_H, restricted to X;

μ^2_X = 2-dimensional Hausdorff measure defined by d_X on X.

Then we get

$$\mu^2_{\text{euc}}(E_k) \leqq c_1 \mu^2_H(E_k).$$

Let $E^*_k = E_k - f_{n_k}(0, z'_k)$. Then $\mu^2_H(E_k) = \mu^2_H(E^*_k)$. Note that $E^*_k \subset X$. Since $d_H \ll d_X$ on X because X is hyperbolically imbedded in Y, it follows that

$$\mu^2_H(E_k) \leqq C_2 \mu^2_X(E^*_k),$$

where μ^2_X is the Hausdorff measure associated with the Kobayashi distance d_X. On the other hand, since

$$f_{n_k} : M\text{-}A = \underline{\underline{D}}^* \times \underline{\underline{D}}^{m-1} \longrightarrow X$$

is distance decreasing from d_{M-A} to d_X, it follows that

$$\mu^2_X(E^*_k) \leqq C_3 \mu^2_{M-A}(\underline{\underline{D}}^*_{1/k}, z'_k).$$

But $\underline{\underline{D}}^*$ is imbedded on $(\underline{\underline{D}}^*, z'_k)$ in M-A, and the imbedding is Kobayashi distance decreasing. Therefore we obtain the final inequality

$$\mu^2_{M-A}(\underline{\underline{D}}^*_{1/k}, z'_k) \leqq C_4\, \mu^2_{\underline{\underline{D}}^*}(\underline{\underline{D}}^*_{1/k}) \longrightarrow 0 \quad \text{as} \quad k \longrightarrow \infty .$$

This contradicts the first inequality in this successive chain, namely that $C_0 \leqq \mu^2_{euc}(E_k)$, and concludes the proof of the theorem in case $A = \underline{\underline{D}}^* \times \underline{\underline{D}}^{m-1}$.

The case when $A = \underline{\underline{D}}^{*p} \times \underline{\underline{D}}^{m-p}$ is then done by induction, since the sequence $\{\bar{f}_n\}$ converges uniformly to $\bar{f}$ on compact subsets of $\underline{\underline{D}}^{*p-1} \times \underline{\underline{D}}^{m-p+1}$. This concludes the proof of the theorem.

The theorem was formulated by Noguchi for sections in a proper family. An analysis of the proof showed that the hypothesis of having sections was irrelevant, and that Noguchi's arguments applied without change to arbitrary maps as formulated above. For the formulation in terms of sections, see below.

Let $\pi : X \longrightarrow Y$ be a proper holomorphic map of complex spaces. We view π as representing a family of complex spaces $\pi^{-1}(y)$ for $y \in Y$. One wants conditions under which the set of sections $\mathrm{Sec}(\pi)$ is locally compact for the compact open topology, that is given any sequence of sections, there exists a subsequence which converges uniformly on every compact subset of Y.

Suppose that $\pi^{-1}(y)$ is hyperbolic for <u>all</u> $y \in Y$. Then for <u>every</u> point $y \in Y$ there is an open neighborhood V which is hyperbolic (in fact some closed subspace of a polydisc chart) and such that $\pi^{-1}(V)$ is hyperbolic. This follows immediately from Brody's criterion for hyperbolicity [Br] which he proves is an open condition. Since a holomorphic map, in this case a section, is Kobayashi distance decreasing, it follows that a family of sections

$$f: V \longrightarrow \pi^{-1}(V)$$

is equicontinuous, and satisfies the hypotheses of Ascoli's theorem, so is locally compact. This is the Kobayashi hyperbolicity version of Grauert's original idea, pursued by Riebesehl in the context where "hyperbolic" means "negative Gauss curvature" in some form, so one could call it "Gauss hyperbolic". Riebesehl falls short of analyzing what happens at "bad" points $y \in Y$, when the fiber $\pi^{-1}(y)$ is not Gauss hyperbolic. Noguchi made substantial progress by proving the following theorem.

<u>NOGUCHI'S THEOREM</u> [No]. <u>Let</u> $\bar{R}$ <u>be a Riemann surface (complex manifold of dimension</u> 1), <u>and let</u>

$$\bar{\pi} : \bar{X} \longrightarrow \bar{R}$$

<u>be a proper holomorphic map</u>. <u>Let</u> S <u>be a discrete subset of</u> $\bar{R}$, <u>and let</u> $R = \bar{R}-S$. <u>Let</u> $X = \pi^{-1}(R)$ <u>and</u> $\pi : X \longrightarrow R$ <u>the restriction of</u> $\bar{\pi}$ <u>to</u> X. <u>Assume</u>:

(a) <u>For all</u> $y \in R$ <u>the fiber</u> $\pi^{-1}(y)$ <u>is hyperbolic</u>.

(b) <u>For each point</u> $y \in \bar{R}-R$ <u>there is an open neighborhood</u> V <u>such that</u> $\pi^{-1}(V - \{y\})$ <u>is hyperbolically imbedded in</u> $\pi^{-1}(V)$.

<u>Then the set of sections</u> $\mathrm{Sec}(\bar{\pi})$ <u>is locally compact</u>.

This follows in the manner described above. Actually, Noguchi formulates a version with a higher dimensional base, and a divisor with normal crossings in the base where the fibers may be bad. He then defines the notion of (X,π,R) being hyperbolically imbedded in $(\bar{X},\bar{\pi},\bar{R})$, and gets the theorem in that case. For simplicity of formulation, in order to rely on the standard definition of hyperbolically imbedded, I limited myself to the case of a one-dimensional base, which is the essential case for applications to the Mordell type problems.

Noguchi also proved in the case of surfaces fibered by curves whose generic member has genus at least 2 that there always exists a model satisfying the hypothesis of the above theorem. The existence of such a model remains open in the case of higher dimensional hyperbolic fibers.

Suppose that for the proper map

$$\pi : X \longrightarrow Y$$

given $y \in Y$, the fiber $\pi^{-1}(y)$ is hyperbolic, or there exists a neighborhood V of y which is hyperbolic, and such that $\pi^{-1}(V - \{y\})$ is hyperbolically imbedded in $\pi^{-1}(V)$. Then the set of sections is locally compact by Noguchi's theorem.

Suppose in addition that X, Y *are projective. Then the set of sections is compact, and the projective degrees of the sections are bounded.*

This is the desired conclusion corresponding to Mordell's conjecture, that the heights of rational points are bounded in the case of number fields, cf. [L 1]. We are thus led naturally into the second part of this paper.

2. ON CARTAN'S THEOREM.

Let F be a field with a family of absolute values $\{v\}$. If $a \in F^*$ then we write $\|a\|_v$ for a suitably normalized multiplicative absolute value, and

$$v(a) = -\log \|a\|_v .$$

We assume that the set of absolute values satisfies the product formula, that is for $a \neq 0$,

$$\prod_v \|a\|_v = 1,$$

or in additive terms,

$$\sum_v v(a) = 0.$$

Let $P = (a_0, \ldots, a_n) \in \underline{\underline{P}}^n(F)$ be a point in projective space, with projective coordinates $a_i \in F$. We define its <u>height</u>

$$h(P) = \sum_v \log \max_i \|a_i\|_v.$$

<u>Examples</u>. (Cf. [La 4], Chapter III, §1 and §3.) Let F be a number field, and let $\{v\}$ be the set of absolute values extending the ordinary absolute value on $\underline{\underline{Q}}$, or a p-adic absolute value. Normalizing $\|a\|_v$ suitably the product formula holds (Artin-Whaples).

Let F be the function field of a projective variety X, which we assume non-singular for simplicity. Say X is defined over an algebraically closed field, or even $\underline{\underline{C}}$. To each irreducible divisor W on X we have the associated order function

$$v_W(f) = \text{ord}_W(f)\deg(W),$$

where ord_W is the order at W, and $\deg(W)$ is the projective degree. Again the product formula holds.

As a third example, we start from Vojta's remark that Jensen's formula is the analogue of the product formula. Thus let r be a fixed positive number, and let F be the field of meromorphic functions on the disc $\underline{\underline{D}}_r$. For each θ we have the absolute value

$$\|f\|_{\theta,r} = |f(re^{i\theta})|, \quad \text{so} \quad v_{\theta,r}(f) = -\log |f(re^{i\theta})|.$$

We allow ∞ to be a value. Given a point $a \in \underline{\underline{D}}_r$ we have the absolute value such that in logarithmic form

$$v_{a,r}(f) = (\text{ord}_a f)\log \left|\frac{r}{a}\right|.$$

The factor $\log |r/a|$ is a normalizing factor, like $\log p$ in number theory when p is a prime number. Let c_f be the leading coefficient of f in the power series expansion at 0. Then we also let

$$v_0(f) = -\log |c_f|.$$

Strictly speaking, v_0 is not quite a valuation, but let that pass. Then

Jensen's formula is the sum formula (fixing r). If $f \neq 0$ then

$$\int_0^{2\pi} v_{\theta,r}(f) \frac{d\theta}{2\pi} + \sum_{a \in \underline{D}_r} v_{a,r}(f) - v_0(f) = 0.$$

The theory of the height (as function of r) associated with this sum formula is, by definition, Nevanlinna theory.

In the number field case, it is easily proved that a set of points of bounded height is finite. In the second example, called the function field case, bounding the height, i.e. the projective degree, implies that the points lie in a finite number of Chow families.

The Mordell conjecture (Faltings' theorem) was (is) that if X is a curve of genus at least 2 defined over a number field F, and say in projective space, then its set of rational points $X(F)$ in F has bounded height - and therefore is finite.

My conjecture is that if X is a projective variety defined over F, and hyperbolic in some imbedding of F in the complex numbers, then again the points of $X(F)$ have bounded height, and so $X(F)$ is finite.

In Nevanlinna theory, given a holomorphic map $f: \underline{C} \longrightarrow X$ into some projective variety, bounding the height of the map implies that the map is constant.

In each case, finding conditions under which the height is bounded is regarded as an end in itself. Such conditions amount conjecturally to conditions of "hyperbolicity", in one form or another.

I shall now discuss the Nevanlinna case at greater length. The height of a map

$$f: \underline{C} \longrightarrow \underline{P}^n$$

is usually denote by T_f, so T_f is a function of r. Suppose that f is represented by coordinate functions $(f_0,\ldots,f_n)$ where f_i are entire functions without common zeros. Then the height T_f is given by

$$T_f(r) = \int_0^{2\pi} \log \max_i \|f_i\|_{\theta,r} \frac{d\theta}{2\pi} - \log \max_i |f_i(0)|.$$

We call this the <u>Cartan height</u>, since Cartan defined it in 1929 [Ca]. It is similar to the Weil height in number theory (also defined by Siegel at about the same time). More generally, if f is represented by meromorphic coordinates $(g_0, g_1, \ldots, g_n)$ then the Cartan height is equal to

$$T_f(r) = \int_0^{2\pi} \log \max_i \|g_i\|_{\theta,r} \frac{d\theta}{2\pi} - \log \max_{i \in M} |c_{g_i}|$$

$$+ \sum_{a \in \underline{\underline{D}}_r} \log \max_i \|g_i\|_{a,r}$$

where the indices $i \in M$ in the term with the leading coefficients c_{g_i} are taken in an appropriate subset of all indices. This height T_f is independent of homogeneous coordinates, because if we change $(g_0, \ldots, g_n)$ by multiplying with a meromorphic function h, then all the max in all the terms of the formula change by $\log \|h\|_v$, where v ranges over v_θ, v_a, and $-\log |c_h|$. The sum of these is 0 by Jensen's formula, i.e. by the sum formula. Cartan also observed that if f is a meromorphic function, and $(1,f)$ is the corresponding map into $\underline{\underline{P}}^1$, then

$$T_{(1,f)} = T_f + O(1),$$

where here T_f is the "characteristic function" defined by Nevanlinna for meromorphic functions. The height satisfies similar properties to those in number theory, better codified in this case than by the analysts. First:

<u>Let</u> f <u>and</u> $g: \mathbb{C} \longrightarrow \underline{\underline{P}}^n$ <u>be two holomorphic maps and</u> $g = A \circ f$ <u>where</u> A <u>is a projective linear transformation. Then</u>

$$T_f = T_g + O(1).$$

We just defined the height of a map into projective space. More generally we can define the height of a map into an arbitrary projective variety. We just state what goes on without proofs.

Let $f\colon \mathbf{C} \longrightarrow X$ *be a holomorphic map into a projective variety. To each Cartier divisor* D *on* X *one can associate a function* $T_{f,D}$ *of real numbers* $\geqq 1$, *depending only on the linear equivalence class of* D, *and uniquely determined up to a bounded function by the following properties.* [*The notation* $O(1)$ *always refers to* $r \longrightarrow \infty$.]

H 1. The map $D \longmapsto T_{f,D}$ *is a homomorphism* mod $O(1)$.

H 2. If E *is very ample, and* $\psi\colon X \longrightarrow \mathbf{P}^n$ *is one of the associated imbeddings into projective space, then*

$$T_{f,E} = T_{\psi\circ f} + O(1),$$

where $T_{\psi\circ f}$ *is the Cartan-Nevanlinna height.*

In addition, the height satisfies further properties as follows.

H 3. For any Cartier divisor D *and ample* E, *we have*

$$T_{f,D} = O(T_{f,E}).$$

H 4. If D *is effective and* f *meets* D *properly, then* $T_{f,D} \geqq -O(1)$.

H 5. The association $(f,D) \longmapsto T_{f,D}$ *is functorial in* (X,D). *In other words, if* $\psi\colon X \longrightarrow Y$ *is a morphism of varieties, and* $D = \psi^{-1}D'$ *where* D' *is a divisor on* Y, *then*

$$T_{f,D} = T_{\psi\circ f,D'}.$$

Other properties of the height can be copied from those of number theory, cf. for instance [La 4], Chapter 4, §1, §2, §3, including the analogue of Weil's theorem, namely Proposition 2.1 of that chapter.

Next we come to the definition of other functions in Nevanlinna theory. Let X be a projective variety. Let D be an effective Cartier divisor on X, represented locally for the Zariski topology by regular functions. Say on a Zariski open set U, D is represented by the function φ. This means $D|U = (\varphi)|U$. A *Weil function* is a function

$$\lambda_D\colon X - \mathrm{supp}(D) \longrightarrow \mathbf{R}$$

which is continuous, and is such that if D is represented by φ on U, then there exists a continuous function $\alpha : U \longrightarrow \mathbb{R}$ such that for all $x \notin \operatorname{supp}(D)$ we have

$$\lambda_D(x) = -\log |\varphi(x)| + \alpha(x).$$

The difference of two Weil functions is the restriction to X-supp(D) of a continuous function on X, and so is bounded. Thus two Weil functions differ by $O(1)$. If L is a line bundle over X and s a holomorphic section whose divisor is D, and ρ is a metric on L, then for instance

$$\lambda_D(x) = -\log |s(x)|_\rho$$

is a Weil function associated with D.

Suppose that $f(\mathbb{C})$ is not contained in D. This is equivalent with the fact that f meets D discretely, i.e. in any disc $\mathbb{D}_r$ there are only a finite number of points $a \in \mathbb{D}_r$ such that $f(a) \in D$. Given a Weil function λ_D associated with the effective divisor D, we define the <u>proximity function</u>

$$m_f(r,D) = \int_0^{2\pi} \lambda_D(f(re^{i\theta}))\,\frac{d\theta}{2\pi}.$$

Then λ_D and $m_f(r,D)$ are additive in D mod $O(1)$.

Let D again be an effective divisor. We let

$$D_f = f^{-1}(D)$$

as a divisor on $\mathbb{C}$, which exists by our basic assumption that f meets D discretely. If D is represented by φ on U, and $f(a) \in U$, then we let

$$\nu_f(a,D) = \nu(a,D_f) = \operatorname{ord}_a(\varphi \circ f) \quad (\geqq 0 \text{ since } D \text{ is effective})$$

Note that $\varphi \circ f$ is a holomorphic map of $\mathbb{C}$ into $\mathbb{P}^1$. Then we define

$$N_f(r,D) = \sum_{\substack{a \in \mathbb{D}_r \\ a \neq 0}} \nu_f(a,D)\log\left|\frac{r}{a}\right| + \nu_f(0,D)\log r.$$

Thus $N_f(r,D)$ can be viewed as the "r-degree" of the divisor D_f, with the weighting factors $\log|r/a|$ similar to $\log p$ in number theory when p is a prime number.

We call $N_f(r,D)$ the *integrated counting function*. It is additive in D and $\geqq 0$ for $r \geqq 1$. It measures the zeros of f in D.

Finally we define the *height*, or *Cartan-Nevanlinna height*, to be

$$T_f(r,D) = m_f(r,D) + N_f(r,D).$$

This is well defined mod $O(1)$.

We extend Weil functions, the proximity function and the integrated counting function to all Cartier divisors by additivity.

FIRST MAIN THEOREM. *If* $D = (\varphi)$ *is linearly equivalent to* 0, *then* $T_f(r,D) = O(1)$, i.e. $T_f(r,D)$ *is a bounded function of* r.

This is proved first for $\mathbf{P}^1$ (Nevanlinna) and in general by functoriality.

The SECOND MAIN THEOREM is today only a conjecture.

CONJECTURE. *Let* D *be an effective divisor with normal crossings on a projective non-singular variety* X. *There exists a divisor* D' *having the following property. Let* $f\colon \mathbf{C} \longrightarrow X$ *be holomorphic such that* $f(\mathbf{C}) \not\subset D'$. *Let* K *be the canonical class. Let* E *be ample. Then*

$$m_f(r,D) + T_f(r,K) \leqq O_{exc}(\log r + \log T_f(r,E)),$$

where O_{exc} *means the usual* O *with the exception of* r *lying in a set of finite Lebesgue measure*.

Only very special cases of this conjecture are known today, essentially for maps into projective space or Grassmanians, in a linear situation. Furthermore, a ramification term is missing in the above inequality, but should be present in the same way that it is present in Nevanlinna's theory in the one-dimensional case. We now state one of the basic known results.

THEOREM 2. *Let* $f = (f_0,\dots,f_n)\colon \mathbf{C} \longrightarrow \mathbf{P}^n$ *be a holomorphic map, with* $f_0,\dots,f_n$ *entire without common zeros. Assume that the image of* f *is not contained in any hyperplane. Let* $H_1,\dots,H_q$ *be hyperplanes in general position with* $q \geqq n+2$ (*meaning any* $n+1$ *are linearly independent*). *Let* $W(f) = W(f_0,\dots,f_n) = W$ *be the Wronskian. Let* H

be any hyperplane. Then

$$\sum_{k=1}^{q} m_f(r,H_k) - (n+1)T_f(r,H) + N_W(r,0) \leqq O_{exc}(\log r + \log T_f(r)),$$

where O_{exc} *has the usual* O *meaning, for* $r \longrightarrow \infty$ *but* r *lying outside some set of finite Lebesgue measure.*

This theorem is due to Cartan [Ca] except for the term $N_W(r,0)$, which constitutes an improvement. Cartan has in its place a term also reflecting the ramification, but depending on the hyperplanes $H_1,\ldots,H_q$. This is the "wrong" structure for this term, which should reflect only the ramification of the mapping f, independently of the divisor D, which in this case is $\sum H_k$. For $n = 1$, the theorem is due to Nevanlinna who started it all.

As an example of a diophantine application, we show how a theorem of Borel is an immediate consequence of Cartan's theorem.

COROLLARY. *Let* $g_1,\ldots,g_n$ *be entire functions without zeros (so units in the ring of entire functions). Suppose that*

$$g_1+\ldots+g_n = 1.$$

Then $g_1,\ldots,g_n$ *are linearly dependent.*

Proof. Let $g\colon \underline{C} \longrightarrow \underline{P}^{n-1}$ be the map $(g_1,\ldots,g_n)$. Let $x_1,\ldots,x_n$ be the homogeneous variables of $\underline{P}^{n-1}$. Let H_k be the hyperplane $x_k = 0$ for $k = 1,\ldots,n$ and $x_1+\ldots+x_n = 0$ for $k = n+1$. Then g does not meet these hyperplanes. Hence $m_g(r,H_k) = T_g(r,H_k) + O(1)$ for $k = 1,\ldots,n+1$. The canonical class is the class of nH for any hyperplane H. If $g_1,\ldots,g_n$ are linearly independent, then by Cartan's theorem

$$(n+1)T_g \leqq nT_g + O_{exc}.$$

Hence $T_g(r) = O_{exc}(\log r)$, and it is then easy to show that all coordinates $g_1,\ldots,g_n$ are polynomials, so are constant since g_i is a unit for each i. This concludes the proof of the corollary.

Cartan's theorem itself is proved according to the Wronskian technique apparently first used by Nevanlinna, also in connection with Borel's theorem.

Vojta [Vo] has translated the Second Main Theorem into a similar conjecture for a height inequality in algebraic number theory. He shows how theorems like those of Roth, Schmidt, Faltings, and conjectures like those of Hall, Lang, Lang-Waldschmidt, and what is neither, like Fermat's Last Theorem, would follow (or more or less follow) from his translation. The "more or less" is due to the presence of an exceptional set, similar to the exceptional set of r of finite Lebesgue measure in Nevanlinna theory. For details, the reader can read Vojta's extensive exposition.

BIBLIOGRAPHY

[Ca] H. CARTAN, "Sur les zéros des combinaisons lineaires de p fonctions holomorphes données", Mathematica 7 (1933) pp. 5-31; announced in CR Acad. Sci. 189 (1929) pp. 521-523, same title

[Br] R. BRODY, "Compact manifolds and hyperbolicity", Trans. AMS 235 (1978) pp. 213-219

[G-R] H. GRAUERT and H. RECKZIEGEL, "Hermitesche Metriken und normale Familien holomorpher Abbildungen", Math. Z. 89 (1965) pp. 108-112

[Gr] P. GRIFFITHS, Entire holomorphic mappings in one and several complex variables, Ann. of Math. Studies, Princeton University Press, 1976

[Ki] P. KIERNAN, "Extensions of holomorphic maps", Trans. AMS 172 (1972) pp. 347-355

[Ko] S. KOBAYASHI, Hyperbolic manifolds and holomorphic mappings, Marcel Dekker, New York, 1970

[Kw] M. KWACK, "Generalizations of the big Picard theorem", Ann. of Math. (2) 90 (1969) pp. 9-22

[La 1] S. LANG, "Integral points on curves", Pub. IHES, 1960

[La 2] S. LANG, "Higher dimensional diophantine problems", Bull. AMS 80 (1974) pp. 779-787

[La 3] S. LANG, "Hyperbolic and diophantine analysis", Bull. AMS April 1986

[La 4] S. LANG, Fundamentals of diophantine geometry, Springer Verlag, 1983

[Le 1] P. LELONG, "Fonctions entières (n variables) et fonctions plurisousharmoniques d'ordre fini dans $\mathbb{C}^n$," J. Analyse Math. 12 (1964) pp. 365-407

[Le 2] P. LELONG, Plurisubharmonic functions and positive differential forms, Gordon and Breach, 1969

[Ma] Y. MANIN, "Rational points of algebraic curves over function fields" (in Russian), Izv. Akad. Nauk 27 (1963) pp. 1395-1440

[No] J. NOGUCHI, "Hyperbolic fiber spaces and Mordell's conjecture", to appear

[Ri] D. RIEBESEHL, "Hyperbolische Complexe Raume und die Vermutung von Mordell", Math. Ann. 257 (1981) pp. 99-110

[Vo] P. VOJTA, Diophantine approximation and value distribution theory, Springer Lecture Notes, to appear. See also Vojta's thesis, Harvard, 1983

Serge Lang
Mathematics Department
Box 2155 Yale Station
New Haven Conn 06520

Vol. 56, No. 1 DUKE MATHEMATICAL JOURNAL © February 1988

THE ERROR TERM IN NEVANLINNA THEORY

SERGE LANG

In the sixties I conjectured that Roth's theorem could be improved to an inequality of type

$$\left|\alpha - \frac{p}{q}\right| \geqq \frac{1}{q^2(\log q)^{1+\varepsilon}}.$$

In other words, given ε and α algebraic, the inequality above holds for all but a finite number of fractions p/q in lowest form. (Cf. [La 6], [La 1], [La 4].) The inequality can also be written in the form

$$-\log\left|\alpha - \frac{p}{q}\right| - 2\log q \leqq (1 + \varepsilon)\log\log q.$$

Osgood ([Os 1], [Os 2]) has pointed out that the 2 in Roth's theorem corresponds to the 2 in Nevanlinna's inequality for meromorphic functions. But Vojta had the extraordinary insight that the second main theorem (SMT) of Nevanlinna theory, in both its one-dimensional and higher-dimensional versions, has an analogous statement in the theory of heights, which gives rise to a major conjecture in diophantine analysis ([Vo 1], [Vo 2], [Vo 3]). It occurred to me to transpose my conjecture about the "error term" in Roth's theorem to the context of this higher-dimensional analytic theory and see whether it held there.

I shall deal here with the Stoll–Carlson–Griffiths theorem in higher dimensions ([St], [C–G], [Gr]), which contains as a special case Nevanlinna's own theorem for $\mathbf{P}^1$, i.e., for meromorphic functions on $\mathbf{C}$. In the dictionary, the Nevanlinna function T_f corresponds to the number-theoretic height $\log q$. I shall prove the analogue of the exponent $3/2 + \varepsilon$ instead of the $1 + \varepsilon$ in this analytic context. (I am unable to get the $1 + \varepsilon$ exactly.) Even in the most classical case of meromorphic functions, I believe this estimate is new and improves both the error term in SMT and Nevanlinna's estimate in the theorem on the logarithmic derivative, where Nevanlinna's constants are replaced by $3/2 + \varepsilon$. However, the problem remains as to whether one has $1 + \varepsilon$ in general, and what is the best possible result in general. Thus the error term has a structure of its own, which should also occur in Vojta's conjecture.

Received May 4, 1987. Revision received August 24, 1987.

In [La 1], [La 4], I defined a **type** for a number α to be a function h such that

$$\left|\alpha - \frac{p}{q}\right| \geqq \frac{1}{q^2 h(q)}$$

for all but a finite number of p/q. One can rephrase my conjecture by saying that an algebraic number has type $\leqq (\log q)^{1+\varepsilon}$. It becomes a problem to determine the type for each algebraic number and for the "classical" numbers. For instance, it follows from Adams's work ([Ad 1], [Ad 2]) that e has type

$$h(q) = \frac{C \log q}{\log\log q},$$

with a suitable constant C, which is much better than the "probabilistic" type $(\log q)^{1+\varepsilon}$. By a theorem of Khintchine and Lebesgue, almost all real numbers have type $\leqq h$ if

$$\sum \frac{1}{q^2 h(q)} < \infty.$$

Thus it is only natural to try out $(\log q)^{1+\varepsilon}$ after the Roth q^{ε}. However, except for quadratic numbers that have bounded type, there is no known example of an algebraic number about which one knows that it is or is not of type $(\log q)^{\kappa}$ for some number $\kappa > 1$.

Transposing to the analytic context, it becomes a problem to determine the "type" of the classical meromorphic functions, i.e., the best possible error terms in SMT with these functions. Such a type measures a refined value distribution of the function. It is classical, for example, that e^z has bounded type, i.e., that the error term in SMT is $O(1)$.

In this paper, I shall point out how to adjust the proofs of [C–G] and [Gr] in order to show that error term in SMT, say in dimension 1, has the form

$$\varepsilon \log r + \tfrac{3}{2}(1 + \varepsilon)\log T_f(r)$$

for all r lying outside the usual exceptional set. The proofs depend on the construction of a singular volume form and rely on two estimates, so that the error term decomposes into two natural terms estimated respectively by

$$\varepsilon \log r + \tfrac{1}{2}(1 + \varepsilon)\log T_f(r) \quad \text{and} \quad \log T_f(r).$$

This decomposition has an independent interest and reinforces the structure inherent in this error term. A remaining problem is whether one can put $1/2$ as a factor of the second term $\log T_f(r)$ to end up with $(1 + \varepsilon)\log T_f(r)$ for the whole error term or eliminate this term altogether.

In this connection, Chern's proof of SMT in a differential geometric context ([Ch]), following the averaging method of Ahlfors, goes through an inequality with an error term in his formula (49). This error term has a factor $1/2$ where we have $3/2$. But formula (49) is not proved, and may be false. The difficulty is that Chern uses the singular volume form

$$\Psi = \frac{\Psi_K}{|s|_{\rho_D}^{2\lambda}}, \qquad 0 < \lambda < 1.$$

He correctly obtains a previous inequality valid outside an exceptional set with what is, from our point of view, an inefficient error term, which is worse than the error term as it is given in Nevanlinna, including F. Nevanlinna's use of a singular volume form ([Ne], chap. 9, §4, or [C–G]). This is because his volume form is not sufficiently hyperbolic (i.e., the curvature tends to 0 along the divisor). When he writes, "Letting $\lambda \to 1$ and using (48) we get (49)," he overlooks the fact that the exceptional set depends on λ, and so he cannot take the limit as $\lambda \to 1$ to get rid of that error term and get (49). It may be that a hidden uniformity in the calculus lemma is used by all authors at this point of the proof, but this uniformity, assuming it exists, has never been studied. Chern's previous inequality is enough, of course, to give the defect relation, which was his main purpose at the time. The error term was not recognized as significant, and so it was not scrutinized carefully either in Chern or, later, in [C–G].

One should also keep in mind that the probabilistic heuristics for the error term may depend on special situations. Vojta has pointed out to me that for the analogue of Roth's theorem over $\mathbf{Q}(\sqrt{-1})$, one gets a factor of $1/2$ in the error term, because $\mathbf{Q}(\sqrt{-1})$ is complex, not real.

§1. The height and the main theorem. Let X be a projective nonsingular variety of dimension n over the complex numbers. Let D be a divisor on X and let $L = L_D$ be a line bundle associated with D. There is a meromorphic section s of L whose divisor is D, which we write

$$(s) = D.$$

Let ρ be a metric on L. We have the first **Chern form** $c_1(\rho)$, given locally outside D by the formula

$$c_1(\rho) = -dd^c \log|s|_\rho^2.$$

As usual,

$$d = \partial + \bar{\partial} \quad \text{and} \quad d^c = \frac{1}{2\pi}\frac{\partial - \bar{\partial}}{2i}.$$

On Euclidean space $\mathbf{C}^n$ we have the form ω, given in terms of the complex coordinates $z = (z_1, \dots, z_n)$ by

$$\omega(z) = dd^c \log\|z\|^2, \quad \text{where } \|z\|^2 = \sum z_j \bar{z}_j.$$

To orient the reader, ω is the pull-back of $c_1(\tau)$, where τ is the Fubini–Study metric on the hyperplane bundle of $\mathbf{P}^{n-1}$, but we won't need this here. We let

$$\mathbf{B}(r) = \text{ball of radius } r \text{ in } \mathbf{C}^n,$$

$$\mathbf{S}(r) = \text{sphere of radius } r \text{ in } \mathbf{C}^n,$$

so $\mathbf{S}(r)$ is the boundary of the ball. Both are centered at the origin. Let

$$f\colon \mathbf{C}^n \to X$$

be a holomorphic map, which we assume to be nondegenerate for simplicity. This means that f is a local holomorphic isomorphism somewhere, and hence that its image is not contained in any proper algebraic subset of X. We define three types of functions as follows. First,

$$\mathbf{t}_{f,\rho}(r) = \int_{\mathbf{B}(r)} f^* c_1(\rho) \wedge \omega^{n-1} \quad \text{and} \quad T_{f,\rho} = \int_0^r \mathbf{t}_{f,\rho}(t) \frac{dt}{t}.$$

We call $\mathbf{t}_{f,\rho}$ the **preheight** and $T_{f,\rho}$ the **height** (rather than the "characteristic function," as in Nevanlinna). More generally, if η is a $(1,1)$-form, we define the **η-heights** (pre and otherwise) by

$$\mathbf{t}_{f,\eta}(r) = \int_{\mathbf{B}(r)} f^*\eta \wedge \omega^{n-1} \quad \text{and} \quad T_{f,\eta}(r) = \int_0^r \mathbf{t}_{f,\eta}(t) \frac{dt}{t}.$$

Note that we are pulling back forms and Chern classes from X to $\mathbf{C}^n$. If η is a form on $\mathbf{C}^n$ itself, then we can define simply

$$\mathbf{t}_\eta(r) = \int_{\mathbf{B}(r)} \eta \wedge \omega^{n-1} \quad \text{and} \quad T_\eta(r) = \int_0^r \mathbf{t}_\eta(t) \frac{dt}{t}.$$

Thus the notation is functorial. If η is a form on X, then

$$T_{f^*\eta} = T_{f,\eta}.$$

Thus, if Z is a divisor on $\mathbf{C}^n$ and ρ is a metric on L_Z, then we write $\mathbf{t}_\rho$ and T_ρ instead of $\mathbf{t}_\eta$ and T_η, where $\eta = c_1(\rho)$.

In practice, we shall also deal with forms η on the complement of some divisor on X, and the pull-back is a form on the complement of a divisor in $\mathbf{B}(r)$, so the integral is defined as an improper integral. We shall do so under conditions when we have absolute convergence of the integral. Actually, the form ω already has a singularity at the origin, so no matter what, we are dealing with improper integrals. If we change the metric, then $T_{f,\rho}$ changes by a bounded function of r. If we are interested in $T_{f,\rho}$ only up to $O(1)$, then we use the notation

$$T_{f,D} = T_{f,\rho} + O(1).$$

Second, let Z be a divisor on $\mathbf{C}^n$. We define $Z(r) = Z \cap \mathbf{B}(r)$, and if $0 \notin Z$, we define the **precounting function** and **counting function** by

$$\mathbf{n}_Z(r) = \int_{Z(r)} \omega^{n-1} \quad \text{and} \quad N_Z(r) = \int_0^r \mathbf{n}_Z(t)\frac{dt}{t}.$$

If $0 \in Z$, then one has to subtract a constant of integration in $\mathbf{n}_Z$, the Lelong number, using [Gr], page 13. We omit this for simplicity. We then define functorially

$$\mathbf{n}_{f,D}(r) = \int_{f^*D(r)} \omega^{n-1} \quad \text{and} \quad N_{f,D}(r) = \int_0^r \mathbf{n}_{f,D}(t)\frac{dt}{t}.$$

Third, we let σ be the standard rotation-invariant form giving the spheres measure 1, that is,

$$\sigma(z) = d^c \log\|z\|^2 \wedge \omega^{n-1}.$$

When $n = 1$, then $\sigma(z) = d\theta/2\pi$.

We assume throughout that $f(0) \notin D$. We then define the **preproximity function** and the corresponding **proximity function** by

$$\mathbf{m}^\circ_{f,D,\rho}(r) = \int_{\mathbf{S}(r)} -d^c \log|s \circ f|^2_\rho \wedge \omega^{n-1} \quad \text{and} \quad m^\circ_{f,D,\rho}(r) = \int_0^r \mathbf{m}^\circ_{f,D,\rho}(t)\frac{dt}{t}.$$

For historical reasons, the notation here is not functorial with respect to the ideas, since I am forced to write a small m for the proximity function. Indeed, M_f classically denotes the sup norm of the log, and Griffiths was adamant about not changing the notation.

If γ is a function on $\mathbf{C}^n$, smooth and > 0 outside some divisor not containing 0, then we define

$$\mathbf{m}^\circ_\gamma(r) = \int_{\mathbf{S}(r)} d^c \log \gamma \wedge \omega^{n-1} \quad \text{and} \quad m^\circ_\gamma(r) = \int_0^r \mathbf{m}^\circ_\gamma(t)\frac{dt}{t}.$$

Thus,

$$\mathbf{m}^{\circ}_{f,D,\rho} = \mathbf{m}^{\circ}_{\gamma} \quad \text{if we put } \gamma = |s \circ f|_{\rho}^{-2}.$$

This expression depends only on D, since sections differ by a multiplicative constant, and its log will disappear under the differentiation by d^c.

LEMMA 1.1. *If all integrals involved are absolutely convergent, then*

$$m^{\circ}_{\gamma}(r) = \frac{1}{2}\int_{\mathbf{S}(r)} (\log \gamma)\sigma - \frac{1}{2}\log \gamma(0).$$

The proof will be recalled in an appendix. We note the extra constant of integration on the right. It is often useful to deal with the integral expression alone, which depends on the section s, so we also define

$$m_{f,s,\rho}(r) = \frac{1}{2}\int_{\mathbf{S}(r)} -\log|s \circ f|_{\rho}^{2}\,\sigma.$$

As for $T_{f,D}$ we let $m_{f,D} = m_{f,D,\rho} + C$, where C can be an arbitrary constant of integration, if this constant is unimportant for certain estimates. In particular, if D is effective, we can choose the constant such that

$$m_{f,D} \geqq 0.$$

The following is sometimes called the **First Main Theorem** of Nevanlinna theory.

THEOREM 1.2. *For any metric ρ on L_D we have*

$$T_{f,\rho} = N_{f,D} + m^{\circ}_{F,D,\rho}.$$

Actually, the theorem holds at the pre-level, that is,

$$\mathbf{t}_{f,\rho} = \mathbf{n}_{f,D} + \mathbf{m}^{\circ}_{f,D,\rho}.$$

At the pre-level, it is proved as a direct application of Stokes's theorem, and the proof will be recalled in an appendix. Note that this is an exact relation, although in [C–G] and [Gr] it is stated only mod $O(1)$. Here I want to clarify the formal structure of the theorem, and I want to distinguish between exact relations and estimates, so I shall be careful in stating the theorems to bring these features out clearly.

Let $D = \Sigma D_j$ be expressed as a sum of irreducible divisors D_j. If each D_j is nonsingular, and if at each point of X there are complex coordinates $z_1, \ldots, z_n$

such that locally near that point D is defined by

$$z_1 \dots z_k = 0 \quad \text{for some } k \leqq n,$$

then we say that D has **simple normal crossings**.

The **Second Main Theorem** with the new error term then reads as follows:

THEOREM 1.3. *Assume that D has simple normal crossings. Let K be a canonical divisor on X. Let* $\operatorname{Ram}(f) = \operatorname{Ram}$ *be the ramification divisor of f (i.e.,* Ram *is defined by the complex Jacobian determinant locally). Assume that* $0 \notin \operatorname{Ram}$ *and $f(0) \notin D$. Given ε and E ample,*

$$m_{f,D} + T_{f,K} + N_{\text{Ram}} \leqq \varepsilon \log r + \frac{n}{2}(1+\varepsilon)\log T_{f,E} + \sum \log m_{f,D_j},$$

except for $r \geqq 1$ lying in a set of finite $r^\delta dr$-measure for some $\delta > 0$.

A set as described in this last sentence will be called an **exceptional set**. Note that an exceptional set has finite Lebesgue measure. Too often the exceptional set is measured by the logarithmic measure dr/r, which gives a weaker result unnecessarily. This confused me for a while, until I checked that the calculus lemma in fact gives the stronger result with the $r^\delta dr$-measure.

Observe that $\log r$ occurs with a factor of ε, which is even better than the constant in front of the height.

Suppose that D has one component and $n = 1$. Then the error term has the form

$$\tfrac{3}{2}(1+\varepsilon)\log T_{f,E},$$

and so conjecturally is not best possible. I shall make comments on this in §4. If D has several components, say q components, then the sum is bounded by $q \log T_{f,E}$, and the problem arises whether one can get rid of this term.

The functions T, m, N are additive in D and, for an effective divisor, they are $\geqq -O(1)$. In particular, $N_{f,D} \leqq T_{f,D} + O(1)$. By adding a sufficiently general hyperplane to D, one then sees that without loss of generality one can assume that $D + K$ is very ample, i.e., is linearly equivalent to a hyperplane section in some projective embedding. Thus we may assume that there is a metric ρ on L_{D+K} such that $c_1(\rho) > 0$.

The pattern of proof is based on Carlson–Griffiths ([C–G]), which has its own origins in F. Nevanlinna's use of a singular form Ψ as in [Ne], chapter 9, §4. A problem is whether the estimate involving the sum $\sum \log m_{f,D_j}$ can be improved by changing the singular volume form used in [C–G]. The formal handling of the proof below will bring out clearly where the difficulty lies. Let Ψ be the singular volume form. Define the **Griffiths function** G_f by

$$\frac{1}{n!} f^* \operatorname{Ric}(\Psi)^n = G_f f^* \Psi,$$

where G_f measures the hyperbolicity of the form. Then we have four main steps.

SMT for Ψ. Let $f^*\Psi = \gamma\Phi$, where Φ is the Euclidean form. Let

$$\mu^{\circ}_{f,\Psi} = m^{\circ}_{\gamma}.$$

Then

$$T_{f,\mathrm{Ric}(\Psi)} + N_{\mathrm{Ram}} = N_{f,D} + \mu^{\circ}_{f,\Psi}.$$

Comparison of heights. We have

$$-O(1) \leqq T_{f,\rho} - T_{f,\mathrm{Ric}}(\Psi) = m^{\circ}_{h_f}.$$

Estimate for $\mu_{f,\Psi}$. Outside an exceptional set we have

$$\mu_{f,\Psi} \leqq n\varepsilon \log r + \frac{n}{2}(1+\varepsilon)\log T_{f,\rho} + m_{1/G_f}.$$

Thus we see the desired error term appearing, together with a sum

$$m_{h_f} + m_{1/G_f} = m_{h_f/G_f}.$$

Estimate for m_{h_f/G_f}. With the Carlson–Griffiths form, $1/G_f = O(1)$, and we find

$$m_{h_f/G_f} \leqq \sum \log m_{f,D_j}.$$

Combining the last two estimates yields the proof of Theorem 1.3.

The problem is essentially a minimizing problem with boundary conditions of some sort, namely:

(a) We must make

$$m_{h_f/G_f} = \frac{1}{2}\int_{S(r)} (\log h_f/G_f)\sigma \leqq \kappa \log m_{f,D},$$

with κ as small as possible.

(b) We must satisfy a condition of Stokes's theorem, say when $n = 1$,

$$\lim_{\delta \to 0} \int_{S_\delta(Z)(r)} d^c \log h = 0,$$

to have Proposition A6 in the appendix. We view (a) as the minimizing problem and (b) as the boundary condition. In (b), $S_\delta(Z)(r)$ is the boundary of a tubular neighborhood of radius δ around the divisor $Z = f^*D$ inside the ball of radius r. The two conditions go in opposite directions and have to be reconciled.

Griffiths's [Gr] formula 3.12 has given a higher-dimensional version of Nevanlinna's lemma on the logarithmic derivative. The error term in this version and in Nevanlinna's original theorem can be improved in the same manner as in the second main theorem. For simplicity, I state here the improved version for the case of maps into $\mathbf{P}^1$, i.e., the classical case of meromorphic functions.

THEOREM 1.4. *Let f be a meromorphic function, viewed as a map of $\mathbf{C}$ into $\mathbf{P}^1$. Let $T_f = T_{f,\infty}$ and $m_f = m_{f,\infty}$. Then*

$$m_{f'/f} \leqq \varepsilon \log r + \tfrac{3}{2}(1+\varepsilon)\log T_f(r)$$

outside the usual exceptional set.

The proof will essentially follow Griffiths's proof, but with appropriate care devoted to the constants. The higher-dimensional case will be stated later, since it requires more terminology. The problem is again to replace 3/2 by 1.

§2. Formal SMT for singular forms. Let Ψ be a volume form on X, or an open subset of X (in practice, the complement of a divisor). Locally, in terms of complex coordinates we can write the volume form as

$$\Psi(z) = \lambda(z)\Phi(z), \quad \text{where } \Phi(z) = \prod \frac{\sqrt{-1}}{2\pi}\, dz_i \wedge d\bar{z}_i,$$

so Φ is the Euclidean volume form and λ is a smooth positive function. There is a unique (1, 1)-form $\mathrm{Ric}(\Psi)$, which is given locally by

$$\mathrm{Ric}(\Psi)(z) = dd^c \log \lambda(z).$$

We recall that a volume form can be viewed as a metric on the canonical line bundle L_K, where K is a canonical divisor.

We observe that the maps

$$\rho \mapsto T_{f,\rho} \quad \text{and} \quad (D,\rho) \mapsto m_{f,D,\rho}$$

are homomorphisms, from the tensor product of metrics (on the tensor product of line bundles) to the additive group of functions on $\mathbf{R}^+$.

Assume that D has simple normal crossings, $D = \sum D_j$. Consider a metric ρ on L_{K+D} and metrics ρ_K on L_K, ρ_j on L_{D_j} such that

$$\rho = \rho_K \otimes \rho_1 \otimes \cdots \otimes \rho_q = \rho_K \otimes \rho_D.$$

Let s_j be a holomorphic section of L_{D_j} with $(s_j) = D_j$. Since s_j vanishes on D_j, after multiplying s_j by a sufficiently small, positive real number, we can assume (for instance) that $|s_j|^2_{\rho_j} < 1/e$, which guarantees that

$$-\log|s_j|^2_{\rho_j} > 0$$

on the complement of D_j. We define the **singular volume form** Ψ by the formula

$$\Psi = \frac{\Psi_K}{\prod |s_j|^2_{\rho_j} h_j}, \quad \text{where } h_j = h\left(|s_j|^2_{\rho_j}\right),$$

h is a suitable function of a real variable t, smooth outside D;
Ψ_K is the volume form on X that corresponds to the metric ρ_K on L_K.
Note that Ψ is smooth on the complement of D.

Carlson–Griffiths uses the function $h(t) = (\log 1/t)^2$. For definiteness the reader may wish to read all that follows by using this function, in which case we call Ψ the **Carlson–Griffiths form**. But we shall carry out as much as possible formally to see precisely where one might obtain an improvement in the estimates eventually to prove the best possible estimate for the error term.

One point of the function chosen is to guarantee the absolute convergence of certain integrals, like

$$\int_0^{1/2} \frac{1}{x^2 h(x)} x\, dx < \infty.$$

So no problem arises in this context with the Carlson–Griffiths form and its h. The convergence of such an integral is similar to the Khintchine convergence in number theory.

Let $f\colon \mathbf{C}^n \to X$ be a nondegenerate holomorphic map. Directly from the definitions we have

$$f^*\mathrm{Ric}(\Psi) = f^*c_1(\rho) - \sum dd^c \log h_j.$$

THEOREM 2.1. *Let ρ be the metric as above on L_{K+D}. Let Ψ be the singular volume form. Then*

$$T_{f,\rho} - T_{f,\mathrm{Ric}(\Psi)} = \sum_j m^{\circ}_{h_j \circ f}.$$

Proof. This comes from Proposition A6 of the appendix and definitions. Note that we have from the definitions explicitly

$$m^{\circ}_{h_j \circ f} = \frac{1}{2}\int_{\mathbf{S}(r)} (\log h_j \circ f)\sigma - \frac{1}{2}\log h_j \circ f(0) \geqq O(1).$$

The constant of integration will not matter in the estimate. We do not yet estimate the sum in Theorem 2.1, since it will combine later with other terms to provide cancellations. See Lemma 4.1.

Proposition A6 is designed to show that in Stokes's theorem, applied to the term corresponding to each $h_j \circ f$, the singularity of $h_j \circ f$ is still sufficiently mild that it contributes nothing to Stokes's formula.

Next, we have a formal SMT for the Ric(Ψ)-height. It is an exact expression. We need to define another term. We let

$$f^*\Psi = \gamma\Phi,$$

where γ could be called the **Carlson–Griffiths function**. Then $\gamma \geqq 0$ and $\gamma > 0$ almost everywhere. We define

$$\mu^\circ_{f,\Psi} = m^\circ_\gamma \quad \text{and} \quad \mu_{f,\Psi} = \frac{1}{2}\int_{\mathbf{S}(r)} (\log \gamma)\sigma = m^\circ_\gamma + \frac{1}{2}\log \gamma(0),$$

so μ° differs from μ only by the constant of integration.

THEOREM 2.2 (**Formal SMT for $T_{f,\mathrm{Ric}(\Psi)}$**). *Let D have simple normal crossings. Let Ψ be the singular volume form. Assume that* $0 \notin \mathrm{Ram}$ *and* $f(0) \notin D$. *Then*

$$T_{f,\mathrm{Ric}(\Psi)} + N_{\mathrm{Ram}} = N_{f,D} + \mu^\circ_{f,\Psi}.$$

This is an exact formula; there is no error term. The proof is by an application of Stokes's theorem to each of the terms in the pulled-back form

$$f^*\Psi,$$

and, of course, Proposition A6 again plays the same role in dealing with the h_j terms. The proof will be given in the appendix. For this theorem we do not need the positivity of $c_1(\rho)$. We merely need absolute integrability of all the integrals concerned and the fact that the function h has sufficiently mild singularities so that Proposition A6 applies.

Having derived the formal results above, we shall now go into estimates of "error terms."

§3. The estimate for $\mu_{f,\Psi}$. We let G be the **Griffiths function**, defined by the equation

$$\frac{1}{n!}\mathrm{Ric}(\Psi)^n = G\Psi.$$

We also let

$$\tau = \mathrm{tr}\,\mathrm{Ric}(\Psi),$$

so, locally, τ is the trace of the Hermitian matrix of the $(1,1)$-form $\mathrm{Ric}(\Psi)$. We let

$$\tau_f = \tau \circ f = \operatorname{tr} f^*\mathrm{Ric}(\Psi) \quad \text{and} \quad G_f = G \circ f$$

in the usual notation.

THEOREM 3.1. *Assume $c_1(\rho) > 0$. Then, for any $\varepsilon > 0$, we have*

$$\mu_{f,\Psi}(r) \leqq n\varepsilon \log r + \frac{n}{2}(1+\varepsilon)\log T_{f,\mathrm{Ric}(\Psi)}(r) + m_{1/G_f}$$

for r outside an exceptional set. Recall that

$$m_{1/G_f}(r) = \frac{1}{2}\int_{\mathbf{S}(r)} (\log 1/G_f)\sigma.$$

Proof. The first part of the argument is standard, and we follow [C–G] in giving another expression for the height based on the following lemma. We shall use the form φ on $\mathbf{C}$ defined by

$$\varphi(z) = dd^c\|z\|^2.$$

Then on the sphere $\mathbf{S}(r)$ we have

$$d^c\|z\|^2 = r^2 d^c\log\|z\|^2 \quad \text{and} \quad dd^c\|z\|^2 = r^2 dd^c\log\|z\|^2.$$

LEMMA 3.2. *If η is a closed $(1,1)$-form, then*

$$r^{2n-2}\int_{\mathbf{B}(r)} \eta \wedge \omega^{n-1} = \int_{\mathbf{B}(r)} \eta \wedge \varphi^{n-1}.$$

Proof. We make a routine application of Stokes's theorem as follows:

$$\begin{aligned} r^{2n-2}\int_{\mathbf{B}(r)} \eta \wedge \omega^{n-1} &= r^{2n-2}\int_{\mathbf{B}(r)} d\left(\eta \wedge d^c\log\|z\|^2 \wedge \omega^{n-2}\right) \\ &= r^{2n-2}\int_{\mathbf{S}(r)} \eta \wedge d^c\log\|z\|^2 \wedge \omega^{n-2} \\ &= \int_{\mathbf{S}(r)} \eta \wedge d^c\|z\|^2 \wedge \varphi^{n-2} \\ &= \int_{\mathbf{B}(r)} \eta \wedge \varphi^{n-1}. \end{aligned}$$

This proves the lemma when η is smooth. In particular, we can apply the lemma to $f^*c_1(\rho)$. Since we also want to have the same formula when

$$\eta = f^*\mathrm{Ric}(\Psi),$$

we need the same relation when $\eta = dd^c\log(h_j \circ f)$ for each j. One uses Proposition A6 of the appendix, which also works with φ^{n-1} instead of ω^{n-1}.

We can further clarify the integral on the right. Let

$$\eta = \sum \eta_{ij} \frac{\sqrt{-1}}{2\pi} dz_i \wedge d\bar{z}_j \quad \text{and} \quad \mathrm{tr}(\eta) = \sum \eta_{ii}$$

be a $(1, 1)$-form and its trace. Since

$$\varphi^{n-1} = (n-1)! \sum \left(\frac{\sqrt{-1}}{2\pi} \right)^{n-1} dz_1 \wedge d\bar{z}_1 \wedge \cdots \wedge \widehat{dz_j \wedge d\bar{z}_j} \wedge \cdots \wedge dz_n \wedge d\bar{z}_n,$$

we get

$$\eta \wedge \varphi^{n-1} = (n-1)!\, \mathrm{tr}(\eta) \Phi.$$

Therefore:

THEOREM 3.3. *If η is a closed $(1,1)$-form, then*

$$\mathbf{t}_\eta(r) = \int_{\mathbf{B}(r)} \eta \wedge \omega^{n-1} = \frac{1}{r^{2n-2}} \int_{\mathbf{B}(r)} (n-1)!(\mathrm{tr}\, \eta) \Phi.$$

Letting $\eta = f^*\mathrm{Ric}(\Psi)$ and integrating once more to get the height, we find

COROLLARY 3.4. *Let Ψ be the singular volume form. Then*

$$T_{f, \mathrm{Ric}(\Psi)}(r) = \int_0^r \left(\int_{\mathbf{B}(t)} (n-1)! \tau_f \Phi \right) \frac{dt}{t^{2n-1}}$$

LEMMA 3.5. *We have $\mu_{f, \Psi} \leqq n m_{\tau_f} + m_{1/G_f}$.*

Proof. Since $\log \gamma = n \log \gamma^{1/n}$, we find

$$\frac{1}{n} \mu_{f, \Psi} = \frac{1}{2} \int_{\mathbf{S}(r)} (\log \gamma^{1/n}) \sigma.$$

But

$$\gamma \Phi = f^*\Psi = G_f^{-1} \frac{1}{n!} f^*\mathrm{Ric}(\Psi)^n = G_f^{-1} (\det f^*\mathrm{Ric}(\Psi)) \Phi,$$

so

$$\gamma = G_f^{-1}\det f^*\mathrm{Ric}\,\Psi.$$

But for a semipositive Hermitian matrix M we have $(\det M)^{1/n} \leqq (1/n)\mathrm{tr}\, M$, so

$$\gamma^{1/n} = G_f^{-1/n}(\det f^*\mathrm{Ric}\,\Psi)^{1/n} \leqq G_f^{-1/n}\frac{1}{n}\,\mathrm{tr}\, f^*\mathrm{Ric}\,\Psi = \frac{1}{n}G_f^{-1/n}\tau_f.$$

We take the log and integrate. We get

$$\frac{1}{n}\mu_{f,\Psi} \leqq \frac{1}{2n}\int_{\mathbf{S}(r)}(\log 1/G_f)\sigma + \frac{1}{2}\int_{\mathbf{S}(r)}(\log \tau_f)\sigma.$$

This proves the lemma.

So we have to analyze the second integral on the right, namely, m_{τ_f}.

Remark. For any sufficiently smooth function α, we have

$$\int_{\mathbf{S}(r)}\alpha\sigma = \frac{1}{r^{2n-1}}\frac{d}{dr}\int_{\mathbf{B}(r)}\alpha\Phi.$$

This is an easy direct verification.

We then use some calculus. For any positive, sufficiently smooth function α, define

$$F_\alpha(r) = \int_0^r\left(\int_{\mathbf{B}(r)}\alpha\Phi\right)\frac{dt}{t^{2n-1}}.$$

LEMMA 3.6. *Given ε, there exists $\delta > 0$ such that for $r \geqq 1$ and outside a set of finite $r^\delta dr$-measure, we have*

$$\log\int_{\mathbf{S}(r)}\alpha\sigma \leqq \varepsilon\log r + (1+\varepsilon)\log F_\alpha(r).$$

Proof. We have, by the standard calculus lemma,

$$\begin{aligned}\log\int_{\mathbf{S}(r)}\alpha\sigma &= \log\left(\frac{1}{r^{2n-1}}\frac{d}{dr}\int_{\mathbf{B}(r)}\alpha\Phi\right) \quad \text{(by the remark)}\\ &= -(2n-1)\log r + \log\left(\frac{d}{dr}r^{2n-1}F_\alpha'(r)\right)\\ &\leqq \varepsilon\log r + (1+\varepsilon)\log F_\alpha(r)\end{aligned}$$

for r outside the exceptional set, as in Nevanlinna ([Ne]), [C–G], or [Gr] (pp. 43–44). This proves Lemma 3.6.

LEMMA 3.7. *For r outside an exceptional set, we have*

$$m_{\tau_f} \leqq \varepsilon \log r + \tfrac{1}{2}(1+\varepsilon)\log T_{f,\rho}.$$

Proof. Let $\alpha = \tau_f$ in Lemma 3.6. Then

$$\begin{aligned} m_{\tau_f}(r) &= \frac{1}{2}\int_{\mathbf{S}(r)} (\log \tau_f)\sigma \\ &\leqq \frac{1}{2}\log \int_{\mathbf{S}(r)} \tau_f \sigma \\ &\leqq \varepsilon \log r + \tfrac{1}{2}(1+\varepsilon)\log F_\alpha(r). \end{aligned}$$

By Corollary 3.4 and Theorem 2.1, we know that

$$F_\alpha = \frac{1}{(n-1)!} T_{f,\mathrm{Ric}(\Psi)} \ll T_{f,\rho} + O(1).$$

This proves Lemma 3.7.

Lemma 3.5 and Lemma 3.7 conclude the proof of Theorem 3.1.

§4. The final step. In Theorems 2.1 and 3.1 we are left with two pieces of the error term, which add up in a natural way to give

$$\sum m_{h_j \circ f} + m_{1/G_f} = m_{1/H_f} \quad \text{where } H = G\prod h_j^{-1}.$$

LEMMA 4.1. *With the Carlson–Griffiths form, if $c_1(\rho) > 0$, then*

$$m_{1/H_f} \leqq \sum \log m_{f,D_j}.$$

Proof. With the Carlson–Griffiths $h(t) = (\log 1/t)^2$ and $t_j = |s_j|^2_{\rho_j}$, the Griffiths function satisfies $G_f \geqq b > 0$ for some constant b, as shown in [C–G], Proposition 2.1. The proof is also given in [Gr], Proposition 2.17. Hence, estimating m_{1/H_f} reduces to estimating each $m_{h_j \circ f}$. We find, as in [C–G],

$$\begin{aligned} m_{h_j \circ f}(r) = \frac{1}{2}\int_{\mathbf{S}(r)} \log\left(-\log|s_j \circ f|^2_{\rho_j}\right)^2 \sigma &\leqq \log \int_{\mathbf{S}(r)} -\log|s_j \circ f|^2_{\rho_j}\sigma \\ &\leqq \log m_{f,D_j}(r). \end{aligned}$$

This proves the lemma.

We have also concluded the proof of Theorem 1.3.

Remarks. Suppose we want to improve on this last estimate. For instance, suppose D has one component, so $H = G/h$. We want to get

$$m_{1/H_f} \leqq \tfrac{1}{2}\log m_{f,D}.$$

For simplicity, suppose we are in dimension $n = 1$. Then

$$\Psi = \frac{\Psi_K}{|s|^2_{\rho_D} h},$$

and directly from the definitions we have

$$\operatorname{Ric} \Psi = c_1(\rho) - dd^c \log h,$$

so that

$$H = G/h = \frac{c_1(\rho) - dd^c \log h}{\Psi_K} |s|^2_{\rho_D}.$$

If we replace h by h^κ with some power κ, then

$$dd^c \log h^\kappa = \kappa\, dd^c \log h,$$

so essentially such a replacement will not give us the desired improvement.

We have the formula

$$-dd^c \log h = -\frac{dd^c h}{h} + \frac{dh \wedge d^c h}{h^2}.$$

The term $(dh \wedge d^c h)/h^2$ is the term that makes the Griffiths function large in [C–G], Proposition 2.1. Thus the presence of h^2 appears unavoidable.

Suppose we write $h = h_0 h_1$ and try for lower-order perturbations. We have

$$-dd^c \log h_0 h_1 = -dd^c \log h_0 - dd^c \log h_1.$$

Attempts to make the Griffiths function large by using $h_1(t) = (\log(-\log t))^\kappa$ and $t = |s|^2_{\rho_D}$ failed to provide the desired estimate.

§5. The theorem on the logarithmic derivative. The next theorem and its proof follow the pattern of Griffiths ([Gr], proof of formula (3.12), pp. 72–73), except that we keep track of the constants carefully. We find an error term similar to that of Theorem 1.3.

THEOREM 5.1. *Again let X be a projective nonsingular variety. Let $K = K_X$ be a canonical divisor, and suppose $-K$ is ample. Let Λ be a meromorphic n-form with no zeros and such that its polar divisor $D = \Sigma D_j$ has simple normal crossings, where D_j are the irreducible components. Let*

$$f^*\Lambda = \lambda_f(z)\, dz_1 \wedge \cdots \wedge dz_n.$$

Define

$$\nu_f(r) = \int_{S(r)} \log^+ |\lambda_f| \sigma.$$

Let $E = -K$. Then, given ε, we have for r outside an exceptional set

$$\nu_f(r) \leqq \varepsilon \log r + \frac{n}{2}(1+\varepsilon)\log T_{f,E} + \sum \log m_{f,D_j}.$$

Proof. Let ρ be a metric on L_E such that

$$\eta = c_1(\rho) > 0,$$

and let ρ_j be a metric on L_{D_j} such that

$$\rho = \rho_1 \otimes \cdots \otimes \rho_q.$$

Let Ψ_K be the volume form corresponding to ρ^{-1}, so that

$$\mathrm{Ric}(\Psi_K) = -\eta.$$

Let s_j be a holomorphic section of L_{D_j} such that $(s_j) = D_j$, and let the Carlson–Griffiths form be as before:

$$\Psi = \frac{\Psi_K}{\Pi |s_j|^2_{\rho_j} h_j} \quad \text{where } h_j = \left(-\log|s_j|^2_{\rho_j}\right)^2.$$

After multiplying s_j by a suitably small constant, we can assume that $h_j \geqq e$. Then the Carlson–Griffiths form satisfies

$$\mathrm{Ric}\,\Psi = \psi - \varepsilon\eta$$

$$\psi > 0 \quad \text{and} \quad \frac{1}{n!}\psi^n \geqq C\Psi,$$

with a suitable constant $C > 0$.

Let

$$f^*\Psi = \gamma\Phi.$$

Then, directly from the definitions, we have

$$\gamma \gg\ll \frac{|\lambda_f|^2}{\Pi(h_j \circ f)}.$$

(The notation $A \ll B$ means $A = O(B)$, and $A \gg\ll B$ means $A = O(B)$ and $B = O(A)$.) Consequently,

$$(1) \qquad m_\gamma(r) = \frac{1}{2}\int_{\mathbf{S}(r)} (\log\gamma)\sigma \leqq \int_{\mathbf{S}(r)} \log^+ |\lambda_f|\sigma = \nu_f(r) + O(1).$$

By SMT, disregarding the ramification term, we get

$$\begin{aligned} T_{f,\mathrm{Ric}\,\Psi} &\leqq N_{f,D} + \nu_f + O(1) \\ &\ll T_{f,E} + \nu_f, \end{aligned}$$

and since $\psi = \mathrm{Ric}\,\Psi + \varepsilon\eta$, we find

$$(2) \qquad T_{f,\psi} \ll T_{f,E} + \nu_f.$$

On the other hand, by Theorem 3.4 we know that

$$T_{f,\psi}(r) = \int_0^r \int_{\mathbf{B}(t)} f^*\psi \wedge \omega^{n-1}\frac{dt}{t} = \int_0^r \left(\int_{\mathbf{B}(t)} ((n-1)!)(\mathrm{tr}\, f^*\psi)\Phi\right)\frac{dt}{t^{2n-1}}$$

and

$$\gamma\Phi = f^*\Psi \ll \frac{1}{n!}(f^*\psi)^n = (\det f^*\psi)\Phi,$$

so

$$(3) \qquad \int_{\mathbf{B}(t)} \gamma^{1/n}\Phi \ll \int_{\mathbf{B}(t)} (\det f^*\psi)^{1/n}\Phi \ll \int_{\mathbf{B}(t)} (\mathrm{tr}\, f^*\psi)\Phi.$$

Let

$$\alpha = |\lambda_f|^{2/n} \quad \text{and} \quad \beta = \frac{1}{\Pi(h_j \circ f)^{1/n}},$$

so that

$$0 \leqq \alpha \quad \text{and} \quad 0 \leqq \beta < 1.$$

Let

$$F_{\alpha\beta}(r) = \int_0^r \left(\int_{\mathbf{B}(t)} \alpha\beta\Phi \right) \frac{dt}{t^{2n-1}}.$$

Since $\gamma^{1/n} \gg \ll \alpha\beta$, we get from (2) and (3)

PROPOSITION 5.2. $F_{\alpha\beta} \ll T_{f,E} + \nu_f$.

We can now start estimating ν_f as follows:

$$\begin{aligned} \frac{1}{n}\nu_f(r) &= \frac{1}{2}\int_{\mathbf{S}(r)} (\log^+ \alpha)\sigma \\ &= \frac{1}{2}\int_{\mathbf{S}(r)} (\log^+ \alpha + \log\beta)\sigma - \frac{1}{2}\int_{\mathbf{S}(r)} (\log\beta)\sigma \\ &= \frac{1}{2}\int \log e^{\log^+\alpha + \log\beta}\,\sigma + \frac{1}{2}\int |\log\beta|\sigma \\ &\leqq \frac{1}{2}\log \int \left(e^{\log^+\alpha + \log\beta}\right)\sigma + \frac{1}{2}\int |\log\beta|\sigma \\ &\leqq \frac{1}{2}\log\int(\alpha\beta)\sigma + \frac{2}{2n}\sum_j \log\int -\log|s_j \circ f|^2\sigma + O(1) \end{aligned}$$

(because $e^{\log^+\alpha + \log\beta} \leqq \alpha\beta + 1$)

$$\leqq \frac{1}{2}\log\int(\alpha\beta)\sigma + \frac{1}{n}\sum \log m_{f,D_j} + O(1).$$

We now apply Lemma 3.6 to estimate the integral on the right-hand side, and we use Proposition 5.2 to conclude the proof of Theorem 5.1.

Remark. The most classical case is when $X = \mathbf{P}^1$ and $\Lambda = dw/w$. In this case, $\lambda_f(z) = f'/f(z)$ and we are in the Nevanlinna situation.

Appendix

In this appendix we give the proofs that have been left out before. (I thought it would enhance the logical clarity of the main steps to postpone them.) Mostly they consist in applying Stokes's theorem with singularities.

The form $\omega(z) = dd^c \log\|z\|^2$ is the pull-back by $\pi\colon \mathbf{C}^n - \{0\} \to \mathbf{P}^{n-1}$ of a form $\omega_{\mathbf{P}}$ on $\mathbf{P}^{n-1}$. We call $\omega_{\mathbf{P}}$ the **Fubini–Study form**. On the hyperplane bundle, if $w_1, \ldots, w_{n-1}$ are coordinates of the affine space not at infinity, then this metric is defined by

$$\tau(w) = 1 + \sum_{j=1}^{n-1} w_j \bar{w}_j.$$

Directly from the definitions, one checks that $\omega_{\mathbf{P}} = c_1(\tau)$.

PROPOSITION A1. $\displaystyle\int_{\mathbf{S}(r)} \sigma = 1 = \int_{\mathbf{P}^{n-1}} \omega_{\mathbf{P}}^{n-1}.$

Proof. We consider the fibering

$$\pi\colon \mathbf{C}^n - \{0\} \to \mathbf{P}^{n-1} \quad \text{or} \quad \pi\colon \mathbf{S}(r) \to \mathbf{P}^{n-1}.$$

In general, if $\pi\colon X \to Y$ is a fibering, $\omega = \pi^* \omega_Y$ is a form on X, and η is a form of the appropriate type and dimension, then we have a globalized version of Fubini's theorem given by

$$\int_X \eta \wedge \omega = \int_{y \in Y} \left(\int_{\pi^{-1}(y)} \eta \right) \omega_Y(y).$$

This is true locally by Fubini's theorem, and extends globally by partitions of unity.

Let I_n be the integral on the left of the proposition, and let J_{n-1} be the integral on the right. For $\pi\colon \mathbf{S}(r) \to \mathbf{P}^{n-1}$ we get

$$\begin{aligned} I_n = \int_{S(r)} \sigma &= \int_{x \in \mathbf{P}^{n-1}} \left(\int_{\pi^{-1}(x)} d^c \log\|z\|^2 \right) \omega_{\mathbf{P}}^{n-1}(x) \\ &= \int_{\pi^{-1}(x)} d^c \log\|z\|^2 \cdot J_{n-1} \\ &= J_{n-1}, \end{aligned}$$

because the first factor is independent of $x \in \mathbf{P}^{n-1}$, and one computes directly in dimension 1, using the formula for d^c given in polar coordinates by

$$\boxed{d^c = \frac{1}{2} r \frac{\partial}{\partial r} \otimes \frac{d\theta}{2\pi} - \frac{1}{4\pi} \frac{1}{r} \frac{\partial}{\partial \theta} \otimes dr.}$$

On the sphere, $dr = 0$, and so a computation involving d^c on the sphere needs only the first term. We may write

$$d^c = \frac{1}{2} r \frac{\partial}{\partial r} \otimes \frac{d\theta}{2\pi}$$

on the sphere. Now we use the fibration $\pi\colon \mathbf{C}^n - \{0\} \to \mathbf{P}^{n-1}$, and we get

$$J_{n-1} = \int_{\mathbf{P}^{n-1}} \omega_{\mathbf{P}}^{n-1} = \int_{\mathbf{C}^{n-1}} \left(dd^c \log\left(1 + \sum w_k \bar{w}_k\right)\right)^{n-1}$$

$$= \lim_{r\to\infty} \int_{\mathbf{B}(r)} \left(dd^c \log\left(1 + \sum w_k \bar{w}_k\right)\right)^{n-1}$$

[by Stokes]

$$= \lim_{r\to\infty} \int_{\mathbf{S}(r)} d^c\left(\log\left(1 + \sum\right)\right) \wedge \left(dd^c \log\left(1 + \sum\right)\right)^{n-2}$$

$$= \lim_{r\to\infty} \int_{\mathbf{S}(1)} d^c \log\left(\frac{1}{r^2} + \sum u_k \bar{u}_k\right) \wedge \left(dd^c \log\left(\frac{1}{r^2} + \sum u_k \bar{u}_k\right)\right)^{n-2}.$$

Taking the limit under the integral shows that $J_{n-1} = I_{n-1}$. Then Proposition A1 follows by induction.

The next proposition generalizes the formula for d^c in higher dimension.

Proposition A2. *For smooth functions* α, *we have*

$$d^c\alpha \wedge \omega^{n-1} | \mathbf{S}(t) = \frac{1}{2} t \frac{\partial \alpha}{\partial t} \sigma \quad \textit{on } \mathbf{S}(t).$$

Proof. Apply the left-hand side as a distribution, i.e., as a functional on smooth functions with compact support. Again use the fibration

$$\pi\colon \mathbf{S}(r) \to \mathbf{P}^{n-1}$$

to reduce the proposition to the case of dimension 1. In that case, we have already given the formula for d^c and thus conclude the proof.

Proof of Lemma 1.1. Put $\alpha = \log \gamma$. Then, by Proposition A2,

$$\int_0^r \left(\int_{\mathbf{S}(t)} d^c\alpha \wedge \omega^{n-1} \right) \frac{dt}{t} = \int_0^r \int_{\mathbf{S}(t)} \frac{1}{2} t \frac{\partial \alpha}{\partial t} \sigma \frac{dt}{t}.$$

Take $\partial/\partial t$ outside the inner integral. Since the integral of a derivative is equal to the function, we get Lemma 1.1 as stated.

Next we come to applications of Stokes's theorem, with singularities. A version is given in [La 5] and [La 2]. The idea is that if the singularities are of dimension one less than the dimensions of the regular points on the boundary of the manifold in question, then Stokes's formula is valid. The reason is that one can apply Stokes to the given forms, multiplied by a smooth function that is 0 near the singularities and 1 slightly farther out from the singularities. One then takes a limit over such functions, depending on a parameter ε, which tends to 0. One has to estimate the derivatives that come into the proof, and all details are given in the references above to carry out this idea. One also needs the improper integrals occurring in the statement of Stokes to be absolutely convergent. For our purposes, we work formally.

Let Z be an effective divisor on $\mathbf{C}^n$, and let g be a holomorphic function with $(g) = Z$. We shall deal with tubular neighborhoods of Z, defined by

$$V_\varepsilon(Z) = \{z \in \mathbf{C}^n \text{ such that } |g(z)| \leqq \varepsilon\}.$$

By Sard's theorem, for almost all ε this neighborhood has a regular boundary. We let its boundary be

$$S_\varepsilon(Z) = \{z \in \mathbf{C}^n \text{ such that } |g(z)| = \varepsilon\}.$$

The boundary of $\mathbf{B}(r) - V_\varepsilon(Z)$ consists of

$$\mathbf{S}(r) - V_\varepsilon(Z) \quad \text{and} \quad S_\varepsilon(Z) \quad \text{(suitably oriented)}.$$

THEOREM A3. *Let α be C^∞ on $\mathbf{C}^n - Z$, and let β be a smooth $(n-1, n-1)$ form. Then*

$$\int_{\mathbf{B}(r)} dd^c\alpha \wedge \beta - \int_{\mathbf{B}(r)} \alpha \wedge dd^c\beta = \int_{\mathbf{S}(r)} d^c\alpha \wedge \beta - \alpha \wedge d^c\beta$$

$$- \lim_{\varepsilon \to 0} \int_{S_\varepsilon(Z)(r)} d^c\alpha \wedge \beta - \alpha \wedge d^c\beta.$$

Proof. We have

$$d(d^c\alpha \wedge \beta) = dd^c\alpha \wedge \beta - d^c\alpha \wedge d\beta$$

$$d(\alpha \wedge d^c\beta) = d\alpha \wedge d^c\beta + \alpha \wedge dd^c\beta.$$

Since β is of type $(n-1, n-1)$, we get $\partial\alpha \wedge \partial\beta = 0$ and $\bar{\partial}\alpha \wedge \bar{\partial}\beta = 0$, so

$$d^c\alpha \wedge d\beta = -d\alpha \wedge d^c\beta.$$

Applying Stokes and subtracting the two expressions above yields the theorem.

The theorem will be used when β is not quite smooth, but in fact

$$\beta = \omega^{n-1}.$$

This form has a mild singularity at the origin, so the general technique of [La 5] or [La 2] can be used to justify this application. Note also that

$$d\omega^{n-1} = 0$$

because $d\omega = 0$. Hence, when $\beta = \omega^{n-1}$, we get the shorter version

$$\boxed{\int_{\mathbf{B}(r)} dd^c\alpha \wedge \omega^{n-1} = \int_{\mathbf{S}(r)} d^c\alpha \wedge \omega^{n-1} - \lim_{\varepsilon \to 0} \int_{S_\varepsilon(Z)(r)} d^c\alpha \wedge \omega^{n-1}.}$$

THEOREM A4. *Let Z be a divisor on $\mathbf{C}^n$, $Z = (g)$ so that g represents a meromorphic section s of L_Z on $\mathbf{C}^n$. Let ρ be a smooth positive function on $\mathbf{C}^n$ defining the metric*

$$|s|_\rho^2 = \frac{|g|^2}{\rho}.$$

Suppose $0 \notin Z$. Then

$$\mathbf{t}_\rho = \mathbf{n}_Z + \mathbf{m}^\circ_{Z,\rho}.$$

Proof. By addivity we may assume that Z is effective and g is holomorphic. We need a lemma.

LEMMA A5.

$$\lim_{\varepsilon \to 0} \int_{S_\varepsilon(Z)(r)} d^c \log|g|^2 \wedge \beta = \int_{Z(r)} \beta.$$

Proof. By a partition of unity we are reduced to the case when the support of β lies in a neighborhood U of a regular point of Z. Then $g = w_1^k$, where

$w_1, \ldots, w_n$ are complex coordinates near that point. Then

$$\lim_{\varepsilon \to 0} \int_{S_\varepsilon(Z) \cap U} d^c\alpha \wedge \beta = \lim_{\varepsilon \to 0} k \int_{|w_1|=1} d^c \log|w_1|^2 \wedge \beta$$

$$= k \int_{w_1=0} \beta = \int_{Z(r)} \beta.$$

This proves the lemma.

Now apply Stokes twice, first with $\beta = \omega^{n-1}$, so $d^c\beta = 0$, and $dd^c \log|g|^2 = 0$. Also, let $\alpha = \log|g|^2$. Then

$$(1) \qquad - \int_{\mathbf{S}(r)} d^c \log|g|^2 \wedge \omega^{n-1} + \int_{Z(r)} \omega^{n-1} = 0.$$

Second, let $D = 0$, $\alpha = \log \rho$, so that $-dd^c \log|g|_\rho^2 = dd^c \log \rho$, and so by Stokes and Lemma A5,

$$(2) \quad - \int_{\mathbf{B}(r)} dd^c \log|g|_\rho^2 \wedge \omega^{n-1} = \int_{\mathbf{B}(r)} dd^c \log \rho \wedge \omega^{n-1} = \int_{\mathbf{S}(r)} d^c \log \rho \wedge \omega^{n-1}.$$

Adding (1) and (2) yields Theorem A4.

The first main theorem, namely, Theorem 1.2, is a special case applied to $Z = f^*D$ and to the pull-back of the metric on L_D by f. Note that we have proved the stronger version with the boldface small functions.

To get the second main theorem, i.e., Theorem 2.2, we use the functoriality, and we observe that

$$\rho \mapsto T_\rho, \qquad (Z, \rho) \mapsto m^\circ_{Z,\rho}, \qquad Z \mapsto N_Z$$

are homomorphisms. Therefore, the second main theorem follows directly from Theorem A3 and the following considerations. We write

$$f^*\Psi = \gamma\Phi,$$

where γ has the form

$$\gamma = \frac{|\Delta|^2 h_k}{\Pi |g_j|_{\rho_j}^2 h_j} \quad \text{with } h_j = h\left(|g_j|_{\rho_j}^2\right),$$

and Δ is an entire function such that $\mathrm{Ram}(f) = (\Delta)$, that is, Δ defines the ramification divisor. On $\mathbf{C}^n$ every divisor is represented by a meromorphic function, say a theta function. Locally, Δ differs from the complex Jacobian determinant of f by an invertible holomorphic function.

Theorem A4 can be applied without change to all the factors entering in γ as above, except for the singular metrics on the trivial line bundle on $\mathbf{C}^n$ defined by the functions h_j. For these, we must go back and reconsider the analogue of Lemma A5. In other words, we want to prove

PROPOSITION A6. *Let* $\rho = h_j$. *Then* $\mathbf{t}_\rho = m_\rho^\circ$ *and* $T_\rho = m_\rho^\circ$, *that is,*

$$\int_{\mathbf{B}(r)} dd^c \log \rho \wedge \omega^{n-1} = \int_{\mathbf{S}(r)} d^c \log \rho \wedge \omega^{n-1}.$$

Proof. The equality to be proved is simply the short version of Stokes's theorem with ω^{n-1} given just before Theorem A3. Thus we let $\alpha = \log \rho$, and what we want to prove is the limit

$$\lim_{\varepsilon \to 0} \int_{S_\varepsilon(Z)(r)} d^c\alpha \wedge \omega^{n-1} = 0.$$

Letting $\lambda = 1/\rho_j$, so that λ is smooth, we are reduced to proving the following lemma, which corresponds to Lemma A5.

LEMMA A7. *Let g be holomorphic, defining the divisor Z, and assume $0 \notin Z$. Let λ be smooth positive such that $\lambda |g|^2 \leqq 1/e$. Let*

$$h = -\log(\lambda |g|^2) \quad \text{and} \quad \alpha = \log h.$$

Then, for any smooth $(n-1, n-1)$ form β, we have

$$\lim_{\varepsilon \to 0} \int_{S_\varepsilon(Z)(r)} d^c\alpha \wedge \beta = 0.$$

Proof. We have

$$d^c\alpha = \frac{d^c \log \lambda + d^c \log |g|^2}{\log(\lambda |g|^2)}.$$

Then $d^c \log \lambda$ is bounded on a neighborhood of Z, and $d^c \log |g|^2$ in the neighborhood of a regular point of Z is like $d\theta/2\pi$. The singular points can be disregarded by the general techniques of Stokes's theorem at singularities. Finally,

$$\log(\lambda |g|^2) \to -\infty$$

in a neighborhood of Z. This proves Lemma A7.

Thus Proposition A6 is also proved, and so is the second main theorem, that is, Theorem 2.2.

References

[Ad 1] W. Adams, *Asymptotic diophantine approximations*, Proc. Nat. Acad. Sci. U.S.A. **55**(1966), 28–31.

[Ad 2] ______, *Asymptotic diophantine approximations and Hurwitz numbers*, Amer. J. Math. **89**(1967), 1083–1108.

[C–G] J. Carlson and P. Griffiths, *A defect relation for equidimensional holomorphic mappings between algebraic varieties*, Ann. Math. **95**(1972), 557–584.

[Ch] S. S. Chern, *Complex analytic mappings of Riemann surfaces I*, Amer. J. Math. **82**(1960), 323–337.

[Gr] P. Griffiths, *Entire Holomorphic Mappings in One and Several Complex Variables*, Annals of Math. Studies, Princeton University Press, New Jersey, 1976.

[La 1] S. Lang, *Asymptotic diophantine approximations*, Proc. Nat. Acad. Sci. U.S.A. **55**(1966), 31–33.

[La 2] ______, *Differential Manifolds*, Springer-Verlag, New York, 1985. (Reprint from Addison-Wesley, Reading, MA, 1972).

[La 3] ______, *Higher dimensional diophantine problems*, Bull. Amer. Math. Soc. **80**(1974), 779–787.

[La 4] ______, *Introduction to Diophantine Approximations*, Addison-Wesley, Reading, MA, 1966.

[La 5] ______, *Real Analysis*, Addison-Wesley, Reading, MA, 1969; 2d ed., 1983.

[La 6] ______, *Report on diophantine approximations*, Bull. Soc. Math. France **93**(1965), 177–192.

[Ne] R. Nevanlinna, *Analytic Functions*, Springer-Verlag, New York, 1970. (Translation of previous editions [1936, 1953]).

[Os 1] C. F. Osgood, *A number theoretic-differential equations approach to generalizing Nevanlinna theory*, Indian J. Math. **23**(1981), 1–15.

[Os 2] ______, *Sometimes effective Thue–Siegel–Roth–Schmidt Nevanlinna bounds or better*, J. Number Theory **21**(1985), 347–389.

[St] W. Stoll, *Value Distribution on Parabolic Spaces*, Springer Lecture Notes **600**, Springer-Verlag, New York, 1973.

[Vo 1] P. Vojta, "A higher dimensional Mordell conjecture," *Arithmetic Geometry*, ed. G. Cornell and J. Silverman, Graduate Texts in Math., Springer-Verlag, New York, 1986, pp. 341–353.

[Vo 2] ______, *Diophantine Approximations and Value Distribution Theory*, Springer Lecture Notes **1239**, Springer-Verlag, New York, 1987.

[Vo 3] ______, *Integral points on varieties*, dissertation, Harvard University, 1983.

Department of Mathematics, Yale University, New Haven, Connecticut 06520

BULLETIN (New Series) OF THE
AMERICAN MATHEMATICAL SOCIETY
Volume 22, Number 1, January 1990

THE ERROR TERM IN NEVANLINNA THEORY. II

SERGE LANG

Nevanlinna theory [Ne] was created to give a quantitative measure of the value distribution for meromorphic functions, for instance to measure the extent to which they approximate a finite number of points. We view a meromorphic function as a holomorphic map $f : \mathbf{C} \to \mathbf{P}^1$ into the projective line. The theory has various higher dimensional analogues, of which we shall later consider maps $f : \mathbf{C}^n \to X$ where X is a projective complex manifold of dimension n.

We first deal with the classical case of Nevanlinna with $n = 1$. Let $a \in \mathbf{P}^1$. By a **Weil function** associated with a we mean a continuous function

$$\lambda_a : \mathbf{P}^1 - \{a\} \to \mathbf{R}$$

having the property that in some open neighborhood of a there exists a continuous function α such that if z is a local coordinate at a, then

$$\lambda_a(z) = -\log|z-a| + \alpha(z).$$

The difference between two Weil functions is a continuous (and therefore bounded) function on $\mathbf{P}^1$. A Weil function roughly measures the distance from a. As usual, for real $x > 0$ define $\log^+(x) = \max(\log x, 0)$. Let z be the standard coordinate on $\mathbf{C}$. Nevanlinna takes the functions

$$\lambda_a(z) = \log^+ 1/|z-a| \quad \text{if } a \neq \infty,$$
$$\lambda_a(z) = \log^+ |z| \quad \text{if } a = \infty.$$

One defines the corresponding **mean proximity function**

$$m_f(\lambda_a, r) = \int_0^{2\pi} \lambda_a(f(re^{i\theta}))\frac{d\theta}{2\pi}.$$

One usually writes $m_f(a, r)$ instead of $m_f(\lambda_a, r)$ since a definite

Received by the editors April 25, 1989 and, in revised form, July 11, 1989.
1980 *Mathematics Subject Classification* (1985 *Revision*). Primary 11J68, 30D35, 32H30.

Weil function has been chosen once for all. As a function of r, $m_f(a, r)$ is defined up to $O(1)$, independently of the choice of Weil function.

For $R > 1$ we define the **normalized zero counting function** by

$$N_f(0, R) = \sum_{\substack{a\in\mathbf{D}(R)\\ a\neq 0,\, f(a)=0}} (\mathrm{ord}_a f)\log\left|\frac{R}{a}\right| + (\mathrm{ord}_0 f)\log R,$$

where $\mathbf{D}(R)$ is the disc of radius R, and ord_a denotes the order of the zero of f at a. We define $N_f(\infty, R) = N_{1/f}(0, R)$ and $N_f(a, R) = N_{f-a}(0, R)$. Thus $N_f(0, R)$ measures the number of zeros of f in the disc of radius R, suitably weighted.

One defines the **height function** associated to f by

$$T_{f,a}(r) = m_f(a, r) + N_f(a, r).$$

Using Jensen's formula, it is easy to prove that $T_{f,a}$ is independent of a modulo $O(1)$. We write T_f instead of $T_{f,\infty}$. We choose λ_a to make $T_{f,a}$ an increasing function of r, and to have certain smoothness properties, by letting for instance

$$\lambda_a(z) = -\frac{1}{2}\log\frac{|z-a|^2}{(1+|z|^2)(1+|a|^2)} \quad \text{for } a,\, z \neq \infty.$$

Write $f = f_1/f_0$ where f_1, f_0 are entire without common zero. Let

$$W(f_0, f_1) = f_0 f_1' - f_0' f_1$$

be the Wronskian. We define the **ramification** counting function

$$N_{f,\mathrm{Ram}}(r) = N_W(0, r).$$

Basic conditions. *Let $a_1, \dots, a_q$ be distinct points of $\mathbf{P}^1$. Suppose for simplicity that $f(0) \neq 0, \infty$, a_j for all j, and $f'(0) \neq 0$.*

Under the basic conditions, Nevanlinna's classical theorem [Ne] is that asymptotically for $r \to \infty$, we have

$$(q-2)T_f(r) - \sum N_f(a_j, r) + N_{f,\mathrm{Ram}}(r) = O(\log r + \log T_f(r))$$

except for r lying in a set of finite Lebesgue measure. Nevanlinna also gives explicit constants in the error term on the right.

Osgood [Os 1, Os 2] noticed a similarity between the 2 occurring in Nevanlinna's theorem above on the left-hand side, and the 2 occurring in Roth's theorem [Ro]. However, Vojta [Vo 1] gave a much deeper analysis by pointing out that the whole theory of heights in algebraic number theory and diophantine geometry is analogous to Nevanlinna theory. Following this analogy, I looked into the error term of Nevanlinna's theorem [La 5], as follows.

Let α be a real, irrational number. In [La 2] and [La 3] I defined a **type** for α to be a positive increasing function ψ such that

$$-\log\left|\alpha - \frac{p}{q}\right| - 2\log q \le \log \psi(q)$$

for all but a finite number of fractions p/q in lowest form, $q > 0$. The height $h(p/q)$ is defined to be $\log\max(|p|, |q|)$. If p/q is close to α, then $\log q$ has the same order of magnitude as the height, so $\log q$ is essentially the height in the above inequality. A theorem of Khintchine states that almost all numbers have type ψ if

$$\sum_{q=1}^{\infty} \frac{1}{q\psi(q)} < \infty.$$

A basic question is whether Khintchine's principle applies to algebraic numbers, although possibly some additional restrictions on the function ψ might be needed. Roth's theorem can be formulated as saying that an algebraic number has type $\le q^{\varepsilon}$ for every $\varepsilon > 0$, and in the sixties I conjectured in line with Khintchine's principle that this could be improved to having type $\le (\log q)^{1+\varepsilon}$. Cf. [La 1, La 3, 4] especially.[1] Thus for instance, we would have the improvement of Roth's inequality

$$\left|\alpha - \frac{p}{q}\right| \ge \frac{C(\alpha, q)}{q^2(\log q)^{1+\varepsilon}}$$

which could be written

$$-\log\left|\alpha - \frac{p}{q}\right| 2\log q \le (1+\varepsilon)\log\log q$$

for all but a finite number of fractions p/q. However, except for quadratic numbers, which all have bounded type (trivial exercise), there is no example of an algebraic number about which one knows that it is or is not of type $(\log q)^k$ for some number $k > 1$. It becomes a problem to determine the type for each algebraic number and for the classical numbers. For instance, it follows from Adams' work [Ad 1], [Ad 2] that e has type

$$\psi(q) = \frac{C\log q}{\log\log q}$$

[1] Unknown to me until much later, similar conjectures were made by Bryuno [Br] and Richtmayer, Devaney and Metropolis [RDM], see [L-T 1] and [L-T 2].

with a suitable constant C, which is much better than the "probability" type, and goes beyond Khintchine's principle: the sum $\sum 1/q\psi(q)$ is divergent.

In light of Vojta's analysis, it occurred to me to transpose my conjecture about the error term in Roth's theorem to the context of Nevanlinna theory.

It becomes a problem to determine the "type" of the classical meromorphic functions, i.e. the best possible error term in the inequality which describes the value distribution of the function. It is classical, and easy, for example, that e^z has bounded type, i.e. that the error term in Nevanlinna's theorem is $O(1)$. But I do not even know an example of a function which does not have bounded type! There are two problems here:

- To determine for "almost all" functions (in a suitable sense) whether the type follows the pattern of Khintchine's convergence principle.
- To determine the specific type for each concrete classical function, using the specific special properties of each such function $\wp$, θ, Γ, ζ, J, etc.

In [La 5] I conjectured a best possible error term, but was not able to prove it exactly. For instance, instead of $1+\varepsilon$ I got only $3/2+\varepsilon$. Using a method from Ahlfors' paper [Ah], P. M. Wong [Wo] obtained the error term with $1+\varepsilon$. I pointed out to him that his method would also prove the desired result with an arbitrary type function ψ satisfying only the Khintchine convergence principle. Thus the result precisely stated is as follows.

Let ψ be a positive (weakly) increasing function such that

$$\int_e^\infty \frac{1}{u\psi(u)}\,du = b_0(\psi)$$

is finite. For any positive increasing function F of class C^1 such that $r \mapsto rF'(r)$ is positive increasing, and for r, $c>0$ we define the **error function**

$$S(f, c, \psi, r) = \log F(r) + \log \psi(F(r)) + \log \psi(crF(r)\psi(F(r))).$$

We let $r_1(F)$ be the smallest number ≥ 1 such that $F(r_1) \geq 1$, and we let $b_1(F)$ be the smallest number ≥ 1 such that

$$b_1 rF'(r) \geq e \quad \text{for } r \geq 1.$$

Theorem 1. *Under the basic conditions, there are constants* $b = b(f, a_1, \ldots, a_q)$ *and* B_q *(depending on* q*) such that for all* $r \geq r_1(T_f)$ *outside a set of measure* $\leq 2b_0(\psi)$ *and all* $b_1 \geq b_1(T_f)$ *we* *have*

$$(q-2)T_f(r)-\sum N_f(a_j, r)+N_{f,\mathrm{Ram}}(r) \le \frac{1}{2}S(B_q T_f^2, b_1, \psi, r)+b.$$

We can take $B_q = 12q^2 + q^3 \log 4$, *and* b *can also be determined explicitly.*

The T_f^2 already occurs in Ahlfors, but in a form with unspecified constants. Since the dominant term in the error term S is essentially a log, the error term amounts to $\frac{1}{2}\log T_f^2 = \log T_f$ in first order approximation. Wong obtained the correct factor $\frac{1}{2}$ by Ahlfors' method, rather than through the singular volume form used previously by other authors.

I shall now describe the result in the higher, equidimensional case, first investigated by Carlson-Griffiths [C-G]. Let

$$f : \mathbf{C}^n \to X$$

be a holomorphic map into a compact complex manifold of dimension n. We assume that f is nondegenerate, in the sense that the derivative of f at some point is nonsingular. Let $z = (z_1, \ldots, z_n)$ be the complex coordinates on $\mathbf{C}^n$, and let $\|z\|$ be the euclidean norm. We define the differential forms on $\mathbf{C}^n$:

$$\omega(z) = dd^c \log \|z\|^2 \quad \text{and} \quad \sigma(z) = d^c \log \|z\|^2 \wedge \omega^{n-1}.$$

Here $d^c = (\partial - \overline{\partial})/4\pi i$ and d is the usual exterior derivative. Note that $\sigma(z)$ in dimension $n = 1$ is $d\theta/d\pi$.

Let D be a divisor on X, so D is locally a hypersurface. We suppose that D is effective, so D can be represented by one holomorphic function φ locally, up to an invertible holomorphic function, and D is defined by $\varphi = 0$. By a **Weil function** associated with D we mean a function

$$\lambda_D : X - \text{support of } D \to \mathbf{R}$$

such that if D is represented by φ on an open set U, there exists a continuous function α on U such that

$$\lambda_D(P) = -\log|\varphi(P)| + \alpha(P) \quad \text{for } P \in U - D.$$

We define the **mean proximity function**

$$m_{f,D}(r) = \int_{\mathbf{S}(r)} (\lambda_D \circ f)\sigma,$$

where $\mathbf{S}(r)$ is the sphere of radius r in $\mathbf{C}^n$. Thus the mean proximity function measures the average approximation of D by the values of f on the spheres. In order to get smoothness, one must select smooth Weil functions λ_D in their class $\bmod O(1)$. This is done as follows.

We let L_D be a holomorphic line bundle over X having a meromorphic section s whose divisor (s) is precisely D. Since X is compact, such a section is well defined up to a constant factor. We let ρ be a hermitian metric on L_D. Then we select

$$\lambda_D = -\log |s|_\rho.$$

We define the **counting function**

$$N_{f,D}(r) = \int_0^r \frac{dt}{t} \int_{(f^*D)\cap \mathbf{B}(r)} \omega^{n-1},$$

where $\mathbf{B}(r)$ is the ball of radius r, and we define the **height function**

$$T_{f,D} = m_{f,D} + N_{f,D}.$$

Strictly speaking, we should write $T_{f,\rho}$, to indicate that T depends on the metric ρ; but a change of ρ changes the height function only by a bounded function. With these differential geometric definitions, $T_{f,D}$ is an increasing function. In fact, if we define the first **Chern form** outside the support of (s) by the formula

$$c_1(\rho) = -dd^c \log |s|_\rho^2$$

then

$$T_{f,\rho}(r) = \int_0^r \frac{dt}{t} \int_{\mathbf{B}(r)} f^* c_1(\rho) \wedge \omega^{n-1}.$$

Let Ω be a volume form on X (i.e. a positive (n, n)-form). Then

$$f^*\Omega = |\Delta|^2 h\Phi,$$

where Φ is the Euclidean volume form on $\mathbf{C}^n$, h is C^∞ real > 0, and Δ is a holomorphic function on $\mathbf{C}^n$, which defines the **ramification divisor** of f. Let Z denote the ramification divisor. We define

$$N_{f,\mathrm{Ram}}(r) = \int_0^r \frac{dt}{t} \int_{Z\cap \mathbf{D}(r)} \omega^{n-1}.$$

We say that a divisor D has **simple normal crossings** if $D = \sum D_j$ is a formal sum of nonsingular irreducible divisors, and locally at each point of X there exist complex coordinates $z_1, \ldots, z_n$ such that in a neighborhood of this point, D is defined by

$$z_1 \cdots z_k = 0 \quad \text{with } k \le n.$$

When $n = 1$, then the property of D having simple normal crossings is equivalent to the property that D consists of distinct points, taken with multiplicity 1. The maximal value of k which can occur will be called the **complexity** of D.

Finally, in higher dimension n, we suppose that $r \mapsto F(r)$ and $r \mapsto r^{2n-1}F'(r)$ are positive increasing functions of r, and we define the **error function**

$$S(F, c, \psi, r) = \log F(r) + \log \psi(F(r)) + \log \psi(cr^{2n-1}F(r)\psi(F(r))).$$

We let $b_1(F)$ be the smallest number ≥ 1 such that

$$b_1 r^{2n-1}F'(r) \geq e \quad \text{for all } r \geq 1.$$

The definition of $r_1(F)$ is the same as for $n = 1$. Then the analogue of Theorem 1 in higher dimension runs as follows.

Theorem 2. *Suppose that $f(0) \notin D$ and $0 \notin \mathrm{Ram}_f$. Let D have simple normal crossings, of complexity k. Let K be a canonical divisor on X. Let $T_f = T_{f,E}$ where E is a hyperplane section in some projective imbedding of X. Then*

$$T_{f,K}(r) + T_{f,D}(r) - N_{f,D}(r) + N_{f,\mathrm{Ram}}(r) \leq \frac{n}{2} S(BT_f^{1+k/n}, b_1, \psi, r) + B'$$

for all $r \geq r_1(T_f)$ outside a set of measure $\leq 2b_0(\psi)$, and some constants $B = B(D, E)$ and $B' = B'(D, E)$ which can be given explicitly.

The general shape of the theorem stems from Carlson-Griffiths [C-G]. I raised the question of a best possible error term in [La 5] but I was not able to prove the conjectured result at that time. By using Ahlfors' method, Wong [Wo] obtained not only the T_f^2 as in Ahlfors, but also the "correct" factor $n/2$. The final improvement with $1 + k/n$ instead of 2 follows from a technical change in Wong's proof at the appropriate moment, and will be given in detail in a forthcoming Springer Lecture Note. I also improved Wong's formulation by using the arbitrary Khintchine type function ψ in the final estimate and by not making any restriction on the divisor other than simple normal crossings. Wong, following some previous authors, assumes unnecessarily that the irreducible components all lie in the same linear system. Otherwise, the general pattern of the proof is due to Wong. It makes use of some ideas of Carlson-Griffiths concerning curvature, but somewhat more efficiently, in a way which should have significance elsewhere in complex differential geometry.

I would conjecture that the exponent $1 + k/n$ is best possible. Thus the error term should be determined by local considerations on the divisor, in terms of the complexity of its singularities. The conjecture can then be transposed to a strengthening of Schmidt's theorem [Sch]. The exponent $1 + k/n$ also applies when there is no divisor D, so $k = 0$, in which case T_f occurs with exponent 1 on the right-hand side. Thus the exponent $1 + k/n$ interpolates very neatly between the two extreme cases $D = 0$ and D defined locally by $z_1 \cdots z_n = 0$.

The same error term can be given in the theorem on the logarithmic derivative, originally stated with a weak error term by Nevanlinna in dimension 1, and proved using a differential geometric method, [Ne], p. 259. A higher dimensional version was formulated and proved by Griffiths [Gr] , p. 70, still with a weak error term. The version I now have runs as follows.

Theorem 3. *Let $f : \mathbf{C}^n \to X$ be holomorphic nondegenerate. Let Ψ be a meromorphic n-form with no zeros on X, and such that its polar divisor D has simple normal crossings. Let*

$$f^*\Psi = L_f(z)\, dz_1 \wedge \cdots \wedge dz_n.$$

Define

$$\nu_f(r) = \int_{\mathbf{S}(r)} \log^+ |L_f| \sigma.$$

Let K be a canonical divisor and assume $-K$ is ample. Then for some constants B, B' we have

$$\nu_f(r) \leq \frac{n}{2} S(BT_f^{1+k/n}, b_1, \psi, r) + B'$$

for all $r \geq r_1$ outside a set of measure $\leq 2b_0(\psi)$.

Note that in dimension 1, taking $X = \mathbf{P}^1$ and $\Psi = dz/z$, then $D = (0) + (\infty)$, and L_f is the logarithmic derivative

$$L_f = f'/f,$$

so Nevanlinna's classical set up is a special case. The proof follows the pattern of Nevanlinna-Griffiths, but using the Ahlfors technique as revived by Wong.

Finally, the same type of error term can be obtained for holomorphic maps

$$f : \mathbf{C} \to X \quad \text{or} \quad f : \mathbf{D}(R) \to X.$$

I shall formulate here one version stemming from Griffiths-King [G-K] and Vojta [Vo 1] Theorem 5.7.2.

Let Y be a complex manifold, and let $f: \mathbf{D}(R) \to Y$ be a nonconstant holomorphic map. If ω is a $(1, 1)$-form on Y then we define the **height**

$$T_{f,\omega}(r) = \int_0^r \frac{dt}{t} \int_{\mathbf{D}(t)} f^*\omega.$$

We write as usual

$$f^*\omega = \gamma_f \Phi, \quad \text{where } \Phi = \frac{\sqrt{-1}}{2\pi} dz \wedge d\bar{z}.$$

Then

$$\gamma_f = |\Delta|^2 h$$

where h is C^∞. If ω is positive, then $h > 0$. The function Δ is holomorphic and defines the ramification divisor Ram_f, which in this case is a discrete set of zeros with multiplicities. We define $\mathrm{Ric}\, f^*\omega = dd^c \log h$.

Theorem 4. *Let Y be a complex manifold (not necessarily compact). Let ω be a positive $(1, 1)$-form on Y and let $f: \mathbf{D}(R) \to Y$ be a holomorphic map. Suppose there is a constant B such that*

$$Bf^*\omega \leq \mathrm{Ric}\, f^*\omega.$$

Assume $0 \notin \mathrm{Ram}_f$. Let $b_1 = b_1(T_{f,\omega})$. Then for $r < R$ we have

$$BT_{f,\omega}(r) + N_{f,\mathrm{Ram}}(r) \leq \tfrac{1}{2} S(T_{f,\omega}, b_1, \psi, r) - \tfrac{1}{2} \log \gamma_f(0)$$

for $r \geq r_1(T_{f,\omega})$ outside a set of measure $\leq 2b_0(\psi)$.

Note that the theorem is formulated for a noncompact manifold, and that the map f is defined on a disc. Under the hypothesis of the theorem, there is no nonconstant holomorphic map of $\mathbf{C}$ into Y. The theorem gives an implicit bound for the radius of a disc on which a holomorphic map is defined.

References

[Ad1] W. Adams, *Asymptotic diophantine approximations to e*, Proc. Nat. Acad. Sci. U.S.A. **55** (1966), 28–31.

[AD2] W. Adams, *Asymptotic diophantine approximations and Hurwitz numbers*, Amer. J. Math. **89** (1967), 1083–1108.

[A-L] W. Adams and S. Lang, *Some computations in diophantine approximations*, J. Reine Angew. Math. **220** (1965), 163–173.

[Ah] L. Ahlfors, *The theory of meromorphic curves*, Acta. Soc. Sci. Fenn. Nova Ser. A **3**(1939/47) no. 4 (1941), 1–31.

[Br] A. D. Bryuno, *Continued fraction expansion of algebraic numbers*, Zh. Vichisl Mat. i Mat. Fiz. **4** nr. 2 (1964), 211–221, translated USSR Comput. Math. Phys. **4** (1964), 1–15

[C-G] J. Carlson and P. Griffiths, *A defect relation for equidimensional holomorphic mappings between algebraic varieties*, Ann. of Math. (2) **95** (1972), 557–584.

[Ch] S. S. Chern, *Complex analytic mappings of Riemann surfaces*. I, Amer. J. Math. **82** (1960), 323–337.

[Gr] P. Griffiths, *Entire holomorphic mappings in one and several complex variables*, Ann. Math. Studies vol. 85, Princeton Univ. Press, Princeton, N. J., 1976

[G-K] P. Griffiths and J. King, *Nevanlinna theory and holomorphic mappings between algebraic varieties*, Acta Math. **130** (1973), 145–220.

[La1] S. Lang, *Report on diophantine approximations*, Bull. Soc. Math. France **93** (1965), 177–192.

[La2] S. Lang, *Asymptotic diophantine approximations*, Proc. Nat. Acad. Sci. U.S.A. **55** (1966), 31–34.

[La3] S. Lang, *Introduction to diophantine approximations*, Addison-Wesley, 1966.

[La4] S. Lang, *Transcendental numbers and diophantine approximations*, Bull. Amer. Math. Soc. **77** (1971), 635–677.

[La5] S. Lang, *The error term in Nevanlinna theory*, Duke Math. J. **56** (1988), 193–218.

[L-T1] S. Lang and H. Trotter, *Continued fractions of some algebraic numbers*, J. Reine Angew. Math. **255** (1972), 112–134.

[L-T2] S. Lang and H. Trotter, Addendum to the above, J. Reine Angew Math. **267** (1974), 219–220.

[Ne] R. Nevanlinna, *Analytic functions*, Springer-Verlag, 1970 (revised translation of the German edition, 1953).

[vN-T] J. von Neumann and B. Tuckerman, *Continued fractions expansion of* $2^{1/3}$, Math. Tables Aids Comput. **9** (1955), 23–24.

[Os1] C. F. Osgood, *A number theoretic differential equations approach to generalizing Nevanlinna theory*, Indian J. Math. **23** (1981), 1–15.

[Os2] C. F. Osgood, *Sometimes effective Thue-Siegel Roth Schmidt-Nevanlinna bounds, or better*, J. Number Theory **21** (1984), 347–389.

[RDM] R. Richtmyer, M. Devaney and N. Metropolis, *Continued fraction expansions of algebraic numbers*, Numer. Math. **4** (1962), 68–84.

[Ro] K. F. Roth, *Rational approximations to algebraic numbers*, Matematika **2** (1955), 1–20.

[Sc] W. Schmidt, *Diophantine approximation*, Lecture Notes in Math., vol. 785, Springer-Verlag, Berlin and New York, 1980.

[Vo1] P. Vojta, *Diophantine approximations and value distribution theory*, Lecture Notes in Math., vol. 1239, Springer-Verlag, Berlin and New York, 1987.

[Vo2] P. Vojta, *A refinement of Schmidt's subspace theorem* Am. J. Math **111** (1989), pp. 489–518.

[Wo] P. M. Wong, *On the second main theorem of Nevanlinna theory* (to appear).

Department of Mathematics, Yale University, New Haven, Connecticut 06520

BULLETIN (New Series) OF THE
AMERICAN MATHEMATICAL SOCIETY
Volume 23, Number 1, July 1990

OLD AND NEW CONJECTURED DIOPHANTINE INEQUALITIES

SERGE LANG *

The original meaning of diophantine problems is to find all solutions of equations in integers or rational numbers, and to give a bound for these solutions. One may expand the domain of coefficients and solutions to include algebraic integers, algebraic numbers, polynomials, rational functions, or algebraic functions. In the case of polynomial solutions, one tries to bound their degrees. Inequalities concerning the size of solutions of diophantine problems are called diophantine inequalities.

During the past few years, new insights have been gained in old problems combined with new ones, and great coherence has been achieved in understanding a number of diophantine inequalities. Some of these results, notably the first section, can be formulated in very simple terms, almost at the level of high school algebra.

Received by the editors July 24, 1989.

1980 *Mathematics Subject Classification* (1985 *Revision*). Primary 11D41, 11D75, 11G05, 11G30.

* Supported by NSF grant, but...: In 1981 the Yale administration turned down the NSF grant to me as principal investigator because I would not sign effort reports. (However the math dept. at Yale passed a resolution to return all effort reports unsigned to the administration, and the Yale administration did nothing about other people's grants.) I was without a grant for 4 years. Then I reapplied for some items (such as travel expenses) in 1985, via someone else's grant, but I have not reapplied for summer salary support since 1981, and have therefore been without such support since. The fact that I am listed by the NSF as being "supported" by the NSF has given rise to misunderstandings concerning the nature of the support I received this last decade, whence it is important to specify the nature of such support. In addition, I find that the shortage of funds affecting the NSF makes it impossible for them to fulfill properly their funding role by following past policies on the allocation of funds. Over the last few years I have found that on the whole, for each person getting a grant, there are several others equally qualified who have been dropped, thus causing chaos in the granting process, and serious misinterpretations as to its implications by university administrators. It is demoralizing for the younger generation of mathematicians when getting a grant becomes a crap-shoot, particularly when university administrators often interpret not having NSF support as making someone unsuitable for tenuring or promoting. As a result of all these factors (plus renewed bureaucratic pressures), I have decided not to apply for any further NSF support in any form whatsoever starting next year.

0273-0979/90 $1.00 + $.25 per page

I shall give a survey starting with these formulations, and ending with more sophisticated applications to elliptic curves. But I have made an attempt to have this article readable by a fairly broad audience by giving basic definitions, and limiting myself to the rational numbers whenever possible.

1. The abc conjecture. This conjecture evolved from the insights of Mason [Ma], Frey [Fr], Szpiro, and others. Mason started one recent trend of thoughts by discovering an entirely new relation among polynomials, in a very original work, as follows. Let $f(t)$ be a polynomial with coefficients in an algebraically closed field of characteristic 0. We define

$$n_0(f) = \text{number of distinct zeros of } f.$$

Thus $n_0(f)$ counts the zeros of f by giving each of them multiplicity one.

Mason's theorem. *Let $a(t)$, $b(t)$, $c(t)$ be relatively prime polynomials such that $a + b = c$. Then*

$$\max \deg\{a, b, c\} \leqq n_0(abc) - 1.$$

In the statement of Mason's theorem, observe that it does not matter whether we assume a, b, c relatively prime in pairs, or without common prime factor for a, b, c. These two possible assumptions are equivalent by the equation $a + b = c$. Also the statement is symmetric in a, b, c and we could have rewritten the equation in the form $a + b + c = 0$.

Mason's theorem is a theorem, not a conjecture, and can be proved as follows. Dividing by c, and letting $f = a/c$, $g = b/c$, we have

$$f + g = 1$$

where f, g are rational functions. Differentiating we get $f' + g' = 0$, which we rewrite as

$$\frac{f'}{f} f + \frac{g'}{g} g = 0,$$

so that

$$\frac{b}{a} = \frac{g}{f} = -\frac{f'/f}{g'/g}.$$

If R is a rational function, $R(t) = \prod (t - \rho_i)^{q_i}$ with $q_i \in \mathbf{Z}$, then

$$R'/R = \sum \frac{q_i}{t - \rho_i}$$

and the multiplicities disappear. Suppose

$$a(t) = \prod (t-\alpha_i)^{m_i}, \qquad b(t) = \prod (t-\beta_j)^{n_j}, \qquad c(t) = \prod (t-\gamma_k)^{r_k}.$$

Then

$$\frac{b}{a} = -\frac{f'/f}{g'/g} = -\frac{\sum \frac{m_i}{t-\alpha_i} - \sum \frac{r_k}{t-\gamma_k}}{\sum \frac{n_j}{t-\beta_j} - \sum \frac{r_k}{t-\gamma_k}}.$$

A common denominator for f'/f and g'/g is given by the product

$$N_0 = \prod (t-\alpha_i) \prod (t-\beta_j) \prod (t-\gamma_k),$$

whose degree is $n_0(abc)$. Observe that $N_0 f'/f$ and $N_0 g'/g$ are both polynomials of degrees at most $n_0(abc) - 1$. From the relation

$$\frac{b}{a} = -\frac{N_0 f'/f}{N_0 g'/g}$$

and the fact that a, b are assumed relatively prime we deduce the inequality in Mason's theorem.

As an application let us prove Fermat's theorem for polynomials. Thus let $x(t)$, $y(t)$, $z(t)$ be relatively prime polynomials such that one of them has degree $\geqq 1$, and such that

$$x(t)^n + y(t)^n = z(t)^n.$$

We want to prove that $n \leqq 2$. By Mason's theorem, we get

$$\deg x(t)^n \leqq \deg x(t) + \deg y(t) + \deg z(t) - 1$$

and similarly replacing x by y and z on the left-hand side. Adding, we find

$$n(\deg x + \deg y + \deg z) \leqq 3(\deg x + \deg y + \deg z) - 3.$$

This yields a contradiction if $n \geqq 3$.

Influenced by Mason's theorem, and considerations of Szpiro and Frey which we shall describe below, Masser and Oesterle formulated the *abc* conjecture for integers as follows. Let k be a non-zero integer. Define the **radical** of k to be

$$N_0(k) = \prod_{p|k} p$$

i.e. the product of the distinct primes dividing k. There is a classical analogy between polynomials and integers. Under that analogy, n_0 of a polynomial corresponds to $\log N_0$ of an integer. Thus for polynomials we had an inequality formulated additively, whereas for integers we formulate the corresponding inequality multiplicatively. Note that if x, y are nonzero integers, then

$$N_0(xy) \leqq N_0(x)N_0(y),$$

and if x, y are relatively prime, then

$$N_0(xy) = N_0(x)N_0(y).$$

The abc conjecture. *Given $\varepsilon > 0$ there exists a number $C(\varepsilon)$ having the following property. For any nonzero relatively prime integers a, b, c such that $a + b = c$ we have*

$$\max(|a|, |b|, |c|) \leqq C(\varepsilon)N_0(abc)^{1+\varepsilon}.$$

Unlike the polynomial case, it is necessary to have the ε in the formulation of the conjecture, and the constant $C(\varepsilon)$ on the right-hand side. To simplify notation in dealing with the possible presence of such constants, if A, B are positive functions, we write

$$A \ll B$$

to mean that there exists a constant $C > 0$ such that $A \leqq CB$. Thus $A \ll B$ means that $A = O(B)$ in the big oh notation. We write

$$A \gg\ll B$$

to mean $A = O(B)$ and $B = O(A)$. In case the functions A, B depend on a parameter ε, the constant C also depends on ε.

The conjecture implies that many prime factors of abc occur to the first power, and that if some primes occur to high powers, then they have to be compensated by "large" primes, or many primes, occurring to the first power. Readers interested in seeing immediately the application to the Fermat problem and similar problems should skip the following remarks, having to do with the equation $2^n \pm 1 = k$. For n large, the abc conjecture would state that k is divisible by large primes to the first power or many primes to the first power. This phenomenon can be seen on the tables of [BLSTW]. The simplest examples showing the need for the constant $C(\varepsilon)$ in the abc conjecture were communicated to me by Wojtek Jastrzebowski and Dan Spielman as follows. We want to show that there is no constant $C > 0$ such that

$$\max(|a|, |b|, |c|) \leqq CN_0(abc).$$

Writing $3 = 1 + 2$, we find by induction that

$$2^n | (3^{2^n} - 1).$$

We consider the relations $a_n + b_n = c_n$ given by

$$3^{2^n} - 1 = c_n.$$

Then

$$N_0(a_n b_n c_n) = 3N_0(c_n) \leqq 3 \cdot 2 \frac{3^{2^n} - 1}{2^n}$$

so there is no constant C such that $c_n \leqq 3CN_0(c_n)$. Other examples can be constructed similarly since the role of 3 and 2 can be played by other integers: instead of 2 we use a prime p, and instead of 3 we use an integer $\equiv 1 \bmod p$.

In line with these examples, we now show that the *abc* conjecture implies a classical conjecture:

There are infinitely many primes p *such that*

$$2^{p-1} \not\equiv 1 \mod p^2.$$

We follow Silverman [Si]. First a remark. Let S be the set of primes such that

$$2^{p-1} \not\equiv 1 \mod p^2.$$

We claim that if n is a positive integer, and p is a prime such that $2^n \equiv 1 \bmod p$ but $2^n \not\equiv 1 \bmod p^2$, then p is in S. Indeed, let d be the period of 2 in the multiplicative group $(\mathbf{Z}/p\mathbf{Z})^*$ of units in $\mathbf{Z}/p\mathbf{Z}$, which has order $p-1$. Then d divides $p-1$ and also divides n. Furthermore $2^n \equiv 1 \bmod p$ but $2^n \not\equiv 1 \bmod p^2$ implies $2^d \not\equiv 1 \bmod p^2$. Hence $2^{p-1} \not\equiv 1 \bmod p^2$, as one sees by writing $p - 1 = dm$, with m prime to p, and $2^d = 1 + pk$ with k prime to p. Then

$$2^{p-1} \equiv 1 + pmk \not\equiv 1 \mod p^2,$$

so p is in S as claimed.

Now suppose that S is finite. Write

$$2^n - 1 = u_n v_n$$

where u_n is the product of primes in S, and all primes dividing v_n are not in S. Then u_n is bounded. If $p | v_n$ then by the claim, p^2 divides $2^n - 1$, so p^2 divides v_n. By the *abc* conjecture applied to the equation

$$(2^n - 1) + 1 = 2^n$$

we conclude that

$$u_n v_n \ll (u_n v_n^{1/2})^{1+\varepsilon} \ll v_n^{(1+\varepsilon)/2}$$

whence v_n is bounded, a contradiction.

Actually, in line with Lang-Trotter conjectures, the probability that $2^{p-1} \equiv 1 + pk (\bmod p^2)$ with a fixed residue class k mod p should be $O(1/p)$, so the number of primes $p \leqq x$ such that $2^{p-1} \equiv 1 \bmod p^2$ should be

$$O\left(\sum_{p \leqq x} \frac{1}{p}\right) = O(\log \log x).$$

Thus most primes should have the property that $2^{p-1} \not\equiv 1 \bmod p^2$.

We now pass to the application of the *abc* conjecture to various diophantine equations. In the case of polynomials, we got an explicit bound for the degree of Fermat's equation satisfied by polynomials which are not all constant. Since there is an unknown constant $C(\varepsilon)$ floating around in the *abc* conjecture, we shall get only an unknown bound for the classical case of Fermat's equation over the integers. Thus we define the **asymptotic Fermat problem** to state that there exists an integer n_1 such that for all $n \geqq n_1$ the equation

$$x^n + y^n = z^n$$

has only a trivial solution in integers, that is with one of x, y, z equal to 0. Of course, for the asymptotic Fermat problem as well as the ordinary one, we may and do assume that x, y, z are relatively prime.

The abc conjecture implies the asymptotic Fermat problem.

Indeed, suppose $x^n + y^n = z^n$ with x, y, z relatively prime. By the *abc* conjecture, we have

$$|x^n| \ll |xyz|^{1+\varepsilon}$$
$$|y^n| \ll |xyz|^{1+\varepsilon}$$
$$|z^n| \ll |xyz|^{1+\varepsilon}.$$

Taking the product yields

$$|xyz|^n \ll |xyz|^{3+\varepsilon},$$

whence for $|xyz| > 1$ we get n bounded. The extent to which the *abc* conjecture is proved with an explicit constant $C(\varepsilon)$ (or say

$C(1)$ to fix ideas) yields the corresponding explicit determination of the bound for n in the application.

We shall now see how the abc conjecture implies other conjectures by Hall, Szpiro, and Lang-Waldschmidt.

> **Hall's original conjecture** *is that if* u, v *are relatively prime*[1] *nonzero integers such that* $u^3 - v^2 \neq 0$ *then*

$$|u^3 - v^2| \gg |u|^{1/2-\varepsilon}.$$

Note that Hall's conjecture describes how small $|u^3 - v^2|$ can be, and the answer is not too small, as described by the right-hand side. Furthermore, if $|u^3 - v^2|$ is small, then $|u^3| \gg\ll v^2$ so $|v| \gg\ll |u|^{3/2}$. The Hall conjecture can also be interpreted as giving a bound for integral relatively prime solutions of

$$v^2 = u^3 + b \quad \text{with integral } b.$$

Then we find

$$|u| \ll |b|^{2+\varepsilon}.$$

More generally, as in Lang-Waldschmidt [La3], p. 212, let us fix nonzero integers A, B and let u, v, k, m, n be variable, with u, v relatively prime and $mn > m + n$. Put

$$Au^m + Bv^n = k.$$

By the *abc* conjecture, we get

$$|u|^m \ll |uvN_0(k)|^{1+\varepsilon}$$
$$|v|^n \ll |uvN_0(k)|^{1+\varepsilon}.$$

If, say, $|Au^m| \leqq |Bv^n|$ then $|u| \ll |v|^{n/m}$. We substitute this estimate for u to get an inequality entirely in terms of v, namely

$$|v|^n \ll |v^{1+n/m}N_0(k)|^{1+\varepsilon} = |v|^{(1+n/m)(1+\varepsilon)}N_0(k)^{1+\varepsilon}.$$

We first bring all powers of v to the left-hand side. We have

$$n - 1 - \frac{n}{m} = \frac{mn - (m+n)}{m}.$$

[1]Actually the original conjecture does not make the assumption that u, v are relatively prime, only that $u^3 - v^2 \neq 0$. The original conjecture also follows from the *abc* conjecture by extracting a common factor, and using the same method as that indicated below. We make the relatively prime assumption to avoid secondary technical complications, and we leave the proof in general to the reader.

We let the reader take care of the extra ε, so we obtain

(1) $$|v| \ll N_0(k)^{m(1+\varepsilon)/(mn-(m+n))} \quad \text{and then also}$$
$$|u| \ll N_0(k)^{n(1+\varepsilon)/(mn-(m+n))}$$

because the situation is symmetric in u and v. Again by the *abc* conjecture, we have

$$|k| \ll |uvN_0(k)|^{1+\varepsilon},$$

so using the estimate for $|uv|$ coming from the product of the inequalities in (1) we find

(2) $$|k| \ll N_0(k)^{mn(1+\varepsilon)/(mn-(m+n))}.$$

The Hall conjecture concerning $u^3 - v^2 = k$ is a special case of (1), after we replace $N_0(k)$ with $|k|$, because $N_0(k) \leqq |k|$.

Again take $m = 3$, $n = 2$ and take $A = 4$, $B = -27$. In this case, we write D instead of k, and we find for

$$D = 4u^3 - 27v^2$$

that

(3) $$|u| \ll N_0(D)^{2+\varepsilon} \quad \text{and} \quad |v| \ll N_0(D)^{3+\varepsilon}.$$

These inequalities are supposed to hold at first for u, v relatively prime. Suppose we allow u, v to have some bounded common factor, say d. Write

$$u = u'd \quad \text{and} \quad v = v'd$$

with u', v' relatively prime. Then

$$D = 4d^3u'^3 - 27d^2v'^2.$$

Now we can apply inequalities (1) with $A = 4d^3$ and $B = -27d^2$, and we find the same inequalities (3), with the constant implicit in the sign $\ll$ depending also on d, or on some fixed bound for such a common factor. Under these circumstances, we call inequalities (3) the **generalized Szpiro conjecture**.

The original Szpiro conjecture was stated for what is called a "minimal discriminant" D. We shall discuss the notion of a minimal discriminant in §4, when we go further into the theory of elliptic curves. Szpiro's inequality was stated in the form

$$|D| \ll N(D)^{6+\varepsilon},$$

where $N(D)$ is a more subtle invariant which is commonly used in the literature, namely the conductor. But for our purposes, it is sufficient and much easier to use the radical $N_0(D)$, since the conductor is harder to define, and its subtleties are irrelevant for what we are doing.

Note that the generalized Szpiro conjecture actually bounds $|u|$, $|v|$ and not just $|D|$ itself in terms of the right power of $N_0(D)$.

The point of D is that it occurs as the discriminant of an elliptic curve. The recent trend of thoughts in the direction we are discussing was started by Frey [Fr], who associated with each solution of $a + b = c$ the elliptic curve

$$y^2 = x(x-a)(x+b),$$

which we call the **Frey curve**. The discriminant of the right-hand side is the product of the differences of the roots squared, and so

$$D = (abc)^2.$$

We make a translation

$$\xi = x + \frac{b-a}{3}$$

to get rid of the x^2-term, so that our equation can be rewritten

$$y^2 = \xi^3 - \gamma_2\xi - \gamma_3,$$

where γ_2, γ_3 are homogeneous in a, b of appropriate weight. The discriminant does not change because the roots of the polynomial in ξ are translations of the roots of the polynomial in x. Then

$$D = 4\gamma_2^3 - 27\gamma_3^2.$$

The translation with $(b-a)/3$ introduces a small denominator. One may avoid this denominator by using the curve

$$y^2 = x(x-3a)(x+3b),$$

so that γ_2, γ_3 then come out to be integers, and one can apply the generalized Szpiro conjecture to the discriminant, which then has an extra factor

$$D = 3^6(abc)^2 = 4\gamma_3^3 - 27\gamma_3^2.$$

The Szpiro conjecture implies asymptotic Fermat.

Indeed, suppose that

$$a = u^n, \quad b = v^n, \quad \text{and} \quad c = w^n$$

with relatively prime u, v, w. Then

$$4\gamma_2^3 - 27\gamma_3^2 = 3^6(uvw)^{2n},$$

and we get a bound on n from the Szpiro conjecture

$$|D| \ll N_0(D)^{6+\varepsilon}.$$

Of course any exponent would do, e.g. $|D| \ll N_0(D)^{100}$ for asymptotic Fermat.

We have already seen that the *abc* conjecture implies generalized Szpiro.

Conversely, generalized Szpiro implies abc.

Indeed, the correspondence

$$(a, b) \leftrightarrow (\gamma_2, \gamma_3)$$

is invertible, and has the "right" weight. A simple algebraic manipulation left to the reader shows that the generalized Szpiro estimates on γ_2, γ_3 imply the desired estimates on $|a|, |b|$. (I found this manipulation a good assignment for my undergraduate algebra class.)

From the equivalence between *abc* and generalized Szpiro, one can use the examples given at the beginning to show that the epsilon is needed in the Szpiro conjecture.

Hall made his conjecture in 1971, actually without the epsilon so it had to be adjusted later. The final setting of the proofs in the simple *abc* context which we gave above had to await Mason and the *abc* conjecture a decade later.

Let us return to the polynomial case and Mason's theorem. The proofs that the *abc* conjecture implies the other conjectures apply as well in this case, so the analog use of Hall, Szpiro and Lang-Waldschmidt are also proved in the polynomial case. Actually, it had already been conjectured in [BCHS] that if f, g are non-zero polynomials such that $f^3 - g^2 \neq 0$ then

$$\deg(f(t)^3 - g(t)^2) \geqq \tfrac{1}{2}\deg f(t) + 1.$$

This (and its analogue for higher degrees) was proved by Davenport [Dav] in 1965, but we now see it as a consequence of Mason's theorem. Both in the case of Hall's conjecture for integers and Davenport's theorem, the point is to determine what lower bound can occur in a difference between a cube and a square, in the simplest case. The result for polynomials is particularly clear since, unlike the case of integers, there is no extraneous undetermined constant $C(\varepsilon)$ floating around, and there is even $+1$ on the right-hand side.

The polynomial case as in Davenport and the Hall conjecture for integers are of course not independent. Examples in the polynomial case parametrize cases with integers when we substitute integers for the variable. Examples are given in [BCHS], one of them due to Birch being:

$$f(t) = t^6 + 4t^4 + 10t^2 + 6 \quad \text{and} \quad g(t) = t^9 + 6t^7 + 21t^5 + 35t^3 + \tfrac{63}{2}t,$$

whence

$$\text{degree}(f(t)^3 - g(t)^2) = \tfrac{1}{2}\deg f + 1 .$$

This example shows that Davenport's inequality is best possible, because the degree attains the lowest possible value permissible under his theorem. Substituting large integral values of $t \equiv 2 \bmod 4$ gives examples of similarly low values for $x^3 - y^2$. A fairly general construction is given by Danilov [Dan]. See also the discussion of similar questions relevant to the size of integral points on elliptic curves in [La2], for instance Conjecture 5.

For those who know the theory of function fields in one variable, it is clear that the abc-property can also be formulated in that case, and can also be proved in that case. In fact, historically it was done that way as in [Ma3]. However, we are interested in algebraic number fields, so we shall now go to the next level of exposition and discuss heights of points in number fields.

So far we have proved the implications and equivalences relating three corners of the following diagram.

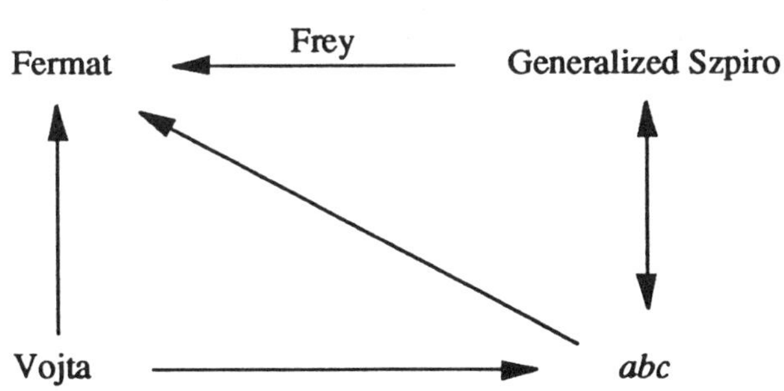

Our purpose is to fill in the Vojta corner, and show how the conjectures follow from the Vojta conjecture. One of the crucial points here is that we cannot stay within the realm of rational numbers, we must look at algebraic numbers. It will be seen how diophantine properties of a *family of curves* over the rational numbers depends on the diophantine properties of a *single curve* but uniformly for solutions in finite extensions of the rationals of bounded degree.

We want to estimate how big solutions of diophantine equations can be. What does "big" mean? Let $x = c/d$ be a rational number expressed in lowest form with relatively prime integers c, d. We define the **height** of x to be

$$h(x) = \log \max(|c|, |d|).$$

Similarly, if $P = (x_0, \dots, x_M)$ is a point in projective space with integer coordinates x_j which are relatively prime, then we define the **height**

$$h(P) = \log \max |x_j|.$$

The abc conjecture thus bounds the height of the point (a, b, c), with relatively prime integers a, b, c which can be seen as representing a point in projective 2-space.

The next two sections which carry out the above program are independent of the final three sections which give applications of the abc conjecture to elliptic curves, via the Szpiro conjecture. The reader may therefore read these in any order.[2]

[2] **Remark for those who know or want to find out about modular curves.** I don't want to discuss modular curves here, I want to confine the discussion to diophantine inequalities. But for those interested, it may be useful to make the following comments, for which we assume that the reader will either know the required definitions or look them up elsewhere.

Frey started the recent train of thoughts concerning Fermat and elliptic curves by pointing out that the elliptic curve associated with a solution of Fermat would have remarkable properties which should contradict the Taniyama-Shimura conjecture that all elliptic curves over the rational are "modular" [Fr]. Frey's idea was to show that the elliptic curve then could not exist, whence Fermat follows. There were serious difficulties in realizing this idea. Serre [Se] pointed out that one needed apparently more than the Taniyama-Shimura conjecture, for instance another conjecture which he had made, concerning the modularity of Galois representations over the rationals. Then Ribet [Ri] proved enough of the Serre conjecture in the modular case to show that the Taniyama-Shimura conjecture suffices for the Fermat application. We do not go into this modular aspect here. On the contrary, in the present discussion, we are pointing in a different direction, namely the direction of diophantine analysis and diophantine inequalities. The Szpiro conjecture can be viewed as lying just in the middle, and is susceptible of being handled or proved in either direction. Thus the abc conjecture has a modular interpretation via the generalized Szpiro conjecture to which it is equivalent. I must also point out that the modular route to Fermat via Ribet-Serre-Taniyama-Shimura would prove Fermat unconditionally. The route via the diophantine inequalities depends

2. The height for algebraic points. In this section we discuss algebraic numbers, and we describe how to define the height for points with algebraic coordinates, which will be used in §3. After that we return to the rational numbers, but we shall use p-adic numbers also.

Let F be a number field, i.e. a finite extension of the rational numbers $\mathbf{Q}$. Then F has a family of absolute values, p-adic and archimedean (at infinity, as one says). We shall now describe these absolute values.

For each prime p there is the **p-adic absolute value** on $\mathbf{Q}$, defined on a rational number $a = p^r c/d$ with $(c, d) = 1$, $p \nmid cd$, by

$$|a|_p = 1/p^r .$$

This p-adic absolute value has a finite number of extensions to F, which are called **p-adic** also. Let v be one of these, extending v_p on $\mathbf{Q}$. Let F_v be the completion of F at v. Then F_v is a finite extension of the field of p-adic numbers $\mathbf{Q}_p$. The absolute value on $\mathbf{Q}_p$ extends uniquely to F_v. Each v is induced by an imbedding of F into the algebraic closure of $\mathbf{Q}_p$:

$$\sigma_v : F \to \mathbf{Q}_p^{\mathrm{a}}$$

and v is induced by the absolute value on $\mathbf{Q}_p^{\mathrm{a}}$. Conversely, given such an imbedding $\sigma : F \to \mathbf{Q}_p^a$ we let v_σ be the induced absolute value.

The rational numbers also have the ordinary absolute value, extending to the real numbers, and denoted by v_∞. An extension of v_∞ to F is said to be **at infinity**, and there is a finite number of such extension. Let v extend v_∞ to F. Then v is induced by an imbedding

$$\sigma_v : F \to \mathbf{R}^{\mathrm{a}} = \mathbf{C}$$

of F into the complex numbers. Thus the situations at infinity and at the ordinary primes are entirely similar. In each case, imbeddings of F into $\mathbf{Q}_v^{\mathrm{a}}$ which differ by an isomorphism of F_v over $\mathbf{Q}_v$ give rise to the same absolute value on F, and conversely. In the case of absolute values at infinity, a pair of complex conjugate imbeddings of F into $\mathbf{C}$ corresponds to an extension of the ordinary absolute value from $\mathbf{Q}$ to F.

---

on the constants in the estimates for these inequalities, and will be as effective as these constants can be made to be effective.

Now consider a point $P = (x_0, \dots, x_n)$ in projective space $\mathbf{P}^n$, with coordinates in F so $x_j \in F$ and not all $x_j = 0$. We define the **height** of P by the formula

$$h(P) = \frac{1}{[F:\mathbf{Q}]} \sum_v [F_v : \mathbf{Q}_v] \log \max_j |x_j|_v$$

where $[F : \mathbf{Q}]$ denotes the degree of the extension, and where the sum is taken over all the above described absolute values v on F. The Artin-Whaples product formula (sum formula in our case) asserts that for $a \in F$, $a \neq 0$ we have

$$\sum_v [F_v : \mathbf{Q}_v] \log |a|_v = 0,$$

and so the height indeed depends only on the point in projective space. An elementary property of the absolute values implies that the height is independent of the field F in which the coordinates $x_0, \dots, x_n$ lie: this is the reason for the normalizing factor $1/[F:\mathbf{Q}]$ in front of the formula defining the height.

Note that if $F = \mathbf{Q}$ and $x_0, \dots, x_n$ are relatively prime integers, then

$$h(P) = \log \max |x_j|,$$

where the absolute value here is the ordinary one on the rational numbers.

We shall need another notion of algebraic number theory. The field F has a subring R, the ring of algebraic integers which are those elements of F satisfying an equation

$$T^n + a_{n-1}T^{n-1} + \cdots + a_0 = 0$$

whose coefficients $a_0, \dots, a_{n-1}$ lie in the ordinary integers $\mathbf{Z}$. This ring R has a basis $\{w_1, \dots, w_N\}$ over $\mathbf{Z}$, where $N = [F : \mathbf{Q}]$. If σ_j $(j = 1, \dots, N)$ ranges over all imbeddings of F into $\mathbf{C}$, then the **logarithmic discriminant** $d(F)$ is defined by

$$\begin{aligned} d(F) &= \frac{1}{[F:\mathbf{Q}]} \log |\det \sigma_i w_j|^2 \\ &= \frac{1}{[F:\mathbf{Q}]} \log |\text{discriminant of } F \text{ over } \mathbf{Q}|. \end{aligned}$$

It is easy to prove that if F_1, F_2 are number fields, then

$$d(F_1F_2) \leqq d(F_1) + d(F_2).$$

Also if $F_1 \subset F_2$ then $d(F_1) \leqq d(F_2)$.

Given a point P in projective space, we let

$$d(P) = d(F(P)).$$

Furthermore, the discriminant of the field $F(P)$ is insensitive to the higher power of primes dividing the coordinates of the point.

EXAMPLE. Suppose $x = a^{1/n}$ where a is a positive integer. Let

$$a = p_1^{\nu_1} \cdots p_r^{\nu_r}$$

be the factorization with $\nu_i \geqq 1$ and with distinct primes $p_1, \ldots, p_r$. Then

$$\log N_0(a) = \sum \log p_i.$$

Furthermore

$$\mathbf{Q}(a^{1/n}) \subset \mathbf{Q}(p_1^{1/n}, \ldots, p_r^{1/n}).$$

If α is a root of $\alpha^n = p$, then the discriminant is the norm of $n\alpha^{n-1}$, so the discriminant is bounded by $n^n p^{n-1}$. Hence

$$d(\mathbf{Q}(\alpha)) \leqq \log n + \frac{n-1}{n} \log p.$$

We apply this estimate to each prime dividing a to get

$$d(\mathbf{Q}(x)) \leqq r \log n + \frac{n-1}{n} \log N_0(a).$$

REMARK. For $N_0(a) \to \infty$ we have

$$r = o(\log N_0(a)).$$

Indeed, if r is bounded this is trivial. If r is unbounded, then $p_1 \cdots p_r \geqq r!$, so our assertion follows from Sterling's formula that $r! \geqq r^r e^{-r}$.

So far we have dealt with notions of algebraic number theory. We now pass to the height in the context of algebraic geometry, in the case of curves. We must assume that the reader is acquainted with the basic notion of an algebraic curve, imbeddable in projective space. Such a curve X will always be assumed irreducible. It is defined by a system of homogeneous polynomial equations when it is so imbedded, and is said to be **defined over** F if the coefficients of these equations lie in F. A **divisor** on X is a formal linear combination of points, with integer coefficients. The **degree** of a divisor is defined to be the sum of these coefficients. The curve X is covered by affine pieces, and a typical affine piece may be defined by one equation $f(x, y) = 0$, for instance. A point will usually be taken with coordinates x, y which lie in some number field, i.e. we deal mostly with algebraic points. The set of points with coordinates in a field E is denote by $X(E)$. Then $X(\mathbf{C})$ is the set of complex points, and if the curve is nonsingular, then $X(\mathbf{C})$ is a compact Riemann surface. The genus g of X may be defined as the genus of this surface, but there are also algebraic definitions. For instance, if X is defined by one homogeneous irreducible equation

$$H(T_0, T_1, T_2) = 0$$

in the projective plane, if X is nonsingular, and H has degree d, then the genus of X is $(d-1)(d-2)/2$.

The genus is also equal to the dimension of the space of regular differential forms. Say over the complex numbers, let ydx be a meromorphic differential form on X where y, x are rational functions on X. Let P be a complex point. Let t be a local uniformizing parameter at P. We write

$$ydx = y(t)\frac{dx}{dt}dt,$$

where y, x are expressible as power series in t. Then we define

$$\mathrm{ord}_P ydx = \text{order of the power series } y(t)\frac{dx}{dt}.$$

We can associate a **divisor to the differential form** ydx by letting

$$(ydx) = \sum_P \mathrm{ord}_P(ydx)(P).$$

Then the degree of this form satisfies

$$\deg(ydx) = \sum_P \mathrm{ord}_P(ydx) = 2g - 2.$$

If f is a rational function on X, then one can also associate a **divisor** (f) to f, namely at the point P we define $\mathrm{ord}_P f =$ order of the power series $f(t)$, in terms of the local parameter t. Then $\deg(f) = 0$ (which corresponds to the product formula). A divisor is said to be **rationally equivalent to** 0 if it is the divisor of a rational function. Equivalence among divisors will always refer to rational equivalence in what follows. In light of the relation $\deg(f) = 0$ we see that the degree function is defined on equivalence classes. The class of divisors of rational differential forms is called the **canonical class**.

Let X be a projective nonsingular curve defined over a number field. To each divisor D on X one can associate a function, the

height,

$$h_D \colon X(\mathbf{Q}^{\mathrm{a}}) \to \mathbf{R}$$

from the set of algebraic points into the reals, satisfying the following properties, and uniquely determined by them modulo bounded functions:

1. $D \mapsto h_D$ is well defined mod $O(1)$ on the rational equivalence class of D and is a homomorphism, i.e. is additive in D.
2. If D is a hyperplane section in a projective imbedding, then $h_D(P)$ is the height of a point P in that projective imbedding.

The uniqueness follows because given a divisor D there exists two projective imbeddings of X with hyperplane sections H_1 and H_2 respectively such that D is equivalent to $H_1 - H_2$ (this is a basic lemma of algebraic geometry).

3. The Vojta conjecture. The conjectures expressed by Vojta [Vo] are basic to the subject. I give here only one of the conjectures having to do with curves.

> **Vojta conjecture** Let X be a projective nonsingular curve defined over a number field. Let K be the canonical class of X. Then given $\varepsilon > 0$,
>
> $$h_K(P) \leqq (1+\varepsilon)d(P) + O_\varepsilon(1) \quad \text{for } P \in X(\mathbf{Q}^{\mathrm{a}}).$$

In the first place, observe that for a curve of genus $\geqq 2$, the Vojta conjecture immediately implies that the set of rational points $X(F)$ is finite (Mordell conjecture—Faltings theorem). Indeed, in that case $d(P)$ is constant, so the height of such points is bounded, and it is easy to show that there is only a finite number of points of bounded degree and bounded height.

We shall now see how the Vojta conjecture implies asymptotic Fermat and also implies the *abc* conjecture. Note that we shall use the Vojta conjecture only for points P of bounded degree over F as in [Vo], *Appendix ABC* p. 84.

Suppose we want to prove Vojta implies asymptotic Fermat. Let

$$X_n \colon x^n + y^n = z^n$$

be the Fermat curve and consider X_4 which has genus 3. For a hypersurface of degree d in $\mathbf{P}^m$ the canonical class is that of $(d-(m+1)))H$, where H is a hyperplane. So for X_n in $\mathbf{P}^2$ the canonical class is that of $(n-3)H$, and for X_4 the canonical class is

$$K = H.$$

Let $u^n + v^n = w^n$ in relatively prime integers. Associate the point

$$P = (x, y, z) = (u^{n/4} : v^{n/4} : w^{n/4}) \quad \text{on } X_4.$$

From the definition of the height, it is immediate that

$$h(P) = \frac{n}{4} \log \max(|u|, |v|, |w|).$$

By Vojta's conjecture we get

$$\frac{n}{4} \log \max |u|, |v|, |w| \ll \log N_0(uvw) + O(1),$$

which gives a bound for n.

Similarly to prove the *abc* conjecture from Vojta, suppose $a + b = c$. Fix n. Associate the point

$$P = (a^{1/n} : b^{1/n} : c^{1/n}) \quad \text{on } X_n.$$

Since $K_n = (n-3)H$ we get by Vojta's conjecture

$$h_K(P) = \frac{n-3}{n} \log \max(|a|, |b|, |c|) \leqq (1+\varepsilon_1) \log N_0(abc) + O_n(1).$$

This IS the log of the *abc* inequality. We let $n \to \infty$ with $\varepsilon = \varepsilon_1 + 4/n$ or whatever.

We shall conclude with sections on elliptic curves, showing how the *abc* conjecture implies diophantine properties of such curves over number fields.

4. Elliptic curves and minimal equations. An elliptic curve for our purposes will be a curve which can be represented by an equation in Weierstrass form

$$y^2 = x^3 - \gamma_2 x - \gamma_3,$$

with coefficients γ_2, γ_3 in some field F. Let A denote the curve. The set of points with coordinates $x, y \in F$ together with the point at infinity is denoted by $A(F)$, and is a group. As before, we shall take F to be mostly a subfield of the complex numbers, in which case $A(F)$ is a subgroup of $A(\mathbf{C})$. Over $\mathbf{C}$, the curve is parametrized by the Weierstrass functions

$$z \mapsto (\wp(z), \tfrac{1}{2}\wp'(z)),$$

giving an analytic isomorphism $\mathbf{C}/\Lambda \to A(\mathbf{C})$ where Λ is a lattice. The lattice points go to the point at infinity under this parametrization. The addition formula for the $\wp$-function defines the group law on $A(\mathbf{C})$, and this addition formula is given by rational functions on the coordinates (x, y), with coefficients in $\mathbf{Q}$, which is why $A(F)$ is a subgroup.

As before we have the **discriminants**

$$4\gamma_2^3 - 27\gamma_3^2 = D \quad \text{and} \quad \Delta = 16D.$$

For any number $c \neq 0$ we can change the elliptic curve by an isomorphism, which under the Weierstrass parametrization maps $z \mapsto cz$. Then the representation of this isomorphism on the equation has the effect:

$$x \mapsto x' = c^{-2}x, \qquad y \mapsto y' = c^{-3}y,$$
$$\gamma_2 \mapsto \gamma_2' = c^{-4}\gamma_2, \qquad \gamma_3 \mapsto \gamma_3' = c^{-6}\gamma_3,$$

so that the isomorphic curve satisfies the equation

$$y'^2 = x'^3 - \gamma_2'x' - \gamma_3'.$$

Suppose the elliptic curve is defined over $\mathbf{Q}$, that is $\gamma_2, \gamma_3 \in \mathbf{Q}$. By an appropriate such isomorphism, we can make $\gamma_2, \gamma_3 \in \mathbf{Z}$. Suppose p is a prime such that p^4 divides γ_2 and p^6 divides γ_3. Then we can change the elliptic curve by an isomorphism, letting $\gamma_2 \mapsto p^{-4}\gamma_2$ and $\gamma_3 \mapsto p^{-6}\gamma_3$. After having done so repeatedly until no further possible, we obtain what is a **minimal model** for the curve over $\mathbf{Z}$, and then Δ is called a **minimal discriminant**.

REMARK. The minimal discriminant is defined up to a factor of ± 1, and is an invariant of an isomorphism class of elliptic curves over $\mathbf{Q}$. We have slightly over simplified the situation in two ways. First by tying ourselves down to the Weierstrass model, we don't quite get the "right" notion for this minimal discriminant, because of the primes 2 and 3. One has to give a more general equation, first studied by Deuring, and carried out systematically in more recent times by Neron and Tate. Also to be absolutely correct, to take care of the primes 2 and 3, we really should consider a more general form of the Weierstrass model, namely:

$$y^2 + a_1xy + a_3y = x^3 + a_2x^2 + a_4x + a_6.$$

One can obtain a minimal model, i.e. one with minimal discriminant, just as in the standard case of the Weierstrass equation. The following statements refer to such a model, but the reader may think of $p \neq 2, 3$ and of the usual Weierstrass equation without harm.

The discussion of isomorphisms above contained the essential aspects of the minimal discriminant. Furthermore, by dealing over the rationals, we avoid all the problems which come from non-unique factorization in number fields, and the existence of a non-trivial group of units in the ring of integers of the number field. A reader acquainted with elementary algebraic number theory will then see that in case of a nontrivial ideal class group, there is no unique minimal model of the curve with minimal discriminant, but there is a finite number of models, with relatively minimal discriminants. These are secondary considerations for what we are doing here, where we want to see clearly the effect of the *abc* conjecture on the diophantine aspects of the curve. These essential aspects are all present for curves over the rationals.

We shall want to apply the Szpiro conjecture, but we must take into account that γ_2, γ_3 may not be relatively prime. We still want to see that for a minimal model, we have $|\Delta| \ll N_0(\Delta)^s$ for some s independent of A. Indeed, by the minimality assumption, if p^m is a common factor of γ_2, γ_3 then $m \leqq 5$. Therefore, if $d = (\gamma_2, \gamma_3)$ is the greatest common divisor, then

$$d \leqq N_0(d)^5.$$

By factoring out a common factor, canceling, and small algebraic manipulations, it is easy to see that for a minimal equation, by the *abc* conjecture we have

$$|\Delta| \ll N_0(\Delta)^{6+\varepsilon} d^4$$

and therefore

$$|\Delta| \ll N_0(\Delta)^s$$

for some low integer s which we may call the **Szpiro exponent**. Without loss of generality, we may then assume that

$$|\Delta| \leqq N_0(\Delta)^s$$

except for a finite number of Δ, and so except for a finite number of elliptic curves. We shall omit these exceptional curves, and *assume in the sequel that* $|\Delta| \leqq N_0(\Delta)^s$.

5. The Szpiro conjecture and torsion points. Let F be a number field, that is a finite extension of the rational numbers. The

Mordell-Weil theorem asserts that the group of rational points $A(F)$ is finitely generated. (Mordell originally proved the theorem over the rationals.) It is a major problem to describe the order of the torsion group and the rank of $A(F)$. We are here concerned with the torsion group, which is finite by the Mordell-Weil theorem. A standard conjecture states:

> *Given the number field F, there exists a positive number C such that for every elliptic curve A defined over F, the order of the torsion group $A(F)_{\text{tor}}$ is bounded by C.*

Over the rational numbers, this was proved in a very strong form by Mazur [Maz], who showed that the order of the torsion group is bounded by 16. For this purpose, Mazur developed a whole theory on modular curves. Here we are concerned with a statement which is weaker in the sense that no specific bound like 16, is given, but stronger in that we want the statement for any number field. Results for number fields have been obtained by Kubert [Ku]. In this section, we show that the *abc* conjecture for number fields implies the uniform boundedness of torsion as stated above by an argument due to Frey which he wrote me in 1986.

For simplicity, we shall work over the rational numbers. A reader acquainted with the basic properties of elliptic curves and number fields will immediately see that the arguments generalize.

As usual we define the isomorphism invariant

$$j = 3^3 4^6 \gamma_2^3 / \Delta .$$

REMARK. The primes which divide the denominator of j play a special role. These primes must also divide Δ, but the converse need not hold. We shall have to use a relatively deep fact:

> *If there exists a rational point in $A(\mathbf{Q})$ of prime order $n \geqq 5$, then γ_2 and γ_3 are relatively prime except for small powers of 2 and 3, and also possibly n itself.*

This fact is hard to prove, and we make some comments for those somewhat acquainted with elliptic curves, or those who may wish to pursue this matter further. Let p be a prime $\geqq 5$. One may reduce the minimal equation of an elliptic curve $\bmod p$. Then the reduced equation defines a possibly reducible curve, called the **fiber**. The group law on the original elliptic curve reduces to a group law on the set of nonsingular points on the fiber, giving rise to the theory of the Néron model. Cf. [Ne], the last section on elliptic curves, and the last table, as well as Artin's exposition (somewhat sketchy) of Néron models [Ar], especially Proposition 1.15. Now suppose that j is p-integral. If p is not a common factor of γ_2 and γ_3 then we say that the elliptic curve is **semistable** at p. If p is a common factor, then the curve reduces $\bmod p$ to $y^2 = x^3$. Let us look at the curve over the p-adic field $\mathbf{Q}_p$. By the theory of Néron models, it can be shown that the nonsingular part of the fiber is an algebraic group, which contains the additive group as a subgroup of index at most 4, and that the kernel of reduction is a p-adic Lie group, which does not contain points of finite order. From this structure, we conclude that the curve is semistable at p when there exists a rational point of prime order $n \geqq 5$, except possibly when $p = n$, because all points on the additive group in characteristic p have order p. For a generalization and examples, see Lenstra-Oort [L-O], especially §3 for examples of large p-torsion p-adically.

Admitting the above fact, we see that except possibly for 2, 3, and n, the primes dividing the denominator of j are precisely the primes dividing Δ.

To prove a theorem about torsion points over the rationals, we must have a model for them which will contain enough algebraic information, even though it may involve some analysis. The classical parametrization $\mathbf{C}/\Lambda \to A(\mathbf{C})$ by the Weierstrass functions is not good enough for this purpose. However, we may take the lattice to have a $\mathbf{Z}$-basis $\{\tau, 1\}$ where τ is in the upper half plane: $\mathrm{Im}(\tau) > 0$. Then we put

$$q_\tau = q = e^{2\pi i\tau}.$$

We can then also represent $A(\mathbf{C})$ as a quotient of the multiplicative group as follows. We have the Fourier series expansion, power series in q, in all standard texts, including [La1] and [La3]:

$$(2\pi i)^{-4}\gamma_2 = \frac{1}{48}\left[1 + 240\sum_{n=1}^{\infty}\frac{n^3q^n}{1-q^n}\right]$$

$$(2\pi i)^{-6}\gamma_3 = \frac{1}{2^5 3^3}\left[-1 + 504\sum_{n=1}^{\infty}\frac{n^5q^n}{1-q^n}\right]$$

$$(2\pi i)^{-12}\Delta = q\prod_{n=1}^{\infty}(1-q^n)^{24}$$

$$j = \frac{1}{q} + 744 + 196884q + \cdots$$

and for a variable $t \in \mathbf{C}^*$:

$$(2\pi i)^{-2}x(t) = \frac{1}{12}\sum_{m\in\mathbf{Z}}\frac{q^m t}{(1-q^m t)^2} - 2\sum_{n=1}^{\infty}\frac{nq^n}{1-q^n} = X(t)$$

$$(2\pi i)^{-3}y(t) = \frac{1}{2}\sum_{m\in\mathbf{Z}}\frac{q^m t(1+q^m t)}{(1-q^m t)^3} = Y(t).$$

We get an analytic isomorphism

$$\mathbf{C}^*/q^{\mathbf{Z}} \xrightarrow{\approx} A(\mathbf{C}) \quad \text{by} \quad t \mapsto (x(t), y(t)).$$

The power series expansions for $X(t)$ and $Y(t)$ have essentially integral coefficients, and as Tate remarked in the late fifties, they can be used to parametrize an elliptic curve in the p-adic domain, where they converge provided that

$$|q| < 1 \text{ or equivalently } p \text{ divides the denominator of } j.$$

This elliptic curve, called the **Tate curve**, is isomorphic to the given curve over a quadratic extension. For simplicity, we shall argue as if this isomorphism is over $\mathbf{Q}_p$ itself.

As with the complex numbers, the absolute value on $\mathbf{Q}_p$ extends uniquely to the algebraic closure $\mathbf{Q}_p^{\mathrm{a}}$ and to the completion of the algebraic closure, which we denote by $\mathbf{C}_p$, and which plays the role of $\mathbf{C}$.

In the p-adic field, using the notation $u \sim v$ to mean that u/v is a p-adic unit, we see that

$$j \sim \frac{1}{q} \quad \text{and} \quad \Delta \sim q.$$

Except for the primes 2 and 3, γ_2 and γ_3 are then p-adic units. Tate normalizes the equation and the power series further to get rid of denominators involving 2 and 3 completely, but we don't go into this here.

Thus $t \mapsto (X(t), Y(t))$ gives a homomorphism

$$\mathbf{C}_p^* \to A(\mathbf{C}_p) \quad \text{inducing} \quad \mathbf{Q}_p^* \to A(\mathbf{Q}_p)$$

from the multiplicative group of the field to the group of points of A in the field. The kernel is the infinite cyclic group $q^{\mathbf{Z}}$. Similarly, if F is a finite extension of $\mathbf{Q}_p$ in $\mathbf{C}_p$ we have a parametrization $F^*/q^{\mathbf{Z}} \to A(F)$. We then obtain a model for the torsion points of A in $A(\mathbf{C}_p)$. The points of order n are parametrized by the group generated by

$$q^{1/n}, \zeta_n \mod q^{\mathbf{Z}}$$

where $q^{1/n}$ is any one of the nth roots of q, well defined modulo an nth root of unity; and ζ_n is a primitive nth root of unity.

We shall first prove that there is only a finite number of primes n such that an elliptic curve over $\mathbf{Q}$ has a rational point of order n, by showing that the minimal discriminants of such curves are divisible by only a finite number of primes.

Suppose that the elliptic curve A over the rationals has a point $P \in A(\mathbf{Q})$ of order precisely n with n prime. For each prime number p dividing the denominator of j, and thus dividing Δ, this point is represented by $q^{1/n}$ or ζ_n. If P corresponds to $q^{1/n}$, since $P \in A(\mathbf{Q}_p)$ it follows that $q^{1/n} \in \mathbf{Q}_p$, so $p^n | q = q_A$. Suppose on the other hand that P corresponds to ζ_n for some primitive nth root of unity. Let (P) be the cyclic group generated by P, and let

$$B = A/(P)$$

be the quotient elliptic curve, which is then also defined over $\mathbf{Q}$. Then we have the minimal discriminants Δ_A and Δ_B, and the two parameters q_A and q_B coming from the Tate parametrization p-adically. In the present case, we claim that $q_B = q_A^n$. Indeed, the system here works just as in the case of a lattice. The map raising to the nth power gives an isomorphism

$$\mathbf{C}_p^*/(q, \zeta_n) \xrightarrow{\approx} \mathbf{C}_p^{*n}/(q^n) = \mathbf{C}_p^*/(q^n).$$

Thus the "period" group of $\mathbf{C}_p^*/(q, \zeta_n)$ is generated by q^n, in case P corresponds to ζ_n. (See the Remark below.) Therefore $q_B = q_A^n$. It follows that in this case, $p^n | q_B$. Hence in both cases,

$$p^n \text{ divides } q_A q_B.$$

In terms of the minimal discriminants, this implies that

$$p^n \text{ divides } \Delta_A \Delta_B.$$

But it is known that Δ_A and Δ_B are divisible by the same primes. By the Szpiro conjecture applied to Δ_A and Δ_B separately, we get

$$|\Delta_A \Delta_B| \ll N_0(\Delta_A)^s N_0(\Delta_B)^s = N_0(\Delta)^{2s}.$$

If $p_1, \dots, p_r$ are the primes dividing Δ and $\neq n, 2, 3$, then we get

$$(p_1 \cdots p_r)^n \ll (p_1 \cdots p_r)^{2s_n 2s},$$

which gives a bound on n as effective as the constant in Szpiro's conjecture, unless there are no primes $p_1, \dots, p_r$. But in that case, Δ is divisible only by n (or 2 and 3). Then for $n > 5$, the curve has good reduction modulo 5, say, and reduction mod 5 induces an isomorphism on the points of order n, which is impossible for n sufficiently large since the cardinality of $A(\mathbf{F}_5)$ is bounded.

It follows that the discriminants of minimal models are divisible by only a finite number of primes. By what we saw at the end of §4, this implies that there is only a finite number of values Δ_A for minimal models A. It remains to be shown that for fixed Δ the equation

$$\Delta = 4\gamma_2^3 - 27\gamma_3^2$$

has only a finite number of solutions γ_2, γ_3 giving minimal models. This is a known theorem, but for our purposes we can deduce it from the Szpiro conjecture. Indeed, the powers of primes occurring as common factors of γ_2, γ_3 are bounded by the minimality condition. Hence the g.c.d. of γ_2, γ_3 is bounded for all elliptic curves with minimal equation over $\mathbf{Q}$. By the generalized Szpiro conjecture, it follows that $|\gamma_2|$ and $|\gamma_3|$ are also bounded, thus proving the desired implication.

REMARK. Concerning the assertion made that $q_B = q_A^n$ when $B = A/(P)$ and P corresponds to the nth roots of unity, the reader should think first in terms of the complex case. Suppose the lattice of the elliptic curve is $[\tau, 1]$ (i.e. the group generated by $\tau, 1$ over $\mathbf{Z}$), so $A(\mathbf{C}) \approx \mathbf{C}/[\tau, 1]$. Let P correspond to $1/n$. Then

$$A(\mathbf{C})/(P) \approx C/[\tau, 1/n] \approx \mathbf{C}/[n\tau, 1],$$

where the second isomorphism is induced by multiplication $n: \mathbf{C} \to \mathbf{C}$. Exponentiating, we see in the complex case that the "q" corresponding to $A/(P)$ is $e^{2\pi i n\tau}$. Now the reader must accept that essentially the same theory of parametrization holds in the p-adic domain, so the argument works the same way, as given in the above proof.

The same method should prove a more general conjecture as follows.

> *Let F be a number field. There exists a positive number C having the following property. If A is an elliptic curve over F, without complex multiplication, and if there exists a cyclic subgroup of order n, invariant under the Galois group over F, then $n \leqq C$.*

The cyclic subgroup is generated by one point P of order n, and the hypothesis means that the extension $F(P)$ is Galois, with group G such that for $\sigma \in G$ we have

$$\sigma P = \chi(\sigma) P,$$

with some element $\chi(\sigma) \in (\mathbf{Z}/n\mathbf{Z})^*$. Thus χ gives a representation of G as a subgroup of $(\mathbf{Z}/n\mathbf{Z})^*$. If n is a prime number, then in particular, the order of G is prime to n.

To follow the same pattern of proof, one needs to start with semistability. Assume for the moment that the curve is semistable, and let $F = \mathbf{Q}$ for simplicity as before. As before, we then get that a prime dividing Δ_A also divides the denominator of j.

In the first step of the proof, supposing that P corresponds to $q^{1/n}$ under the Tate parametrization, we note that either the polynomial $X^n - q$ is irreducible over $\mathbf{Q}_p$ or it has a root in $\mathbf{Q}_p$, by an elementary criterion of field theory. Therefore, if q is not an nth power in $\mathbf{Q}_p$ then

$$\mathbf{Q}_p(q^{1/n}) = \mathbf{Q}_p(P)$$

has degree n over $\mathbf{Q}_p$. Since the Galois group has order prime to n, this cannot happen and therefore q is an nth power in $\mathbf{Q}_p$. Thus we are in the same situation as before, and exactly the same arguments using the Szpiro conjecture show that n is bounded.

However, under the weaker hypothesis of the Galois invariant cyclic subgroup instead of a rational point of sufficiently high order, it is not clear how to reduce the general case to the semistable case, so at this time, the above arguments apply only to the family of semistable curves.

6. The height on elliptic curves. Let A be an elliptic curve over the rationals again, and let Δ_A be its minimal discriminant. We let $y^2 = x^3 - \gamma_2 x - \gamma_3$ be a minimal equation, with integer coefficients. A fundamental problem is to estimate the absolute values of x, y as function of γ_2 and γ_3 for integral solutions in $\mathbf{Z}$, and to estimate the height $h(x(P))$ for rational solutions. Since $A(\mathbf{Q})$ is finitely generated, $A(\mathbf{Q})$ modulo its torsion group is a free abelian group, finitely generated. It is a problem to give an upper bound for the heights of free generators for this group. For conjectures see [La2]. The height $h(x(P))$ has a great deal of structure. On the group $A(\mathbf{Q})$ modulo torsion, according to a fundamental theorem of Néron-Tate, there exists a positive definite quadratic form

$$h_A\colon A(\mathbf{Q})/A(\mathbf{Q})_{\mathrm{tor}} \to \mathbf{R}$$

such that

$$h_A(P) = \tfrac{1}{2}h(x(P)) + O(1).$$

In the next section we shall formulate a fundamental conjecture about this height. Here, we describe a number of basic properties which allow us to compute it, and which will be used to deal with that conjecture.

In the preceding section, we used certain explicit formulas parametrizing the functions $(\gamma_2, \gamma_3, \Delta, j, x, y)$. We now need similar formulas to express the height. It turns out that the height can be given as a sum

$$h_A(P) = \sum_v \lambda_v(P),$$

where λ_v is a function (the **Néron function**) given by an analytic expression on $A(\mathbf{Q}_v)$ for each absolute value v. We shall now describe these functions. As a matter of notation, for any absolute value v, we define

$$v(a) = -\log|a|_v.$$

For instance, if v is p-adic, then $v(p^m) = m\log p$, so $v(a)$ is the order m of a at p, times a normalizing factor $\log p$ which is used to get global formulas putting all the absolute values together. As a approaches 0 we want the p-adic order to approach ∞, and similarly for an absolute value at infinity.

The height h_A is a quadratic form, but the local functions λ_v cannot be quadratic: there has to be some extraneous term appearing in the quadratic relations locally, and only after taking the sum over all v does this term disappear. For elliptic curves, a neat characterization for the Néron functions was given by Tate as follows.

> *Let F be a p-adic field, or the complex numbers. Let A be an elliptic curve defined over F. There exists a unique function $\lambda_v\colon A(F) - \{0\} \to \mathbf{R}$ satisfying the following conditions:*
>
> (i) *λ_v is continuous and bounded outside every neighborhood of 0.*
>
> (ii) *Let z be a local uniformizing parameter at 0. Then there exists a bounded continuous function α on an open neighborhood of 0 such that for all P in that neighborhood, $P \neq 0$, we have*
>
> $$\lambda_v(P) = v(z(P)) + \alpha(P).$$
>
> (iii) For all $P, Q \in A(F)$ such that P, Q, $P \pm Q \neq 0$ we have
>
> $$\lambda_v(P+Q) + \lambda_v(P-Q) = 2\lambda_v(P) + 2\lambda_v(Q) + v(x(P) - x(Q)) - \tfrac{1}{6}v(\Delta).$$

Without the last two terms, the third relation would be the relation defining a quadratic function. The final constant involving $v(\Delta)$ is a conveniently normalized term. When applying the relation of (iii) globally, the sum over all v of the last two terms will vanish by the product formula, when $x(P)$, $x(Q)$, are rational numbers, say; so summing over all v yields a quadratic function on the group of rational points. It is not difficult to show that this function is the quadratic height.

We shall now give explicit formulas due to Tate for the Néron functions λ_v in order to be able to estimate the height from below later. In each case, it is not difficult to verify that the formulas we give satisfy the desired quadratic relation. The other conditions (i) and (ii) simply express that the Néron function has a logarithmic singularity at the origin, and these conditions are also immediately verified in each case. We shall omit the verification.

$\boldsymbol{v = v_\infty}$.

In dealing with torsion points on elliptic curves, we have already remarked that we have a complex analytic isomorphism

$$\mathbf{C}/[\tau, 1] \to A(\mathbf{C})$$

where τ lies in the upper half plane, and $[\tau, 1]$ is the lattice generated by τ, 1 over $\mathbf{Z}$. We can change τ by any element of $SL_2(\mathbf{Z})$, and in particular, we can take τ to be in the standard fundamental domain for $SL_2(\mathbf{Z})$. If we let

$$q = q_\tau = e^{2\pi i \tau},$$

then

$$(*) \qquad \operatorname{Im} \tau \geqq \tfrac{1}{2}\sqrt{3} \quad \text{and so} \quad |q_\tau| \leqq e^{-\pi\sqrt{3}}.$$

We let $u = u_1\tau + u_2 \in \mathbf{C}$ $(u_1, u_2 \in \mathbf{R})$, and we let

$$t = q_u = e^{2\pi i u}$$

be a variable in $\mathbf{C}^*$. We let

$$g_0(t) = g_0(q, t) = (1 - t)\prod_{n=1}^{\infty}(1 - q^n t)(1 - q^n/t).$$

Then we have a functional equation for g_0, namely

$$g_0(qt) = -t^{-1}g_0(t) = g_0(t^{-1}).$$

Let $\mathbf{B}_2$ be the second Bernoulli polynomial,

$$\mathbf{B}_2(T) = T^2 - T + \tfrac{1}{6}.$$

We define the **Néron function**

$$\lambda_v(u, \tau) = \tfrac{1}{2}\mathbf{B}_2(u_1)v(q) + v(g_0(q_u)) = \tfrac{1}{2}\mathbf{B}_2(u_1)v(q) + v(g_0(t)).$$

Immediately from the functional equation, we see that $\lambda_v(u, \tau)$ is even in u, that is

$$\lambda_v(-u, \tau) = \lambda_v(u, \tau).$$

The term involving $\mathbf{B}_2$ serves the purpose of making the function as we have defined it periodic, with periods $1, \tau$. This can be verified directly and simply from the functional equation for g_0. The function λ_v is real analytic, except for a logarithmic singularity at the lattice points, as one sees directly from its definition.

Proposition 6.1. *For $v = v_\infty$ there exists a constant $C_\infty > 0$ having the following property. Let $\operatorname{Im}\tau > \sqrt{3}/2$ and let $|u_1| \leqq 1/6$. Then*

$$\lambda_v(u, \tau) \geqq -C_\infty.$$

Proof. By the periodicity and the fact that $\lambda_v(u, \tau)$ is even in u, we can normalize a representative for a point by the condition

$$(**) \quad 0 \leqq u_1 \leqq \tfrac{1}{2} \text{ whence } |q^{1/2}| \leqq |q_u| \leqq 1 \quad \text{and} \quad |q^{1/2}/q_u| \leqq 1.$$

In that case, we conclude that the Néron function has the value

$$\lambda_v(u, \tau) = -\tfrac{1}{2}\mathbf{B}_2(u_1)\log|q| - \log|1 - q_u| - O(1),$$

and it is easy to compute an explicit value for $O(1)$, independent of u and τ. Namely, for $n \geqq 1$, we have estimates

$$|q^n q_u| \leqq e^{-n\pi\sqrt{3}/2}$$

and

$$|q^n / q_u| = |q^{n-1/2} q^{1/2} / q_u| \leqq |q^{n-1/2}| \leqq e^{-(n-1/2)\pi\sqrt{3}/2}.$$

These inequalities show that the term $-O(1)$ is independent of u and τ. In addition, the choice of $|u_1| \leqq 1/6$ was made so that $\mathbf{B}_2(u_1) \geqq 0$, and hence the term $-\mathbf{B}_2(u_1)\log|q|$ is actually $\geqq 0$ since $|q| \leqq 1$. Finally $|q_u| \leqq 1$, so

$$\log|1 - q_u| \leqq \log 2.$$

The uniform lower bound of Proposition 6.1 for $\lambda_v(u, \tau)$ follows at once from these estimates.

We have an analytic isomorphism

$$\mathbf{C}/[\tau, 1] \xrightarrow{\approx} A(\mathbf{C}) \text{ by } u \mapsto P_u.$$

We also write $u = u(P)$. The function $\lambda_v(u, \tau)$ being periodic in u, we use it to define the v-component of the height, namely for a rational point P corresponding to u we let

$$\lambda_v(P) = \lambda_v(u(P), \tau).$$

Proposition 6.1 can be restated in part by saying that if P is sufficiently close to the origin, then $\lambda_v(P)$ has a uniform lower bound as in Proposition 6.1.

We shall apply this to multiples of an arbitrary point P which lie close to the origin, thus giving us a lower bound for the component of the height at infinity.

Proposition 6.2. *Let C_∞ be the constant of Proposition 6.1. Let A be an elliptic curve over $\mathbf{Q}$, and let $P \in A(\mathbf{Q})$. Given an integer $M \geqq 1$, there exists an integer b satisfying*

$$1 \leqq b \leqq 6M$$

such that for $v = v_\infty$ we have

$$\lambda_v(mbP) \geqq -C_\infty \quad \text{for } 1 \leqq m \leqq M.$$

Proof. We get a homomorphism $A(\mathbf{Q}) \to \mathbf{R}/\mathbf{Z}$ by mapping

$$Q \mapsto u_1(Q) \mod \mathbf{Z}.$$

Let n be a suitably large integer. Partition $\mathbf{R}/\mathbf{Z}$ into n small intervals of length $1/n$. The multiples of the point P given by

$$0, P, 2P, \dots, nP$$

map to $n+1$ elements of $\mathbf{R}/\mathbf{Z}$. Hence there exist integers $0 \leqq n_1 < n_2 \leqq n$ such that $n_1 P$ and $n_2 P$ lie in the same small interval. Let $b = n_2 - n_1$. Then $0 < b \leqq n$ and bP lies in the small interval containing the origin. Then

$$|u_1(bP)| \leqq \frac{1}{n}$$

and therefore

$$|u_1(mbP)| \leqq \frac{M}{n} \quad \text{for } 1 \leqq m \leqq M.$$

Thus if we let $n = 6M$ we get $|u_1(mbP)| \leqq 1/6$, and we can apply Proposition 6.1 to conclude the proof.

Next we deal with the absolute values associated with prime numbers, and we have to distinguish cases, depending on whether j is p-integral or not.

$v = v_p$ for p prime, and j is p-integral.

Given a prime p, one may reduce the minimal equation of an elliptic curve mod p. Then the reduced equation defines a possibly reducible curve, called the fiber. The group law on the original elliptic curve reduces to a group law on the set of nonsingular points on the fiber.

Proposition 6.3. *Let $v = v_p$. Then for all points P whose reduction* mod p *is nonsingular on the fiber, the Néron function is defined by*

$$\lambda_v(P) = \frac{1}{2}\max\{0, \log|x(P)|_v\} + \frac{1}{12}v(\Delta) \geqq \frac{1}{12}v(\Delta).$$

This proposition is proved essentially by brute force, cf. for instance Theorem 4.4 and Theorem 6.1 of [La3, Chapter III] for the standard Weierstrass equation.

The following result is crucial to know how far a point is from having singular reduction on the fiber.

Proposition 6.4. *Assume that we are dealing with a minimal model, and that j is p-integral. Then*

(i) *For every point $P \in A(F)$ the point $12P$ has nonsingular reduction on the fiber.*

(ii) *Furthermore, $\lambda_v(12mP) \geqq 1/12v(\Delta)$ for all positive integers m.*

Proof. The proof of the first statement is by computation, using the Néron-Kodaira classification of all possible cases of degeneracy which can occur. See Néron's last chapter on elliptic curves ([Ne], the final table), and for more recent insights, Tate's algorithm [Ta2]. It is a very tedious matter. To replace such computations by theoretical arguments takes very heavy machinery. The second statement follows by applying Theorem 6.3.

For our purposes the factor 12 is not important, all that matters is that there is some universal integer which can be used to multiply a point and land it into the nonsingular part of the fiber.

$v = v_p$ with p prime, and j is not p-integral.

Finally we must give a description of the Néron function for v_p when p divides the denominator of j, i.e. j is not p-integral. This case is a p-adic analogue of the case over the complex numbers, with essentially the same formulas. We deal here with the Tate curve as in §4.

Let $\mathbf{C}_p$ be the completion of the algebraic closure of $\mathbf{Q}_p$ and let t be a variable in $\mathbf{C}_p^*$. Define

$$g_0(t) = (1-t)\prod_{n=1}^{\infty}(1-q^n t)(1-q^n/t).$$

The functional equation was formal, and so we have again

$$g_0(qt) = g_0(t^{-1}) = -t^{-1}g_0(t),$$

valid for all $t \in \mathbf{C}_p^*$. Define

$$u(t) = \frac{v(t)}{v(q)} = \frac{\mathrm{ord}_p t}{\mathrm{ord}_p q}.$$

Proposition 6.5. *Let p divide the denominator of j. Let P_t be the image of t in $A(\mathbf{C}_p)$ under the Tate parametrization. Then*

$$\lambda_v(P_t) = v(g_0(t)) + \tfrac{1}{2}\mathbf{B}_2(u(t))v(q).$$

Thus we see the analogy with the complex case. From the functional equation of g_0 we conclude again that $\lambda_v(P_t)$ is periodic with period q, and so is defined on the elliptic curve.

Now let F be a finite extension of $\mathbf{Q}_p$ in $\mathbf{C}_p$, and suppose $q \in F^*$. For every $t \in F^*$ we can find a representative $\bmod q^{\mathbf{Z}}$ which we denote by t_P, and which is uniquely determined by the condition

$$0 \leqq u(t) < 1 \text{ or equivalently } |q|_v < |t|_v \leqq 1 .$$

For such a representative we see from the formula for g_0 that $v(g_0(t_P)) \geqq 0$. It is therefore convenient to define the periodic function

$$\mathbf{B}: \mathbf{R}/\mathbf{Z} \to \mathbf{R} \quad \text{by} \quad \mathbf{B}(u) = \mathbf{B}_2(u) \text{ if } 0 \leqq u \leqq 1 ,$$

and $\mathbf{B}$ is extended by periodicity to all of $\mathbf{R}$. One can also write

$$\mathbf{B}(u) = \{u\}^2 - \{u\} + \tfrac{1}{6}$$

where $\{u\}$ is the fractional part of u, and we obtain:

Proposition 6.6. *Let p divide the denominator of j, and $v = v_p$. Then*

$$\lambda_v(P) \geqq \tfrac{1}{2}\mathbf{B}(u(t_P))v(q) .$$

Just as Proposition 6.4 gave us a simple criterion to get a nice formula for the height of points whose reduction is nonsingular on the fiber, we can formulate a similar criterion which we can apply to Proposition 6.6. Indeed, if $u(t_P) = 0$ then we get the simple inequality

$$\lambda_v(P) \geqq \tfrac{1}{12}v(q) = \tfrac{1}{12}v(\Delta)$$

just as in Proposition 6.3. Suppose however that P is an arbitrary point in $A(\mathbf{Q}_p)$. Let

$$b = \operatorname{ord}_p(q) = \operatorname{ord}_p(\Delta) .$$

Then from the definitions, $t_P^b \in q^{\mathbf{Z}}$, and therefore t_{bP} is a p-adic unit, and so $u(t_{bP}) = 0$. Combining the two cases of Proposition 6.4 and 6.6, we find:

Proposition 6.7. *Given a positive integer n_0 there exists an integer $b > 0$ having the following property. For all elliptic curves A over $\mathbf{Q}$, and non-torsion rational point $P \in A(\mathbf{Q})$, if $v = v_p$ is such* *that j is p-integral or* $\operatorname{ord}_p(\Delta) \leqq n_0$ *then*

$$\lambda_v(bP) \geqq \tfrac{1}{12} v(\Delta).$$

7. The Szpiro conjecture implies the minimal height conjecture. Let A be an elliptic curve defined over $\mathbf{Q}$, and let h_A be the Néron-Tate height. In [La1] I conjectured that this height satisfies a minimum condition as follows. We let Δ_A be the minimal discriminant.

> *There exists constants $C_{\mathbf{Q}}$ and $C'_{\mathbf{Q}} > 0$ such that for all elliptic curves A over $\mathbf{Q}$ and a non-torsion point $P \in A(\mathbf{Q})$, we have*
>
> $$h_A(P) \geqq C_Q \log|\Delta_A| - C'_{\mathbf{Q}}.$$

Note that given an integer $\Delta \neq 0$ there is only a finite number of elliptic curves A over $\mathbf{Q}$ whose minimal discriminant is Δ. Consequently we really did not need to mention the constant $C'_{\mathbf{Q}}$ in the stated inequality, because for $\log|\Delta_A|$ sufficiently large, the right-hand side is $\geqq C''_{\mathbf{Q}} \log|\Delta_A|$ for some $C''_{\mathbf{Q}} > 0$, and by picking $\log|\Delta_A|$ sufficiently large, we are omitting only a finite number of values of $h_A(P)$. Hence by shrinking the constant $C_{\mathbf{Q}}$ suitably, we obtain the stated inequality without a $C'_{\mathbf{Q}}$.

On the other hand in [La2] I also conjectured an upper bound for the height of suitable free generators of $A(\mathbf{Q})/A(\mathbf{Q})_{\text{tor}}$, and I showed that the two conjectures are not independent: the lower bound conjecture is used to motivate the upper bound conjecture, and to carry out certain steps in trying to prove it. The essential point is as follows. Let $\langle\ ,\ \rangle$ be the symmetric bilinear form giving rise to the quadratic form h_A. Then from L-series considerations, conjecturally one gets a bound for the determinant $\det\langle P_i, P_j\rangle$ of a basis for $A(\mathbf{Q})$ mod torsion. It is possible to construct a basis $\{P_1, \dots, P_r\}$ which is "almost" orthogonalized. "Almost" is because we are over $\mathbf{Z}$, not over $\mathbf{R}$. For a precise definition, see [La2] or [La3]. We order the points by ascending height, so

$$h_A(P_1) \leqq \cdots \leqq h_A(P_r).$$

Then we get a conjectural upper bound for the product

$$h_A(P_1) \cdots h_A(P_r).$$

In order to get an upper bound for $h_A(P_r)$ itself, we can then divide by the first $r-1$ terms, and this is where one needs a lower bound for $h_A(P_1)$.

It occurred to Marc Hindry and to me independently that the Szpiro conjecture should imply this minimum, and the implication was proved by Hindry-Silverman [H-S]. I shall describe their proof in this section.

In the first place, Silverman [S] had proved my conjecture in the case of integral j-invariant, and so the problem was how to expand his arguments so that they could apply to the non-integral case.

As before, we let s be the Szpiro exponent, so that we have

$$|\Delta| \leqq N_0(\Delta)^s$$

except for a finite number of elliptic curves, which we disregard. We always take the elliptic curve in minimal form, so Δ is the minimal discriminant.

For the subsequent proof, it will be convenient to decompose the minimal discriminant into a product. We let

$$\Delta = \Delta_1 \Delta_2$$

where:

Δ_1 is the product of the prime powers which occur with an exponent $< 2s$

Δ_2 is the product of those prime powers which occur with an exponent $\geqq 2s$.

Then

$$(*) \qquad \log|\Delta_1| \geqq \frac{1}{2s} \log|\Delta|.$$

This follows immediately from the inequality $N_0(\Delta_2) \leqq \Delta_2^{1/2s}$, the inequality

$$|\Delta| \leqq N_0(\Delta)^s \leqq |\Delta_1|^s N_0(\Delta_2)^s \leqq |\Delta_1|^s |\Delta_2|^{1/2},$$

and by substituting $\Delta_2 = \Delta/\Delta_1$ on the right-hand side.

The primes dividing Δ_2 are the ones which caused trouble in extending Silverman's proof. To cancel their negative contribution to the height, due to a term with the Bernoulli function, Hindry-Silverman use an averaging process, based on a simple inequality of analysis, which we extract here. As in the previous section, we let $\mathbf{B}$ be the periodic function obtain from the Bernoulli polynomial $\mathbf{B}_2$.

Lemma 7.1. *For all $u \in \mathbf{R}/\mathbf{Z}$ and all positive integers M we have*

$$\sum_{m=1}^{M}\left(1-\frac{m}{M+1}\right)\mathbf{B}(mu) \geqq -\frac{1}{12}.$$

Proof. We have the Fourier expansion

$$\mathbf{B}(u)=\frac{1}{2\pi^2}\sum_{n\neq 0}\frac{1}{n^2}e^{2\pi inu}.$$

But also

$$\frac{1}{M+1}\left(\sum_{n=1}^{M+1}z^n\right)\left(\sum_{k=1}^{M+1}z^{-k}\right)=\sum_{m=1}^{M+1}\left(1-\frac{m}{M+1}\right)(z^m+z^{-m})+1.$$

Therefore

$$\begin{aligned}
&\sum_{m=1}^{M}\left(1-\frac{m}{M+1}\right)\mathbf{B}(mu)\\
&\quad=\frac{1}{2\pi^2}\sum_{m=1}^{M}\sum_{n=1}^{\infty}\left(1-\frac{m}{M+1}\right)\frac{1}{n^2}(e^{2\pi inmu}+e^{-2\pi inmu})\\
&\quad=\frac{1}{2\pi^2}\sum_{n=1}^{\infty}\frac{1}{n^2}\left(\frac{1}{M+1}\left|\sum_{m=1}^{M+1}e^{2\pi imnu}\right|^2-1\right)\\
&\quad\geqq -\frac{1}{12}\ \text{because}\ \zeta(2)=\pi^2/6
\end{aligned}$$

as desired.

Hindry and Silverman place the lemma in the context of an inequality for Fourier transforms by Blanksby-Montgomery [B-M]. The lemma and its proof also fall under the pattern used by Elkies in estimating Green's functions on Riemannian manifolds, cf. [La4], Chapter VI, Theorem 6.1.

All that will be used of Lemma 7.1 is that a linear combination of $\mathbf{B}(mu)$ with suitable positive coefficients is uniformly bounded from below, and the sum of the coefficients tends to infinity. The specific form of the coefficients is not important for the applications we shall make.

We now give the proof of the minimal height conjecture assuming the Szpiro conjecture. First, by taking $n_0 = 2s$ (for instance), we can apply Theorem 6.7. Since for all points $P \in A(\mathbf{Q})$ we have

$$h_A(bP)=b^2h_A(P),$$

it suffices to prove the lower bound for the height of points P which satisfy the inequality

$$\lambda_v(P) \geqq \frac{1}{12} v(\Delta_1)$$

for all $v = v_p$ such that $p|\Delta_1$. From now on, we assume that the point P has this property. We pick M sufficiently large as a function of s. For instance, $M = 4s$ suffices, as will become apparent from the following arguments. We let

$$c_m = 1 - \frac{m}{M+1} \quad \text{whence} \quad \sum_{m=1}^{M} c_m = \frac{M}{2} \quad \text{and} \quad C_M = \sum_{m=1}^{M} c_m m^2 .$$

We select b as in Proposition 6.2. Then

$$\begin{aligned} C_M b^2 h_A(P) &= \sum_{m=1}^{M} c_m m^2 b^2 h_A(P) \\ &= \sum_{m=1}^{M} c_m h_A(mbP) = \sum_{v} \sum_{m=1}^{M} c_m \lambda_v(mbP) . \end{aligned}$$

We shall give a lower bound for the v-contribution $\sum c_m \lambda_v(mbP)$ for each v. We partition the absolute values v into four sets:

$v = v_\infty$;
$v = v_p$ with $p \nmid \Delta$;
$v = v_p$ with $p|\Delta_1$;
$v = v_p$ with $p|\Delta_2$.

We note that for every non-zero integer d we have trivially

$$\sum_{p|d} v_p(d) = \log |d| .$$

If $v = v_\infty$ then using Proposition 6.2 and the b of that proposition, we conclude that the v-contribution is $\geqq -MC_\infty$. If $v = v_p$ with $p \nmid \Delta$, then the v-contribution is $\geqq 0$ by Proposition 6.3. Therefore by Proposition 6.6 and Lemma 7.1 for $p|\Delta_2$, and Proposition 6.7 for $p|\Delta_1$, we get

$$\begin{aligned} C_M b^2 h_A(P) &\geqq -MC_\infty + \frac{1}{12}\frac{M}{2} \log|\Delta_1| - \frac{1}{24} \log|\Delta_2| \\ &\geqq -MC_\infty + \frac{1}{48s} M \log|\Delta| - \frac{1}{24} \log|\Delta| \end{aligned}$$

using (*) and the trivial fact that $|\Delta_2| \leqq |\Delta|$. We may now pick for instance $M = 4s$. Then the right-hand side is

$$\geqq -4sC_\infty + \frac{1}{24} \log|\Delta| ,$$

which concludes the proof.

REFERENCES

[Ar] M. ARTIN, *Néron models*, Arithmetic Geometry (G. Cornell and J. Silverman, eds.), Springer-Verlag, Berlin and New York, 1986, pp. 213–230.

[BCHS] B. BIRCH, S. CHOWLA, M. HALL, AND A. SCHINZEL, *On the difference* $x^3 - y^2$, Norske Vid. Selsk. Forrh., **38** (1965), pp. 65–69.

[B-M] P. BLANKSBY AND H. MONTGOMERY, *Algebraic integers near the unit circle*, Acta Arith., **18** (1971), pp. 355–369.

[BLSTW] J. BRILLHART, D. H. LEHMER, J. L. SELFRIDGE, B. TUCKERMAN AND S. S. WAGSTAFF, JR., *Factorization of* $b^n \pm 1$, $b = 2, 3, 5, 6, 7, 10, 11$ *up to high powers*, Contemporary Mathematics Vol. 22, AMS, 1983.

[Dan] L. V. DANILOV, *The diophantine equation* $x^3 - y^2 = k$ *and Hall's conjecture*, Mat. Zametki., **32** No. 3 (1982), pp. 273–275.

[Dav] H. DAVENPORT, *On* $f^3(t) - g^2(t)$, K. Norske Vid. Selsk. Forrh. (Trondheim), **38** (1965), pp. 86–87.

[Fr1] G. FREY, *Links between stable elliptic curves and certain diophantine equations*, Annales Universitatis Saraviensis, Series Mathematicae, **1** (1986), pp. 1–40.

[Fr2] ———, *Links between elliptic curves and solutions of* $A - B = C$, J. Indian Math. Soc., **51** (1987), pp. 117–145.

[Ha] M. HALL, *The diophantine equation* $x^3 - y^2 = k$, Computers in Number Theory (A. O. L. Atkin and B. J. Birch, eds.), Academic Press, London, 1971, pp. 173–198.

[H-S] M. HINDRY AND J. SILVERMAN, *The canonical height and integral points on elliptic curves*, Invent. Math., **93** (1988), pp. 419–450.

[Ku1] D. KUBERT, *Universal bounds on the torsion of elliptic curves*, Proc. London Math. Soc., vol. XXXIII (1976), pp. 193–237.

[Ku2] ———, *Universal bounds on the torsion of elliptic curves*, Compositio Math., **38** (1979), pp. 121–128.

[La1] S. LANG, *Elliptic functions*, Addison-Wesley, 1973, reprinted by Springer-Verlag, Berlin and New York, 1987.

[La2] ———, *Conjectured diophantine estimates on elliptic curves*, Arithmetic and Geometry, Volume dedicated to Shafarevich, Vol. I, edited by M. Artin and J. Tate, Birkhauser, 1983, pp. 155-171.

[La3] ———, *Elliptic Curves: Diophantine Analysis*, Springer-Verlag, Berlin and New York, 1978, pp. 212-213.

[La4] ———, *Introduction to Arakelov Theory*, Springer-Verlag, Berlin and New York, 1988.

[L-O] H. LENSTRA AND F. OORT, *Abelian varieties having purely additive reduction*, J. Pure and Applied Algebra, **36** (1985), pp. 281–198.

[Ma1] R. C. MASON, *Equations over function fields*, Springer Lecture Notes **1068** (1984, 149–157), in Number Theory, proceedings of the Noordwijkerhout, 1983.

[Ma2] ———, *Diophantine equations over function fields*, London Math. Soc. Lecture Note Series, vol. 96, Cambridge University Press, United Kingdom, 1984.

[Ma3] ——, *The hyperelliptic equation over function fields*, Math. Proc. Cambridge Philos. Soc., **93** (1983), pp. 219–230.

[Maz] B. Mazur, *Modular curves and the Eisenstein ideal*, Inst. Hautes Études Sci. Publ. Math. (1978).

[Ne] A. Néron, *Modèles minimaux des variétés abéliennes sur les corps locaux et globaux*, Inst. Hautes Études Sci. Publ. Math., **21** (1964), pp. 361–482.

[Ri] K. Ribet, *On modular representations of* Gal($\overline{\mathbf{Q}}/\mathbf{Q}$) *arising from modular forms* (to appear).

[Se] J. P. Serre, *Sur les représentations modulaires de degré* 2 *de* Gal($\overline{\mathbf{Q}}/\mathbf{Q}$), Duke Math. J., **54** (1987), pp. 179–230.

[Si] J. Silverman, *Lower bound for the canonical height on elliptic curves*, Duke Math. J., **48** (1981), pp. 633–648.

[Ta1] J. Tate, *The arithmetic of elliptic curves*, Invent. Math., **23** (1974), pp. 179–206.

[Ta2] ——, *Algorithm for determining the type of a singular fiber in an elliptic pencil*, Modular Functions in One Variable IV, Lecture Notes in Math., vol. 476, Springer-Verlag, Berlin and New York, (Antwerp Conference).

[Ve] P. Vojta, *Diophantine approximations and value distribution theory*, Lecture Notes in Math., vol. 1239, Springer-Verlag, Berlin and New York, 1987.

Department of Mathematics, Yale University, New Haven, Connecticut 06520

ISBN 0-387-98800-9

EAN

9 780387 988009